MICROBIAL HAZARD IDENTIFICATION IN FRESH FRUIT AND VEGETABLES

MICROBIAL HAZARD IDENTIFICATION IN FRESH FRUIT AND VEGETABLES

Edited by

Jennylynd James, PhD
Dublin, Ireland

A JOHN WILEY & SONS, INC., PUBLICATION

Published by John Wiley & Sons, Inc., Hoboken, New Jersey.
Published simultaneously in Canada.

Library of Congress Cataloging-in-Publication Data:

Microbial hazard identification in fresh fruit and vegetables / edited by Jennylynd James,
p. cm.
Includes bibliographical references and index.
ISBN-13: 978-0-471-67076-6
ISBN-10: 0-471-67076-6 (cloth)
1. Fruit–Microbiology. 2. Vegetables–Microbiology. 3. Fruit–Contamination. 4. Vegetables–Contamination. 5. Fruit trade–Safety measures. 6. Vegetable trade–Safety measures. I. James, Jennylynd.

QR122.M53 2006
664′.8001579–dc22

2005056741

Printed in the United States of America

10 9 8 7 6 5 4 3 2 1

This book is dedicated to my parents Kenneth and Gloria James, and my daughter Tiffany – Thanks for your patience and encouragement.

CONTENTS

CONTRIBUTORS

Belem Avendaño, Facultad de Economica de la Universidad, Autonoma de Baja California, Mexico

Christine Bruhn, Center for Consumer Research, University of California, Davis, California, USA

Linda Calvin, USDA-Economic Research Service, 1800 M Street, N.W. Washington, District of Columbia, USA

Elmé Coetzer, EUREPGAP, c/o FoodPLUS GmbH, Spichernstr. 55, Koeln (Cologne), Germany

Tom Ford, Ecolab, Greensboro, North Carolina, USA

Brett Gardner, Raley's Supermarket, Sacramento, California, USA

Maria Isabel Gil, Research Group on Quality, Safety and Bioactivity of Plant Foods, Dept. Food Science and Technology, CEBAS-CSIC, Campus de Espinardo, Murcia, Spain

Toni Hofer, Raley's Supermarket, Sacramento, California, USA

Jennylynd James, Food Consultant, Dublin, Ireland

LeeAnn Jaykus, Departments of Food Science and Microbiology, North Carolina State University, Schaub Hall, Raleigh, North Carolina, USA

Lynette Johnston, Food Science Department, Box 7624, North Carolina State University, Raleigh, North Carolina, USA

Lise Korsten, Plant Pathology Division, University of Pretoria, New Agriculture Building, Pretoria, South Africa

Kristina Mena, Environmental Sciences, University of Texas Health Science Center at Houston, El Paso Campus, El Paso, Texas, USA

Barry Michaels, B. Michaels Group Inc. Consulting, Palatka, Florida, USA

Christine Moe, Department of International Health, The Rollins School of Public Health, Emory University, Atlanta, Georgia, USA

Debbie Moll, National Center for Environmental Health, Centers for Disease Control and Prevention, Atlanta, Georgia, USA

Victoria Selma, Research Group on Quality, Safety and Bioactivity of Plant Foods, Dept. Food Science and Technology, CEBAS-CSIC, Campus de Espinardo, Murcia, Spain

Rita Schwentesius, Center for Economic, Social, and Technological Research on Agroindustry and World Agriculture, Universidad Autonoma de Chapingo, Chapingo, Mexico

Ewen Todd, National Food Safety and Toxicology Center, Michigan State University, East Lansing, Michigan, USA

Devon Zagory, Davis Fresh Technologies, Davis, California, USA

PREFACE

A healthy, balanced diet would not be complete without fresh fruit and vegetables. The United States Produce for Better Health Foundation recommends five to nine servings of fruits and vegetables daily for better health. Several government agencies have also introduced initiatives with the aim of increasing consumption of fresh fruits and vegetables. As the public has become more health conscious over the years, a number of widely publicized foodborne outbreaks associated with fresh produce have caused some concerns to the industry and consumers alike. Even though statistics show an increased trend in produce-related foodborne outbreaks, it is possible this trend is directly related to an improvement in monitoring programs. Not all countries are equipped to monitor outbreaks and conduct traceback investigations. Thus, a lack of information on outbreaks in many countries does not signify the absence of outbreaks.

Knowledge of microbial hazards in fresh fruit and vegetables, from the farm to the table, will help in providing wholesome, healthy food for consumers. Only if hazards are identified can adequate control measures be implemented to reduce risk of product contamination. Government agencies have proposed mitigation measures, guidelines, and codes of best practice for the industry to reduce contamination of fresh produce. In addition, trade associations have provided food-safety guidelines in farm production and fresh-cut processing operations. Scientific research, if carefully directed, would give growers, packers, and shippers the necessary tools to create preventive programs. Many guidelines are based on the potential for product contamination and not actual scientific data. The paucity of data opens the doors to many areas of future research for the fresh produce industry to understand the impact of product contamination.

All participants in the food chain, from the farm to the fork, should take responsibility for the safety of the food supply. Because vegetables and fruits may be eaten raw, growers, shippers, processors, and retailers have the added responsibility of safeguarding the product and supporting consumer confidence in the industry.

This book is a comprehensive reference for the fresh fruit and vegetable industry. It focuses on the major stages in growing and handling of produce. Possible hazards in production, harvesting, packing, distribution, retail, and consumer handling are identified. This book also covers a case study of a foodborne outbreak associated with fresh produce and the actual costs to the industry because of this outbreak.

The text will be particularly useful to advanced undergraduate and graduate students interested in postharvest biology and food-safety issues affecting horticultural crops. The text would also be useful in organizing short courses through University Extension programs catering to research and extension workers, consultants, quality management staff, and other people involved in managing food-safety programs for the fresh produce industry.

I would like to thank each author for participating in this project. The efforts of these authors, as well as contributions and encouragement by other individuals, have made this publication possible.

JENNYLYND JAMES

CHAPTER 1

Overview of Microbial Hazards in Fresh Fruit and Vegetables Operations

JENNYLYND JAMES

Food Consultant, Dublin, Ireland

Microbial Hazard Identification in Fresh Fruit and Vegetables, Edited by Jennylynd James

1.1 INTRODUCTION

Scientists recommend that everyone eat five to nine servings of fruit and vegetables every day in order to promote good health. The improved availability of fresh produce year round and increased choices of items on the supermarket shelves should certainly help consumers to meet this target of fresh produce consumption. Raw fruit and vegetables, however, have the potential of becoming contaminated with microorganisms, including human pathogens. Several widely publicized foodborne outbreaks in recent years have been associated with sprouted seeds, minimally processed produce, unpasteurized vegetable and fruit juices, as well as intact products. However, the proportion of fresh-produce-related outbreaks is low when compared to the number of foodborne outbreaks per year.

Fruits and vegetables normally carry nonpathogenic, epiphytic microflora. During production on the farm and all stages of product handling from harvest to point of sale, produce may be contaminated with pathogens (Beuchat, 1996; Beuchat and Ryu, 1997). Possible microbial hazards on the farm include the use of raw manure and contaminated soil amendments, dirty irrigation water, wild animals and birds, and dirty farming equipment. At harvest, employee health and hygiene is critical. In addition, farm tools, utensils, and packaging could possibly contaminate the product. Packhouses pose a risk when using water to wash product or convey product in water flumes. The water quality plays a key role in determining the final quality and safety of the product. Employee hygiene and food contact surfaces have the potential to affect product in the packhouse. In addition, transportation and distribution practices determine product quality and safety for future use. When product is displayed at retail and handled in food service operations, there is the potential for contamination. The end user or consumer also plays a critical role in maintaining product safety as produce items are taken from the store, preserved, and prepared in the home.

Major stakeholders in the fresh produce chain have introduced measures to prevent product contamination (FDA/CFSAN, 2001a). At the farm level, Good Agricultural Practices (GAPs) and documentation of these practices were introduced. Government guidelines for the industry help in promoting safe practices and large retailers encourage the use of these guidelines by demanding results of audits of practices (FDA, 1998a, b). Retailers feel assured that the product presented to consumers has been handled safely when farms and packhouses are audited to guidelines and standards.

Minimal processing of fruits and vegetables presents unique challenges, because cutting and slicing remove the natural protective barriers of the intact plant. Thus, implementing Hazard Analysis and Critical Control Points (HACCP) programs in high care facilities adds more assurance of food safety. More research is needed in the fresh produce chain to prove the effectiveness of mitigation measures. Many monitoring programs are based on assumptions that contamination can occur. Scientists have used surrogate organisms to imitate the survival of microbial pathogens in fresh produce and these studies could provide significant insight into controlling the spread of pathogens in the industry (FDA/CFSAN, 2001b).

Produce is moved globally to supply year-round demands and improvement in traceability methods would help in epidemiological investigations. Outbreak data have been limited to just a few industrialized nations because only these countries have active surveillance systems for monitoring. Thus, the generation of more epidemiological data on produce-related foodborne illness worldwide will help determine true levels of illness.

1.2 PATHOGENS AND OUTBREAKS ASSOCIATED WITH THE FRESH PRODUCE INDUSTRY

Biological hazards are of great concern to the fresh produce industry. They can be classified into spore-forming bacteria, non-spore-forming bacteria, viruses, and parasites. Certain bacteria form spores to withstand environmental stress conditions such as high heat on freezing. Spore-forming organisms can attach to vegetables grown near the soil. Examples of these organisms include *Bacillus cereus*, *Clostridium perfringens*, and *Clostridium botulinum.* Maintaining refrigeration temperatures at less than 5°C and promoting oxygen in packaging would reduce the risk of vegetative cell formation and the production of dangerous toxins that cause illness (Linton, 2003). Non-spore-forming bacteria such as enterotoxigenic and enterohemorrhagic *Escherichia coli*, *Campylobacter jejuni*, *Listeria monocytogenes, Salmonella, Shigella* spp., *Staphylococcus aureus*, and *Vibrio* spp. could contaminate fresh produce by cross contact with humans or animals carrying these organisms. The fecal–oral route is possibly the main mechanism of transfer. All of these bacteria have been associated with publicized fresh produce foodborne outbreaks of public health significance. The transfer of these organisms could be controlled by practising good personal hygiene, cleaning food contact surfaces, and always using potable water when water is required.

Foodborne viruses require a living host in which to grow and reproduce. Viruses tend to move from one food to another, from water supply to food, or from food handler to food. Hepatitis A, Norwalk virus, and rotavirus are viruses of public health significance. Hepatitis A has been isolated in vegetables washed with nonpotable water. A food worker can carry the organism virus for up to six weeks and contaminate food and other workers without any knowledge of signs and symptoms. The Norwalk virus and rotavirus have been associated with many foodborne infections. Raw fruits and vegetables washed with contaminated water were implicated in some outbreaks. These viruses are transmitted by person-to-person contact and by fecal contamination. Practising good personal hygiene and controlling staff carrying the virus are measures that could possibly eliminate foodborne illness.

Parasitic protozoa include *Cyclospora cayetanensis*, *Giardia lamblia*, and *Cryptosporidium parvum.* They are single-cell microorganisms that must live on or inside a host to survive. These parasites may be transmitted via contaminated water, by person-to-person contact, and by fecal contamination. Use of potable water for operations is critical.

In the last 15 years, knowledge of foodborne disease epidemiology evolved while the fresh fruit and vegetable industry was undergoing notable changes. Factors increasing the risk of foodborne illness associated with fresh produce include the following:

1. Modifications in agronomic practices, processing and packaging technologies;
2. Global marketing strategies allowing fresh produce supply to consumers with a wide variety of products, year round;
3. Changes in population demographics; and
4. Changes in food consumption patterns.

Increased awareness because of unique epidemiologic surveillance programs and increased media attention has contributed to better documentation of foodborne illness. Numerous pathogens have been isolated from a wide variety of fresh fruits and vegetables. It is important to note that the number of samples in each study varied substantially. Although not all of the pathogens have been associated with produce-related foodborne disease outbreaks, they are all capable of causing illness.

In the United States, a specific etiologic agent was identified for 187 produce-associated outbreaks during the years 1990–2002. Among these outbreaks, 102 (55%) were caused by bacteria, 68 (36%) were caused by viruses, and 17 (1%) were caused by parasites. Among the bacterial agents, *Salmonella* accounted for 60% of outbreaks, and pathogenic *E. coli* was responsible for 25% of bacterial outbreaks. Norovirus caused a majority of viral outbreaks, accounting for over 80% of cases. The apparent prevalence of norovirus has increased possibly as a result of improved surveillance and detection methods. *Cyclospora* caused the majority (65%) of parasitic produce-associated outbreaks. Over 40% of the outbreaks were caused by salad items (including lettuce and tomatoes), whereas fruit and fruit salads comprised 13% of the outbreaks. Melons, including cantaloupe, honeydew, and watermelon, represented 12% of produce-associated outbreaks, and sprouts comprised 10% of the outbreaks (CDC, 2004).

In spring of 1996, CDC and Health Canada were alerted to over 1465 cases of foodborne illness caused by *Cyclospora* in the United States and Canada. The source of illness was incorrectly identified as California strawberries. At the peak of the California strawberry season, this mistake cost the industry $16 million in lost revenue (Calvin et al., 2002). It was later discovered that the illness was caused by Guatemalan raspberries and the United States stopped all imports of this commodity. Another incident occurred when 200 school children and teachers in Michigan contracted Hepatitis A in 1997. This outbreak was traced to frozen California strawberries. The fresh strawberry industry, through the California Strawberry Commission, was quick to alert the public of the subtle difference with the fresh strawberry market, thus limiting the financial impact and decline of fresh market sales as in the previous year (Calvin, 2003). Japan had the world's largest reported vegetable outbreak in 1996 when over 11,000 people were affected and 6000 were culture confirmed. Three school children died from this outbreak, which was caused by *E. coli* O157:H7 (Ministry of Health and Welfare of Japan, 1997). In the United States, *Salmonella* traced back to Mexican cantaloupes caused outbreaks in 2000, 2001, and 2002. The investigation was time-consuming and caused a decline in California cantaloupe sales following the Mexican growing season.

Growers and shippers were the focus of U.S. government investigations in the mid-1990s for outbreaks of food illness linked to fresh produce (Tauxe, 1997). Epidemiology of foodborne outbreaks will be discussed extensively in a following chapter. The lack of strong traceability details and poor reporting systems for outbreaks limits a thorough evaluation of the role of fruits and vegetables as a source of foodborne infections (European Commission, 2002).

1.3 POTENTIAL HAZARDS IN THE FOOD CHAIN/POINTS OF CONTAMINATION

1.3.1 Hazards in Production

Ranch History and Adjacent Land Use. The ground where product is grown plays a vital role in safety of the product. If the area has a history of use for chemical waste or the processing of biosolids, this would present a potential source of contamination for crops. It is important to know the land history and the time required for the area to lay fodder, thus reducing the level of contaminants in the soil. Adjacent land use also affects the safety of the crop grown. If fruit and vegetables are grown next to an animal-rearing operation, there is a potential for product to become contaminated by animals. These animals may physically enter fields. Waste, high winds, and run-off from the animal operation may contaminate crops. The decision to grow next to a potentially hazardous location should be followed up with a risk assessment and the implementation of preventive measures to control risks identified. Sloping land from an adjacent field could be curbed by digging a ditch along the full length of the field to catch run-off water. Physical barriers and trenches may also prevent unwanted animal entry into fresh produce operations.

The importance of adjacent land was demonstrated in the first documented outbreak of *Escherichia coli* O157:H7 infection associated with a treated municipal water supply in Canada in May–June 2000. This was the largest multibacterial waterborne outbreak in Canada at that time (Public Health Agency Canada, 2000). The Walkerton residents who became ill numbered approximately 1286. Researchers confirmed that a well was subject to surface water contamination and elevated turbidity. Environmental testing of 13 livestock farms within a 4 km radius of the three wells identified human bacterial pathogens in animal manure on all but two farms. On nine farms, *Campylobacter* spp. were identified, on two farms both *E. coli* O157:H7 and *Campylobacter* spp. were found; this included a farm adjacent to the affected well. The evidence suggested that the pathogens that entered the well were likely to have originated from cattle manure on this farm (Public Health Agency Canada, 2000).

Animals. Fruit and vegetable growers and packers are discouraged from keeping animals because they represent a source of product contamination. Domestic animals such as chickens, dogs, and horses can contaminate crop with fecal droppings if they pass through growing areas. Nonfarm animals such as deer, other mammals, and birds can serve as reservoirs for pathogens (Dingman, 2000; Moncrief and Bloom, 2005). Wild animals present a unique hazard to fruit and vegetable growers and these animals are difficult to control. Droppings from birds, deer, and other wild animals present a risk. It is not economical in some instances to fence large production areas, but small farms may be fenced to keep out wild animals. Other physical barriers such as mounds, diversion berms, vegetative buffers, and ditches, provide protection from animals. Growers use scarecrows, reflective strips, and gunshots to ward off birds and pests from crop. Mechanical traps have been used to catch fieldmice in lettuce-growing operations in Yuma, Arizona. Growers implemented this emergency measure to alleviate a problem of mice in lettuce bound for the prepared salad industry. This measure was simple, yet effective in preserving product integrity.

Manure and Soil Amendments. An increased demand for organically grown produce promotes the use of alternative measures to protect plants from pests, mites, and fungi. In addition, organic fertilizers such as animal manure could introduce fecal pathogens to fresh produce if manure is not aged and treated before application. Treated manure or biosolids can be beneficial soil amendments. Growers have to manage these amendments very well to minimize contamination of fruits and vegetable with pathogenic fecal microorganisms. Published literature shows the presence and isolation of zoonotic pathogens in manure and on the surface of fresh produce (Dingman, 2000; Moncrief and Bloom, 2005). However, more research is needed to show a direct link from a manure application to finding pathogens in fresh produce. This research could help in assessing the level of risk. Active composting uses three Ts to treat manure: time, temperature, and turnings.

Composting can effectively reduce pathogens and parasites commonly found in manure as well as those that have mutated into different strains with new abilities, like surviving in acidic environments. Millner (2003) estimates that once certain time and temperature criteria are achieved, *E. coli* and *Salmonellae* in the compost are nearly eliminated (99.9999 percent kill rate). The pathogen-reduction criteria include a temperature of at least 131°F for three consecutive days in an aerated pile or 131°F for two weeks in the hot zones of a windrow pile with five turnings. This process can kill nearly all pathogenic microbes and still maintain populations of beneficial ones.

Applying raw manure to a vegetable-growing operation could cause product contamination. In addition, manure piles stored next to growing operations may also be a problem because of run off. Organic growers use a great deal of treated manure as fertilizer and soil amendment. Organic growers must be vigilant not to use fresh manure, because this would increase the potential for product contamination (Blaine and Powell, 2004; FDA, 1998b). Manure application should be carried out well in advance of harvest so there is sufficient time for potential pathogens to break down naturally.

Water. Water may be used throughout the growing and harvesting of fresh fruit and vegetables. Irrigation water and water used for application of plant protection product, fertilizer, and frost protection should not introduce risk of pathogens to product. The source of agricultural water could determine the final safety of the food product. Municipal water or wells are generally safe. In the United States, municipal water is usually treated and tested routinely to monitor microbial contamination. Wells with sound casing and absence of leaks should provide water free of pathogens. Many growers also draw water from open water systems in areas where water is scarce. If effluent water from sewage plants is used in hydroponics plant production, the quality of this water is a concern for introduction of pathogens in edible food. The microflora of these systems changes daily and it is unknown what quality of water exists from one day to the next. Irrigation practices dictate product safety when using open systems like rivers and canals. Drip irrigation or furrow irrigation would bring water to the root of the plant without making contact with leaves or other edible portions. In the early growing stages, it is possible to use sprinklers to wet young plants. The time interval between planting and harvesting is sufficiently long to expect pathogens present to die off.

Greenhouse production is generally well protected from the environment, but is seen to favor the survival of human pathogens (Nguyen-the and Carlin, 2000). Special issues of control of pathogens in greenhouses concern the fact that all nutrients are delivered to plants in liquid form. If water used for mixing agricultural chemicals becomes contaminated in any way, the implications for the greenhouse are extensive. The pathogen would spread over a wide range of product on which it was applied. The conditions in the greenhouse – warm temperature ($>21°C$), moisture (>70 percent relative humidity), and light – would encourage the growth of pathogens (Moncrief and Bloom, 2005).

1.3.2 Contamination in the Field at Harvest

Harvesting is a critical period in the fresh produce chain, with potential for most product contamination. Employees in farming operations play a key role in maintaining the integrity of product grown. Their safe and hygienic practices – from land preparation, planting, weeding, and pruning, to harvesting – could influence whether product becomes contaminated. Personal hygiene, including hand washing all along the food chain, is critical in reducing or eliminating contamination with fecal pathogens. Ill employees harvesting by hand could contaminate product bound for the consumer if care is not taken in basic hygiene. Growers may use portable toilet facilities that follow workers in large fields to provide sanitary facilities. As part of Good Agricultural Practices (GAPs), companies train field employees in the basics of using toilet facilities instead of the field, where product may be contaminated. Employees are trained in proper hand-washing techniques. In the United States, the Occupational Health and Safety Act (29) CFR 1928.110, subpart I, regulates the use of portable toilets. The Act specifies that the following practices should be adopted:

1. One toilet should be provided for every 20 workers.
2. The field sanitation unit should be placed within one quarter mile of the harvest crew.
3. Sanitary tissue, and soap and potable water for hand washing should be provided.
4. A means of hand drying, preferably disposable paper towels, should be provided.
5. Sanitary facilities in the field must be managed effectively so they do not become a source of contaminated product. A professional company should be used to empty and clean sanitary facilities.
6. Growers should have a containment plan in place in the unlikely event of a spillage.

Wounds are hosts to pathogenic bacteria and direct food contact should be avoided. Workers who receive minor cuts while harvesting should only continue to work if the cut is well protected using a plaster/bandage, and further covered with gloves. If gloves are used, they should not cause product contamination. Thus, gloves should be changed regularly or when damaged.

Equipment used for harvesting may also be a source of product contamination. Utensils should be cleaned and sanitized regularly to avoid cross-contamination. Some growers now use a sanitizing dip in the field during harvest so knives and other tools can be cleaned during the break and before start of harvest. Field containers should be washed routinely to avoid a build-up of debris. Some packhouses are equipped with crate washers that comprise a tunnel full of powerful water jets and in some cases an initial spray of foam cleaning detergent. Because produce crates are taken to the field daily, it is important not to create any instance of cross-contamination of produce by using dirty packaging.

Minimal processing of produce items like romaine hearts and head lettuce may be carried out in the field. Product is harvested, rinsed in the field on harvest rigs, and then packed in consumer packages and boxes. The harvest rigs should be cleaned and sanitized daily, because these products are advertised as "ready-to-eat". Any wash water used on harvest rigs should be potable and suitable for its application. Utensils and vehicles used to transport packaged fruit and vegetables should be dedicated for this purpose, because cross-contamination could occur if other products are transported in the same vessels. A precautionary measure would be to wash and sanitize containers before use.

1.3.3 Post-Harvest Handling of Fresh Produce

Packhouses. After product is harvested, it should be protected to prevent any cross-contamination. Packing facilities should be cleaned and well maintained to reduce the introduction of harmful microorganisms to product. Some growers move product from the field in large bins, which are taken to the packhouse for selection, grading, and repacking. No matter what method of packing is used, care must be taken with product. Cartons and other empty containers should not be stored uncovered in fields, because dust and animals could contaminate product. Rigs and utensils used for packing should be cleaned and monitored daily. Packhouses, whether open or enclosed, should be cleaned and protected to deter pest entry and possible product contamination. Harvest storage facilities, containers, or bins should be cleaned regularly, based on a set program.

Poor sanitation in the packhouse could lead to the formation of biofilms. Biofilms are layers of bacteria that attach to surfaces like stainless steel and plastic, and also attach to each other with the help of polymeric materials. The biofilms trap other bacteria, debris and nutrients. Poor sanitation programs cause biofilms to build up and become established. Nonpathogenic and pathogenic bacteria can form biofilms. Organisms in the film tend to be resistant to cleaners and sanitizers, as well as heat treatment. Thus, in the food industry a sound sanitation program is needed so that biofilms do not become entrenched on food contact surfaces.

Good sanitation practices enhance a company's food-safety program. An important step is to provide training in sanitation to a wide base of employees, even those outside the sanitation department (Redemann, 2005). In daily sanitation programs, seven steps could be followed to ensure clean equipment:

1. Dry cleaning to remove gross debris from equipment and floors;
2. Prerinsing to remove debris from surfaces;
3. Using soap and scrubbing equipment on surfaces and floors;
4. Postrinsing to remove soap;
5. Removing standing water and reassembling equipment if necessary;
6. Inspecting cleaned area and recleaning with detergent if necessary;
7. Sanitizing equipment and floors in high-care facilities.

Many companies use chlorine-based cleaning chemicals. For effective cleaning, some chemicals could require a specific residence time on equipment. Sanitizers include halogen-based compounds and quaternary ammonium-based compounds. For a review on food plant cleaning and sanitizing, the reader is referred to Redemann (2005).

Pest management could prevent product contamination, recall, and other loss of productivity. A company should have a preventive program of pest control so that problems never develop. The packhouse manager should assess the risk in the plant to determine the level of prevention needed. Completely closed packhouses may have sophisticated pest control programs using a certified contract company. Windows and air vents should be screened and facilities kept free of debris. Any pest-control program implemented should be monitored and documented regularly to protect product.

Water Flumes. Where water is used to transport produce from one part of the packhouse to the next, the quality of that water will determine the quality and safety of the product being packed. Primary water for rinsing to remove dirt may be of agricultural grade. However, potable water should be used for subsequent rinsing steps. Fresh-cut rinsing operations may use automated chlorination and acidification systems to control and monitor water quality. Flume channels used for sorting and grading product should be cleaned according to a planned sanitation program, thus preventing the build-up of debris from recycled water and avoiding product contamination. Where water is recycled, it should be treated to reduce the build up of microorganisms. Many companies use chlorine or other disinfecting chemicals to control the microbial load. Packers and processors use chlorine dioxide, chlorine (hypochlorous acid), UV light, ozone, hydrogen peroxide, peroxyacetic acid, and other sanitizers. The standard sanitizer used in packhouses in the United States is chlorine. This could be in the form of sodium hypochlorite granules, tablets, or liquid.

Water has been demonstrated to enter tomatoes through the stem end in a water dunk tank. This may be caused by the temperature differential of the produce item harvested in the field and the cooler water in the packhouse. At intake, any microorganisms in the water would also enter the produce item. Thus, it is imperative that any wash water in the plant be potable water. Any pathogens in water could become the source of a foodborne outbreak. A multi-State outbreak of *Salmonella enterica* Serotype Newport infection was linked to mango consumption in December 1999. Traceback of the implicated mangoes led to a single Brazilian farm, where hot water treatment was identified as a possible point of contamination. Hot water treatment was a new process introduced to prevent importation of an agricultural pest, the Mediterranean fruit fly. Contaminated water caused product contamination. This outbreak highlighted the potential global health impact of foodborne diseases and newly implemented food processes and the vulnerability of product placed in contaminated water (Sivapalasingam et al., 2003).

Employees. Packhouse employees should be trained in safe food-handling practices. Product safety of final consumer packs is directly influenced by the handling practices in the packhouse. Employees should receive training on the proper use of toilets, hand-washing procedures, use of protective clothing, and headgear to avoid product contamination. Many foodborne outbreaks have been linked to a sick employee who transferred pathogens via the fecal–oral route. Employees shedding pathogens in diarrhea may not wash their hands adequately before handling food. This is one possible route of contamination of fresh produce.

1.3.4 Storage and Distribution

Refrigeration temperatures in storage and distribution are crucial to maintaining product quality. These temperatures also reduce the proliferation of human pathogens if they are present on produce items. Refrigeration units are thought to spread mold throughout warehouses, and routine servicing of air filters and refrigeration systems is required. As cold air systems blow mold spores into the air, there is also the risk that pathogens may be spread along with the spores from one pallet to the other. More research is needed in the area of air quality in storage and distribution in order to implement effective control measures. This topic will be discussed extensively in Chapter 7.

Pest control programs are necessary at any storage facility. A basic rodent control program would reduce the presence of pests that harbor harmful microorganisms forming a potential hazard to food. A contract company or trained individual should monitor traps or bait stations regularly and document any pest activity observed.

Vehicles and containers used to transport fresh produce could also be sources of potential contamination. Vehicles used to transport fresh produce should be clean and free of odors, dirt, and debris before loading. The ideal situation would be to use dedicated containers and vehicles for each application; that is, produce, meat, or refuse should be transported by separate means. The produce containers and vehicles should also be cleaned routinely as part of Good Hygienic Practices (GHPs) to prevent contamination between loads. Good hygienic and cleaning practices ensure product safety when loading or during inspections. The temperature of transport would also determine the potential for growth of pathogens. Thus, refrigeration temperatures are used to transport many produce items. The cold temperature helps to preserve product quality as well as safety.

1.3.5 Fresh-Cut Fruit and Vegetables: Potential Hazards

The fresh-cut fruit and vegetable industry in the United States grew from supplying quick-serve restaurant chains to providing consumer-size convenience products. From the 1980s to the 2000s, bagged salads and other convenience items experienced tremendous growth in demand. Sales of bagged salads topped $3 billion in 2004, while whole peeled carrots reached about $1 billion (Gorny, 2005).

Fresh-cut processors have worked with growers, retailers, and food service operations to ensure raw material and finished product are handled safely. When fresh produce is washed, cut, and sliced, the natural defense mechanisms on the plant material are removed. The high level of handling increases the potential for product to be contaminated by microorganisms in the environment. Because there is no kill step in fresh-cut processing, using HACCP is imperative in maintaining food safety standards with preventive programs (Nguyen-the and Carlin, 1994, 2000). Growers implement Good Agricultural Practices (GAPs) reducing risks during growing and harvesting (FDA, 1998a). Retailers and food service operations implement the principles of HACCP to manage food safety in their operations. In the United States, the International Fresh Cut Produce Association (IFPA) published Food Safety Guidelines for fresh-cut food processors. Guidance documents produced include a model HACCP plan, best practice guidelines for activities, a model food allergen plan, and a Sanitary Equipment Buying Guide and Development Checklist. The IFPA has also developed a GAP program in conjunction with the Western Growers' Association. The industry is thus equipped with proactive guidance to produce safe food.

1.3.6 Retail and Food Service Operations

The primary cause of foodborne illness in the United States is thought to be the mishandling of food during preparation in food service operations or in the home (Gorny, 2005). Statistics collected by the Center for Disease Control and Prevention (CDC) provide a strong link between foodborne illness outbreaks and the end users in restaurants and retail (CDC, 2004). Consumers could be a source of fresh produce contamination in retail outlets. Consumers touch fruit and vegetables as they make a decision on whether to purchase product. If a person's hands are contaminated because of improper hygiene, this product could be affected. The consumer may also place bare fruits and vegetables into shopping carts, which are generally unwashed. Cross-contamination by microorganisms present in the shopping cart onto fresh produce items could present a food safety hazard, especially if the cart was used to transport meat, fish, or poultry. The final step of bagging produce items could present a food safety hazard if fruits and vegetables are placed in the same bag as raw meat or fish.

When the consumer places grocery bags in the car, vehicle temperature and time to cooling determine the potential for present pathogens to multiply. Linton (2003) describes a "temperature danger zone" of 5–60°C for food in general. A rule of thumb in the industry is that it takes four hours in the temperature danger zone for pathogens to multiply enough to cause illness. For this reason, hot food should be stored at temperatures over 60°C and cold foods at less than 5°C for optimum safety (Linton, 2003).

Retail operations and restaurants should take a proactive role in carrying out a risk assessment of their operations to understand the different types of hazards. When these potential problems are identified using the principles of HACCP, they may be controlled. Salads and juices are sometimes prepared in the fresh produce section of a retail operation. Workers should be trained in good food-handling practices to ensure food quality and safety are maintained:

1. Ready-to-eat foods should be kept at safe temperatures.
2. Food contact surfaces in food-preparation areas should be cleaned and sanitized regularly.
3. Display counters and shelves should be cleaned regularly to prevent cross-contamination.
4. Water used to wash produce should be potable, because poor-quality water is a major source of contamination.
5. Color-coded chopping boards should be used to distinguish boards for cutting produce from those used for cutting meat.

6. Employees bagging product at the cashier's desk should be trained to avoid cross-contamination by placing raw meat in separate bags from fresh produce.
7. Regular hand washing should be enforced to prevent product contamination.

All employees in a retail operation should thus have a basic knowledge of food safety and their responsibility of protecting the public.

Outreach and education of the consumer would be the most effective means of explaining the dangers of mishandling food to the ultimate customers. Retailers may also assume the role of educating the public so the food-safety message reaches into people's homes. Brochures, videos, and posters at display counters have been used effectively for consumer education in retail operations. Modern technology and a bit of creativity on the part of retailers could go even further to enlighten the public and promote safe handling of food.

Restaurant chains represent one important point in the handling of fresh fruits and vegetables before consumption. Food-service operations have been linked to several foodborne outbreaks involving fresh produce (CDC, 1999; CNN, 1996, 2003). Adequate training of food-service employees and implementation of a HACCP program to control food-safety in kitchens and food-display counters are the best ways to control food handling.

The "Serve Safe" program developed by the International Food Safety Council of the National Restaurant Association Educational Foundation has become the industry standard in food safety training. Serve Safe is accepted in almost all U.S. jurisdictions that require employees' certification. The Serve Safe program provides accurate, up-to-date information for all levels of employees on all aspects of handling food, from receiving and storing to preparing and serving. The International Food Safety Council promotes food-safety education to the restaurant and food service industry, and also conveys the industry's food-safety commitment to the public. The Serve Safe program has introduced a number of control measures for food-service operations similar to those mentioned above for retailers handling minimally processed product. In the United States, food service operations may also be subject to food-safety inspections, depending on the laws in each State. Only a very strict food-safety program in food-service operations would reduce the risk of pathogen contamination and illness to the ultimate consumer. The topic of U.S. retailer programs reducing food-safety risks in stores and educating the public will be presented in Chapter 8.

1.3.7 Consumer Handling of Fruits and Vegetables

Fruit and vegetables, besides being perishable items, could be the source of mishandling by consumers, eventually leading to foodborne illness. Consumers sometimes mishandle produce by cross-contamination with meat items being placed in the same bag or cart. Temperature abuse in a hot car would promote the growth of microorganisms, if present. In the home, food-safety practices such as hand washing before handling fresh produce, or after handling meat, may not be observed. Most consumers store produce in the refrigerator; however, some items are stored at room temperature (Li-Cohen and Bruhn, 2002). Room-temperature storage is good for tomatoes, bananas, and unripe climacteric fruits. However, refrigeration extends the freshness and slows bacterial growth if a produce item contains harmful microorganisms. The home refrigerator may also harbor harmful organisms if it is not cleaned routinely. Meat drippings on shelves could cross-contaminate fresh produce placed on the same shelves. Consumers do not generally have a cleaning schedule for refrigerators, but this is a critical point in home hygiene in the kitchen. Refrigerator surfaces should be cleaned and sanitized to remove potential pathogens every two weeks, or more often, depending on contamination.

Another reservoir for pathogenic microorganisms is the kitchen sink. Consumers may place fresh produce items in the sink without washing or sanitizing the area. This causes cross-contamination from items previously placed in the sink. Sinks should be cleaned and sanitized before being used to wash fresh fruits and vegetables. In addition, mixing of utensils would invite cross-contamination, leading to foodborne illness. Consumers may not always wash fruits and vegetables, but even the simplest washing with running water is sufficient to cause one $\log_{10}$ CFU/g reduction in microbes. Fresh-cut fruit and vegetables should always be stored in refrigeration, first to extend shelf life, and then to reduce the potential growth of pathogens, which are usually mesophilic organisms. Chopping boards carry microorganisms in the form of biofilms if they are not cleaned and sanitized thoroughly. Consumers should implement good cleaning programs when using chopping boards for fresh produce, and preferably use different utensils than those used for meat.

Consumer attitudes to handling of fresh produce and statistical data collected in surveys of consumer behavior will be presented in Chapter 9.

1.4 MITIGATION MEASURES

1.4.1 Improvements in Produce Handling and Research Efforts

Scientists at the Agricultural Research Service, USDA, proposed research programs in 1999 to investigate new technologies for decontaminating fresh produce containing human pathogens. The presence and survival of human pathogens in unpasteurized juices, sprouts, melons, lettuce, and berries is known. These commodities have been implicated in a number of foodborne illness outbreaks. Typical methods of washing and sanitizing produce are not effective in removing pathogens, and increasing the knowledge of sources of microbial contamination requires more research. Thus, scientists have tried to identify sources of human pathogen contamination and develop interventions to prevent contamination and remove or inactivate pathogens on fresh and minimally processed produce (Beuchat, 1998; Jaquett et al.,

1996; Zhang and Farber, 1996). These studies provide a foundation for the development of treatments to assure the microbiological safety of fresh produce.

Sapers et al. 2000 showed that washing with hypochlorite or commercial surfactant formulations, as a means of decontaminating apples inoculated with *E. coli*, had limited success with only 1 to 2-logs (90–99 percent) reduction. However, experimental washing with hydrogen peroxide resulted in a 3-log (99.9 percent) reduction. Annous et al. (2001) demonstrated the inability of brush washing to decontaminate apples in field tests. It was thought that bacteria in inaccessible sites, growth of bacteria in skin punctures, and possible infiltration of bacteria into calyx and core tissues prevented their destruction.

In work with apples, mature and immature fruit from various locations were examined for evidence of internal or external bacterial contamination. A decay-causing fungus, *Glomerella cingulata*, permitted growth of *E. coli* O157:H7 in inoculated apples, probably because of a decrease in acidity that resulted from fungal growth. This clearly demonstrates the potential risk of using decayed apples for production of unpasteurized cider (Riordan et al., 2000). Sapers et al. (2001) researched the microbiology of fresh and fresh-cut cantaloupe, providing knowledge about attachment and survival of bacteria, including surrogates of human pathogens on external melon surfaces. They studied the efficacy of washing treatments in decontaminating cantaloupe melons, and effective treatments for extending the shelf-life of fresh-cut cantaloupe. Cantaloupe melons were artificially contaminated with nonpathogenic *E. coli* and *Salmonella stanley*, a human pathogen, and survival of these bacteria on melon rind during washing and their transfer to the melon flesh during fresh-cut processing was measured. Researchers found that chlorine and hydrogen peroxide solutions became progressively less effective in reducing contaminant populations as the time between contamination and washing increased. Surviving bacteria were transferred to the melon flesh and growth occurred (Sapers et al., 2001).

Chlorine dioxide is one sanitizer approved by the U.S. Food and Drug Administration (FDA) and the Environmental Protection Agency (EPA). Chlorine dioxide is said to be as powerful as peroxyacetic acid and more economical. It has less of a negative impact on the environment than quaternary ammonium salts, chlorine, or bromine (Stier, 2005). Ozone and peroxyacetic acid have been introduced to decrease the microbial content of fresh-cut produce. Chlorine bleach and bromine create carcinogenic trihalomethanes that enter the environment through drains. Chlorine dioxide is also thought to destroy biofilms and prevent their formation. Chlorine continues to be the most widely used sanitizer in the produce industry in the United States; however, certain countries in the European Union have banned its use. Heat pasteurization is one innovation receiving a great deal of attention in disinfecting fresh-cut apples and melons (Gorny, 2005). The outside of the fruit is pasteurized before cutting, thus reducing the entry of surface microorganisms to the flesh.

In fresh-cut processing operations, ATP bioluminescence is used to monitor food contact surface contamination. It is a rapid means of informing sanitation crews if bacteria and organic material remain on food contact surfaces. This method of testing is, however, very expensive and finds application mainly in large processing operations.

US Food Safety Programs: GAP, GMP, and HACCP. Use of GAP, GMP, and HACCP is voluntary in the fresh produce sector. However, retailers have demanded that their suppliers implement structured food-safety programs under various schemes and these practices are now subject to auditing by a third party.

Good Agricultural Practices (GAPs) describe preventive measures implemented in farming operations to reduce product contamination. The U.S. government in 1998 introduced a "Guide to Minimize the Microbial Contamination of Fresh Fruit and Vegetables". This Guide was used in the United States, as the basis for developing GAPs (FDA, 1998a). The major risk factors identified as needing control included water, manure, worker health and hygiene, sanitary facilities, pest control, and traceback. GAPs provide guidance for food-safety practices in the field. Implementing GAPs provides some assurance to the retailer that product is safe from the farmer's gate. A small number of foodborne illnesses have been associated with contamination at the farm through an ill harvester, or use of unsanitary water.

Current GMPs (FDA, 2003) were made law in U.S. food-processing and handling facilities and this code has been reviewed. The code provides guidance on the safe handling of product in buildings used for food production. In the fresh produce sector, packhouses should follow guidance set out in cGMPs in order to put out wholesome food product.

The principles of HACCP, a system developed by the NASA space program in the 1950s, are now used as the basis for risk assessment and initiating food-safety measures throughout the industry (FDA, 2001). Although most fresh produce farming and packing operations do not have critical control points, the HACCP program forces operations to go through each area in a packhouse, conducting a risk assessment. The assessment may also indicate where GMPs are failing and help in improving GMPs. The use of GAPs, GMPs, and HACCP in the fresh fruit and vegetable industry provide the basic framework for safe products for the consumer.

1.4.2 Improvements in Distribution and Retail

To aid the integration of the supply chain from the source to the end consumer, automated systems are being developed. These data-based systems would provide complete and constant visibility of product, package, purchasing, and distribution of food. Automated warehouse management systems using robotics could increase accuracy of the picking and assembling of pallets for dispatch. With laws restricting the weight of product lifted at food plants, automated systems could also aid manufacturers in compliance with labor laws. As customer demands for fresh product increases, smaller, more frequent deliveries will be required. Automated warehouse management systems would again improve accuracy of product rotation and picking for small deliveries. Automated systems would improve inventory control and real-time tracking of products that pass through a warehouse.

The advent of barcoding has revolutionized the way product movement is managed. A unique barcode is prepared for product harvested on the farm. This barcode stores information on the grower, farm lot number, day harvested, and crop harvested. A barcode placed on a bin of fruit after harvest could then be scanned at the packhouse at intake. This code stays with the bin until it is repacked into consumer size items. If product were stored before repackaging, the warehouse management system would assist in stock rotation based on information received from the barcode. When product is repacked into consumer size packages, a new barcode is placed on the product to assist in tracing finished product back to the field in case of an emergency.

The Uniform Code Council (UCC) takes a global leadership role in establishing and promoting multi-industry standards for product identification and related electronic communication. The purpose of UCC/EAN-128 is to establish a standard way of labeling a package with more information than just a product code. It provides supplemental information such as batch number and "use before" dates. The symbology for this is the UCC/EAN-128, which can be scanned to capture automatically the details related to the location (EDI, 2005). International standards in barcoding include the use of a Global Location Number or GLN. This is a number that identifies any legal, functional, or physical location within a business or organizational entity such as

1. Legal entities – whole companies, subsidiaries, or divisions such as supplier, customer, bank, forwarder;
2. Functional entities – a specific department within a legal entity, such as the accounting department;
3. Physical entities – a particular room in a building, such as a warehouse or warehouse gate, delivery point, or transmission point.

Such GLNs are governed by strict rules to guarantee that each one is unique worldwide. They always have the following features:

1. They are numeric.
2. They have a fixed length of 13 digits, which must be processed in their entirety.
3. They end with a check digit.

The GLNs can be presented in barcode format and physically marked onto trade units to identify the parties involved in the transaction (buyer, supplier), for example (1) transport units (consignor, consignee), (2) physical locations (place of delivery, place of departure).

X-ray technology is another method with great potential in identifying physical contamination in fresh fruits and vegetables. The widely used metal detectors are not able to detect wood chips, plastic, and other nonmetal contaminants. Thus, machine vision systems could help in eliminating foreign contaminants in food. The cost of purchasing equipment and adapting it to fresh-produce operations far exceeds the benefits from using this technology.

Although some in the produce industry are still grappling with the introduction of barcoding systems for traceability programs, other sectors in the food industry have moved on to the new technology of radio-frequency identification (RFID) systems, a technology that uses grain-sized computer chips to track items at a distance. Each tiny chip is connected to an antenna that picks up electromagnetic energy beamed at it from a reader device. When it picks up the energy, the chip sends back its unique identification number to the reader device, allowing the item to be remotely identified. "Spy chips," as they are sometimes called, can beam back information anywhere from a few inches to 30 ft ($\sim$10 m) away. This RFID will be used in the future to trace product. Retailers such as Walmart have started demanding this technology for packaged goods. Within time, this technology could be used to assist in stock rotation and distribution of fresh produce.

Radio frequency identification is not necessarily better than barcodes. The two are different technologies and have different applications, which sometimes overlap. The big difference between the two is that barcodes are a line-of-sight technology. In other words, a scanner has to see the barcode to read it. This means that people usually have to orient the barcode towards a scanner for it to be read. Radio frequency identification, by contrast, does not require line of sight. RFID tags can be read as long as they are within range of a reader. All existing RFID systems use proprietary technology, but many of the benefits of tracking items come from tracking them as they move from one company to another and even one country to another. RFID readers could cost as much as $1000, making it too expensive for small companies. This technology is still cost-prohibitive, but many large companies are creating proprietary supply-chain information systems to manage product flow and traceability. Another challenge would be sending radio waves through high water content fresh produce items (Gorny, 2005).

1.4.3 Traceability for the Industry

Traceability is defined as the "ability to trace the history, application or location of that which is under consideration." The Codex Alimentarius Commission of the Food and Agriculture Organization defines traceback is the ability to track product from the consumer point of purchase back to the grower (Calvin, 2003). The objectives of a good traceability system should be to improve supply management and aid traceback of food-safety and quality parameters. Some companies use their traceability programs as a marketing tool to differentiate them from competitors because of the benefits of the program. In most cases, a well-run traceability program can generate revenue for a company because it improves efficiency at all levels. The source of a safety or quality problem can be isolated and managed using a detailed traceability program. Benefits of traceability also include the fact that unsafe products are not shipped, and the risk of bad publicity, lawsuits, and recalls is reduced. A detailed tracing system would speed corrective actions when food-safety and quality problems are identified. Many large restaurant chains and retailers now expect their suppliers to create traceability systems. These systems are verified through third-party audits of overall food-safety systems. If a product recall or foodborne illness occurs, an investigation is well controlled if a company has a means of retrieving information. This would help in identifying a problem in the production system and point to immediate corrective actions.

Under the FDA Recording and Reporting Rule of the Bioterrorism Act 2002, food companies must provide the FDA with source data, including immediate previous source (IPS) and immediate subsequent recipient (ISR) for every component of a food item (FDA, 2002). U.S. and international food processors, transportation, and distribution companies must keep records if food or biological products are imported or exported from the United States. Companies have developed innovative software databases in information systems with the ability to store detailed crop history and distribution data. Guidelines have to be set so that meaningful data can be stored and retrieved when needed. Systems usually include information on the shipper, date harvested, field, and picker.

The basic requirements of information databases are

1. To quickly isolate affected product in the event of a recall;
2. To track spray intervals to harvest and pesticide application records, if pesticide use is a concern;
3. To keep historical data on fields where crops were grown.

The U.S. Government has developed systems to monitor the incidence of foodborne illness. FoodNet (www.cdc.gov/foodnet/) and PulseNet (http://www.cdc.gov/pulsenet/) are surveillance systems that increase the capability of the entire food supply chain to respond to food-safety problems before more consumers are affected. FoodNet – Foodborne Diseases Active Surveillance Network (FoodNet) – is a collaborative project of the CDC, ten Emerging Infections Program (EIP) sites, and the U.S. Department of Agriculture. PulseNet is a national network of public health and food regulatory agency laboratories coordinated by the Centers for Disease Control and Prevention (CDC). The network consists of state health departments, local health departments, and federal agencies (CDC, USDA/FSIS, and FDA). PulseNet participants perform standardized molecular subtyping (or "fingerprinting") of foodborne disease-causing bacteria by pulsed-field gel electrophoresis (PFGE), which can be used to distinguish strains of organisms of public health significance.

Scientists can trace fresh produce microbial contaminants using genetic data. The global food-safety testing market is expected to reach as much as $415.6 million by the year 2009 (Food Safety Magazine, 2005). The increase in pathogen test sales continues to rise because of many factors, including

1. Consumer demand for safe food;
2. Evolution of bacteria and emergence of new strains;
3. Increased government regulation.

Food testing globally focuses on pathogens, pesticide residues, genetically modified organisms (GMOs), and other contaminants. In the United States, pathogen testing far exceeds pesticide testing in foods. This is different in the European Union, where testing focuses on pesticide residues. The European Union has also required increased documentation of GMO-containing foods and testing for identity preservation may increase in the future.

1.4.4 Effectiveness of Auditing

Customers are reassured when information supplied by companies is verified by third-party certification. Different forms of auditing include self-audits, second-party audits, and third-party audits. Self-audits are used by companies to monitor the efficiencies of in-house programs: food safety, GMP, HACCP, and so on. Self-audits may be used as a tool to monitor execution of QA and QC programs. One setback would be the familiarity of the auditor with the operation and his inability to view errors in the system objectively. A company verifying practices of its suppliers conducts second-party audits. Some bias is removed, because the company would want the best-quality product as raw material and packaging from vendors. Third-party audits are conducted by neutral auditing firms that carry out inspections on behalf of a retailer or other party of interest. The auditing firm does not usually have a stake in the business and so the assessment is felt to be unbiased. In order to be effective, third-party auditors should be knowledgeable about the business they are auditing and well versed in the standard against which they are auditing.

Different auditing and certification protocols exist in different parts of the world. Each protocol may be applied in a specific area of the food chain, but there is no one correct auditing scheme. Nations have not yet united all schemes under one accreditation body. Good Agricultural Practices (GAPs) refer to food-safety guidelines for field operations and each auditing firm has its own verification system for monitoring GAPs. Another scheme for farm safety, EurepGAP, has been developed in Europe. This protocol provides many allowances for environmental and social issues in the growing operation. European retailers who helped develop the EurepGAP auditing scheme ask growers and shippers to subscribe to this auditing program.

Good Manufacturing Practices (GMPs) have been clearly defined by the U.S. FDA for food processing operations (FDA, 2003). Different auditing firms in the United States would interpret GMP guidelines and audit packhouses and processing plants based on their own checklist. The U.K. Food Safety Act 1990 states that all persons involved in supplying food have an obligation to take all reasonable precautions and exercise all due diligence in avoiding failure when presenting food products to the consumer. Retailers are thus obligated to verify technical performance at food-production sites. In 2003, the British Retail Consortium (BRC) proposed a Food Technical Standard applying to enclosed areas for food production. The standard has been updated with version 4, the BRC Global Standard – Food, in June 2005.This standard for food establishments is seen to be a major international standard specifying safety, quality, and operational criteria within a food manufacturing operation. Accredited auditing bodies inspect food establishments against this standard. Schemes such as the BRC's Global Food Standard provide training for auditors based on the set standard. Training ensures uniformity in auditing.

Another example of a certification protocol is the International Organization for Standardization's ISO 9000 series, or "EN 29000," in Europe. In 1987, ISO published a series of five international standards (ISO 9000, 9001, 9002, 9003, and 9004), developed by ISO Technical Committee (TC) 176 on quality systems. This series, together with the terminology and definitions contained in ISO Standard 8402, provides guidance on the selection of an appropriate quality management program (system) for supplier operations. Conformance to ISO 9000 standards is required in purchasing specifications by some companies. The ISO 9000 Standard Series was adopted in the United States as the ANSI/American Society for Quality Control (ANSI/ASQC Q 9000 series). In Europe, the series was adopted by the European Committee for Standardization (CEN) and the European Committee for Electrotechnical Standardization (CENELEC) as the European Norm (EN) 29000 Series. Many countries have national standards that are identical or equivalent to the ISO 9000 Standard Series (Breitenberg, 1992).

Growers use results of third-party audits of GAPs as a tool to demonstrate that effective due diligence programs are in place. When the FDA introduced GAPs for the produce industry, auditing firms used these guidelines to develop extensive auditing programs for the fruit and vegetable industry. Retailers rely on auditing firms to provide them with the assurance they need from their suppliers. However, fruit and vegetable growers in the United States have the following concerns:

1. There is no government regulation of third-party auditing companies.
2. Food auditors from the food-processing industry attempt to apply the same standards of auditing in open fields.
3. Cost of auditing fees is seldom supported by an increase in revenue for growers.
4. Different retailers may demand different auditing programs, because there is no industry standard.

A major retailer, Tesco, has developed the Tesco Nature's Choice standard for growing and packing operations. Walmart has developed the Walmart Global

Procurement Group Audits. Safe Quality Food (SQF 1000), a scheme owned by the U.S. Food Marketing Institute, verifies food safety as well as quality of product. Retailer auditing programs are not necessarily similar and may increase costs for those growers that supply different stores. Microbial food safety is the focus of United States based produce industry audits. However, there are insufficient data to establish actual risks of microbial contamination in the industry. Thus, auditing is based on many assumptions of risk. Even companies that passed audits have been associated with foodborne outbreaks (Calvin, 2003). This shows that food-safety programs allow for a minimization of risk, but not complete elimination of microbial contamination.

The consolidation and globalization of the retail industry has brought a new dimension to the auditing process. It is hoped that the market will consolidate its basic food safety requirements. The U.S. culture of regulatory control of the food industry is different to the European Union, where a culture of private certification exists. In addition, niche markets may make higher demands of suppliers.

Auditing schemes could be made more effective by the following steps:

1. Developing data to assist in objectivity of decision-making in auditing;
2. Testing and evaluating management tools to add objectivity to audits;
3. Introducing controls on accreditation schemes for auditors in all countries;
4. Harmonizing international auditing schemes so they cover major quality and food-safety concerns;
5. Developing product-specific and region-specific auditing schemes;
6. Introducing an international accreditation body using set standards to verify programs of all auditing bodies, thus making the auditing process uniform.

The Global Food Safety Initiative (GFSI) was created in May 2000 at the initiative of a group of major retailers, including METRO AG, Rewe, Edeka, and Globus. The GFSI is a working group within the CIES (The Food Business Forum), an international network of companies in food retailing and food production. In total, 250 major food retail companies are represented in CIES. GFSI is a separate legal entity governed by an Advisory Group. The mission of GFSI is to strengthen consumer confidence in the food bought in retail outlets. GFSI is to develop product-related standards for food safety that cover all links of the food chain. It wants all suppliers to be audited and certified according to a uniform international standard. This will be achieved by implementing and maintaining a scheme to benchmark food-safety standards (for private label products) as well as farm assurance standards. GFSI will facilitate mutual recognition between standard owners and ensure worldwide integrity in the quality and the accreditation of food safety auditors (CIES, 2005). EurepGAP is also leading discussions to harmonize farm auditing schemes worldwide.

1.4.5 Educating the Public

The retail industry drives the introduction of GHPs in the food industry. As the last point of contact in the fresh-produce chain, retailers could play a proactive role in educating the public on safe food-handling practices.

Programs retailers may use to educate the public include the following:

1. Use of membership cards and frequent shopper cards to store data on buying patterns of regular customers;
2. Tailor-made food safety information to send to customers;
3. Posting signs on the sides or handles of shopping carts to deliver food-safety information to the customer before they enter the store;
4. Posting point-of-sale signs, wall hangings, and floor markers;
5. Distributing fliers with information in shopping bags to keep the customer informed and aware of food-safety issues concerning the item purchased;
6. Posting food-safety information on retailer websites;
7. Sponsoring radio/TV and other media programs promoting food safety;
8. Using videos in store to impart knowledge of safe food-handling practices;
9. Using demonstration stations in the store to sample new food items while incorporating food-safety education.

In the United States, Federal Governmental agencies (the FDA, USDA, and CDC) are responsible for setting and monitoring food-safety standards, as well as communicating key food-safety knowledge to consumers. Local government agencies have been busy educating and informing consumers in the areas of food safety. "Fight Bac!" is one government program that was a success and can serve as a model for conveying food-safety knowledge to the consumer (www.fightbac.org). Educational material for consumers must be presented in a convenient way. According to Li-Cohen and Bruhn (2002), consumers who were interested in receiving information found brochures in supermarkets informative. Consumers also preferred it if information was presented in easy-to-follow picture form, instead of wordy material. Information should also be targeted toward children as part of school studies and with special activities. For example, safe food handling could be a required component of any social studies curriculum. It is important to channel food-safety education to school-aged children, the consumers of the future. Educating children in safe food-handling practices would perhaps reduce foodborne illnesses for future consumers.

In 1994, the National Restaurant Association Educational Foundation's (NRAEF) International Food Safety Council created National Food Safety Education Month® as an annual campaign to heighten awareness about the importance of food-safety education. Each year a new theme and training activities are created for the restaurant and food-service industry to reinforce proper food-safety practices and procedures. Training material is accessible on the NRAEF website (www.foodsafety.gov/~fsg/september.html). In addition, promotional materials are available to make it easier to spread the word about the importance of food-safety education.

1.4.6 U.S. Government Intervention

In the United States, Federal Agencies with roles in food-safety include the Food Safety and Inspection Service (FSIS) of the U.S. Department of Agriculture (USDA), the Environmental Protection Agency (EPA), and the Food and Drug Administration (FDA). These agencies monitor imports and have, in the past, surveyed production in foreign countries. The FDA oversees all import inspections, testing for pesticide residues or microbial contamination.

In 1997, President Clinton introduced the U.S. Food Safety Initiative to monitor the safety of all food in the United States. Under this program in 1998, the FDA produced the "Guide to Minimize Microbial Contamination of Food" (FDA, 1998a). This guide was intended to assist domestic and foreign growers, packers, and shippers of fresh produce by increasing awareness of potential hazards and providing suggestions to minimize these hazards. The basic principle in creating this guide was that prevention of microbial contamination was more effective than relying on corrective measures after contamination had occurred. The "Guide" provided broad-based voluntary measures and identified Water, Manure and Municipal Biosolids, Worker Health and Hygiene, Field Sanitation, Packing Facilities, Transportation, and Traceback, as risk factors for product contamination.

In March 1999, the FDA initiated a 1000-sample survey focused on high-volume imported fresh produce. Broccoli, cantaloupe, celery, cilantro, culantro, loose-leaf lettuce, parsley, scallions (green onions), strawberries, and tomatoes were collected and analyzed for *Salmonella* and *E. coli* O157:H7. All commodities except for cilantro, culantro, lettuce, and strawberries were analyzed for *Shigella.* Twenty-one countries were represented in the collection and sampling of fresh produce. Of 1003 samples collected, 4 percent were contaminated with *Shigella*, *Salmonella*, and/or *E. coli* O157:H7. Of the 44 contaminated samples, 35 (80 percent) were contaminated with *Salmonella*, 9 (20 percent) were contaminated with *Shigella*, and none was contaminated with *E. coli* O157:H7 (FDA/CFSAN, 2001c).

Because of the presence of pathogens in samples, 21 companies were placed on detention without physical examination (DWPE). Seven companies were placed on DWPE because *Shigella* was present in one composite, whereas 14 firms were placed on DWPE because of *Salmonella* in two composite samples. Another company was placed on DWPE for the presence of both *Shigella* and *Salmonella.* The FDA decided to remove the product, shipper, grower, or importer from DWPE only if evidence showed unsanitary conditions had been resolved.

In March 2000, the FDA initiated a 1000-sample survey focused on high-volume domestic fresh produce (the Domestic Produce 1000 Sample Survey, FDA/CFSAN, 2003). Cantaloupe, celery, cilantro, loose-leaf lettuce, parsley, scallions (green onions), strawberries, and tomatoes were collected and analyzed for *Salmonella* and *E. coli* O157:H7. In addition, cantaloupe, celery, parsley, scallions, and tomatoes were analyzed for *Shigella.* This survey was the domestic complement to the FY 1999 Imported Produce Survey. Of 1028 domestic samples that were collected and analyzed, 99 percent were not contaminated with *Shigella*, *Salmonella*, or *E. coli* O157:H7. Eleven samples (1 percent of the total number sampled) were contaminated. Of the 11 contaminated samples, six (55 percent) were contaminated with *Salmonella* and five (45 percent) were contaminated with *Shigella.* For regulatory follow-up, domestic samples that were found to contain pathogens were reported to the collecting district and the Center for Food Safety and Applied Nutrition's (CFSAN) Case Processing Contact. The domestic produce was either reconditioned or destroyed. Firms with contaminated produce were encouraged to conduct voluntary recalls. The FDA conducted follow-up investigations at three farms to determine potential sources of contamination.

In 2001, Congress provided the U.S. Department of Agriculture (USDA) Agricultural Marketing Service (AMS) with $6.235 million to implement a Microbiological Data Program (MDP) (USDA/AMS, 2003). Congress has followed this with appropriations of $6 million annually to continue the program. The aim of this program was to collect data on the incidence and identification of targeted foodborne pathogens and indicator organisms on fresh fruits and vegetables in the United States. The AMS formed a partnership with ten States, and in 2002 the MDP program tested five commodities: cantaloupe, celery, leaf lettuce, romaine lettuce, and tomatoes for *E. coli* and *Salmonella* sp. because of their public health significance. Of the 10,317 samples screened, MDP identified only 0.62 percent with virulence factors and three *Salmonella* spp. isolates from domestic leaf lettuce (0.03 percent of overall rate). The exercise, though not conclusive, showed new trends in fresh produce contamination. No significant differences were found in contamination rates between imported and domestic produce.

The FDA, in October 2004, developed the "Produce Safety Action Plan" to further minimize foodborne illness associated with the consumption of fresh produce (FDA, 2004). The plan was aimed at minimizing microbial food-safety hazards (bacteria, viruses, and parasites) in products from the United States and abroad. The Action Plan focused on all steps in the food chain from the farm to the home. Specific areas included farm production, packing, processing, transportation, distribution, and preparation of minimally processed product. The major objective of the plan was to reduce the number of illnesses per outbreak associated with fresh produce.

Forty-eight percent of CFSAN's accomplishments in 2004 focused on a concern for food security and safety (Carrington and Cianci, 2005). The Bioterrorism Act of 2002 was signed into law on June 12, 2002. In 2004, two segments of this law were implemented – Administrative Detention and Prior Notice. Final rules for Food Facility Registration, Prior Notice of Imported Foods, and Record Keeping were published subsequently. When negotiating access to other markets with different standards, the United States uses the rules set out by the Agreement on the Application of Sanitary and Phytosanitary (SAPS) Measures, established by the World Trade Organization in 1995.

The FDA continues to review food regulations such as cGMPs to ensure they are up to date. In 2001, the FDA approved the Final Rule, making HACCP mandatory for all juice processing operations. A critical control point (CCP) of the pasteurization step was considered the most effective means of destroying potential pathogens in juice (FDA, 2001).

1.4.7 Use of Surrogates and Indicators in Food Safety Research

Researchers have used surrogates and indicator microorganisms as tools in assessing fresh fruit and vegetable safety and the effectiveness of microbial control measures. FDA/CFSAN (2001b) defines an indicator organism as "a microorganism or group of microorganisms that indicate a food has been exposed to conditions posing an increased risk that food may be contaminated with a pathogen or held under conditions conducive for pathogen growth." Indicator analyses are sometimes used to determine the effectiveness of sanitation programs. An indicator could be a specific microorganism, a metabolite, fragment of DNA, or other substance. For a detailed review of indicators, the reader is referred to Ray (1989), Mossel et al. (1995), Jay (1996, 2000), Smoot and Pierson (1997), and Buchanan (2000). The indicator should be present in the food when the target pathogen is present, should have similar growth patterns as the target, and be easy to detect and quantify.

Surrogate organisms are a unique type of indicator. They mimic growth and survival patterns of a pathogen and can help in studying what occurs with a pathogen during handling and storage. Surrogates are used to study the effects and responses to specific processing treatments. In the fresh fruit and vegetable industry, surrogates could be used to determine the effectiveness of a produce wash. They are usually prepared in a laboratory and introduced onto the produce item. They may also be microorganisms occurring naturally on the produce item (FDA/CFSAN, 2001b). Indicators are naturally occurring, whereas surrogates are introduced artificially on the food item for the study. Surrogates are frequently used for research as opposed to pathogens, because scientists need to prevent the release of harmful organisms in a production environment or, in the case of fresh produce, into a field. However, isolated pilot facilities have used actual pathogens for controlled research.

Mitigation measures in the fresh produce chain are mostly based on risk assessment and making an educated guess as to what measures would reduce the risk. Indicators and surrogates could be used to assure safety for individual produce items in all stages of the fresh produce chain: growing, harvesting, processing, handling, storage, distribution, and retail. These microorganisms may be used to determine the effectiveness of a water treatment or composting on pathogen reduction. If an indicator is in low concentration or not present, this could mean a treatment would be effective in reducing growth of the target organism. It is difficult to select the right target organisms, as well as an indicator organism that would relate to the target (Busta et al., 2003).

A surrogate should have the following main characteristics (FDA/CFSAN, 2001b):

1. It should be a nonpathogenic strain of the target organism.
2. It should demonstrate similar behavior as the target organisms in the environment studied.

3. It should be easy to prepare and stable.
4. It should exhibit enumeration, sensitivity, and inexpensive detection means.
5. It should attach to the produce surface at a similar rate as the target organism.
6. It should exhibit genetic stability to allow the test to be reproduced independently.

The use of surrogate organisms started in the low-acid canning industry. The thermophilic organism *Bacillus stearothermophilus* was used as a surrogate to establish processing conditions to destroy *Clostridium botulinum* spores (FDA/CFSAN, 2001b). Gram-positive and gram-negative bacteria, viruses, and protozoan cysts are pathogens associated with produce and foodborne outbreaks. Strains of *Listeria innocua* have served as nonpathogenic surrogates of the pathogen *Listeria monoytogenes* (Corte et al., 2004).

Worobo (1999) used a nonpathogenic surrogate microorganism, *E. coli* ATCC 25922, with the same UV sensitivity/resistance as three pathogenic strains of *E. coli* O157:H7, to evaluate effectiveness of UV irradiation to sanitize apple cider. Numerous strains of microorganisms were tested and differences in the response to ultraviolet light were observed. Because ATCC 25922 showed almost identical UV sensitivity to *E. coli* O157:H7, it was used as the surrogate microorganism to test additional production units. This surrogate was used to validate each *CiderSure* unit (a measure of UV irradiation) to ensure its compliance with the regulated 5-log reduction of microorganisms for juice pasteurization.

Use of indicators and surrogates can be the most scientific means of generating data to validate the effectiveness of microbial control measures. Additional research is needed to identify specific surrogates and indicators for ready-to-eat and fresh-cut produce, thus helping to validate decontamination steps in the process. It is unknown if pathogens would develop increased resistance as a stress response when subjected to intervention steps. In addition, the country of origin and packing environment may induce a stress response (FDA/CFSAN, 2001b). Protocols have to be developed to accurately measure and retrieve indicator microorganisms after a stress is introduced. For an extensive list of research needs in the use of indicators and surrogates, the reader is referred to FDA/CFSAN (2001c).

Scientists raise concerns that researchers are not using established or comparative studies to validate their surrogates or indicator strains with the appropriate pathogen. The number of published works on attenuated pathogens has increased in several laboratories. However, researchers presume that elimination of virulence traits or entire plasmids has no impact on stress tolerance, phenotypes, competitiveness, or environmental persistence. These concerns must be addressed in future research.

1.5 CHALLENGES FOR THE GLOBAL FRESH PRODUCE INDUSTRY

1.5.1 Trends in Consumption and Globalization of the Industry

All countries share responsibility for food safety with the increase in international food trade. Consumers now benefit from worldwide trade with year-round supplies, good quality, lower prices, and a wide variety of food (Buzby and Unnevehr, 2003). Food safety standards may sometimes present technical barriers to trade, but improvements in standards are likely to extend worldwide with the spread of private and governmental control. In the past, foodborne outbreaks were localized, but now, with increase in international trade, low-level contamination could possibly spread to many regions and even cross borders (Tauxe, 1997). Trade volumes are increasing for fresh, minimally processed, or high-value food, with an increase in imports from developing countries needed to fill the gaps (Hooker and Caswell, 1999; Lin et al., 1999). No scientific evidence has shown imported fresh produce to pose any greater food safety threat than U.S. grown products. In addition, less developed countries are acquiring higher food-safety standards to assist in entering new markets. Both Government agencies and private retailers and importers are driving this effort to increase food-safety standards. The higher standards do not, however, give complete protection from disease or injury.

Branded products, if perceived to be safe, could improve a firm's image and international competitiveness (Buzby and Unnevehr, 2003). Private retailers and importers are strongly motivated to prevent a food-safety crisis, because their companies may suffer from loss of reputation, plants closing for cleanup, reduced stock prices, lawsuits, or higher insurance premiums. A crisis could affect an entire industry, as seen in the outbreak associated with Guatemalan raspberries (Calvin et al., 2002). Large organizations are now promoting vertical integration, controlling commodities at two or more stages of production (Martinez and Reed, 1996; Caswell and Henson, 1997). Private companies have promoted the following:

1. Self-regulation;
2. Vertical interaction;
3. Third-party certification;
4. Use of HACCP principles in risk assessment;
5. Application of voluntary GAPs in growing operations.

Government regulation also plays an important role in addition to the private sector. Industrialized nations saw the development of new food-safety regulation in the 1990s with improvements in science and increased awareness of food-safety risks. Regulatory agencies in industrialized nations were seen to follow main trends, forming agencies to focus only on food safety. Risk analysis was used to design guidelines for the industry and a "farm-to-table" approach has been used to address food-safety hazards. Some government agencies have adopted a HACCP-based system for new regulation of microbial pathogens in food. Newly identified hazards have encouraged adoption of more stringent standards and market performance has improved through the government providing the industry with more food safety information (Buzby and Unnevehr, 2003).

Trade disputes over food safety may be persistent, but are manageable. Food safety challenges vary with each commodity. Concerns differ, depending on the perishability of the product, the nature of human health risks, the possible link of the food-safety issue with productivity, and the extent of vertical coordination and cooperation among stakeholders.

In the United States, fresh produce growers and shippers form many trade networks to promote year-round sales of once seasonal items. Because of improvements in storage technology, transportation, and communication, fresh produce is obtained from many new areas. Tariffs on many agricultural products have been eliminated, for example, the Caribbean Basin Initiative and the Andean Trade Preference Act (Calvin, 2003). Nontraditional suppliers now have increased U.S. market share – for example, asparagus from Peru (increased from 10 percent in 1990 to 47 percent in 2001). Cantaloupe from Costa Rica and Guatemala increased from 17 percent in 1990 to 60 percent in 2001 of all cantaloupe imports (Calvin, 2003).

Consumers are more frequently eating at food-service establishments. Even food-production practices have changed, leading to a host of new food-safety risks for the industry to manage. A marked increased in cut, ready-to-eat fruit and vegetables increases the challenge of preparing safe food. Consumers are demanding convenience food items to suit the hectic pace of the life we now lead. Washing, cutting, and preparing noncooked vegetables increase the potential for product contamination. This could increase the potential for more foodborne outbreaks because food handling is increased. Introduction of mandatory HACCP in juice processing in the United States was one move of the government to control food safety in the sector (FDA, 2001). Catering operations are monitored by food inspectors, whereas fresh-cut processing operations are audited at the request of retailers and restaurant chains. Food-safety programs in fresh-cut operations, however, are voluntary and not under mandatory regulation by government bodies.

Food-safety disputes could arise in trade with the appearance of new hazards or increase in volumes from new sources. Rising standards or changing food-safety regulation in industrialized countries would create new challenges for trade from developing nations. Different countries would have unique ideas with respect to consumer risk preferences. Nonscience issues also play a role in regulatory decision-making and this would vary from one country to the next (Buzby and Unnevehr, 2003). Growth in demand and increased access to markets will continue to expand food trade. Improvements in food safety are essential to consumer welfare and both producers and consumers have a stake in seeing the industry make great strides in this area.

1.5.2 International Oversight: WHO, European Commission on Health and Consumer Protection Directorate, Codex Alimentarius

The Codex Alimentarius Committee on Food Hygiene developed two codes of hygienic practice for fruits and vegetables. One focuses on primary production and the other on precut, ready-to-eat fruits and vegetables. The code of "Hygienic Practices for Fresh Fruits and Vegetables" includes an Annex on "Ready-to-Eat Fresh Pre-cut Fruits and Vegetables" and on "Sprout Production". The World Health Organization (WHO) was also active in developing guidance for the fruit and vegetable industry with the preparation of the document WHO/FSF/FOS/98.2 on surface decontamination of fruits and vegetables eaten raw (Beuchat, 1998; WHO, 2002). The European Commission on Health and Consumer Protection Directorate called on a Scientific Committee on Food to develop a risk profile on the microbiological contamination of fruits and vegetables eaten raw. This Committee developed a paper indicating that the proportion of reported food-borne outbreaks associated with fruits and vegetables is low (European Union Commission, 2002).

EU production of fruits and vegetables in 1998 was estimated at 76 million tons. Spain and Italy were the main producing countries, with 20 million tons each per annum, followed by France and Greece. The European Union also imports more fruits and vegetables than other nations through internation partnerships, as well as importing tropical fruits into the region. Cumulative impacts of food-related health crises have shaken consumer confidence in the safety of food products. The European Union has approached this challenge by putting in place a comprehensive strategy to restore people's belief in the safety of their food "from the farm to the fork." The European Union adopted general principles of food safety in a Regulation called the General Food Law introduced in 2002: Regulation (EC) No. 178/2002. This Law revised EU food safety legislation, with emphasis on feed because feed contamination had been at the root of major food scares. Under this Law, it became compulsory from January 1, 2005, for food and feed businesses to guarantee that all foodstuffs, animal feed, and feed ingredients be traceable right through the food chain. The General Food Law was supplemented by targeted legislation on other food-safety issues, such as use of pesticides, food supplements, colorings, antibiotics, and hormones in food production. In addition, the Law outlined stringent procedures on release, marketing, labeling and traceability of crops and foodstuffs containing genetically modified organisms (GMOs). Updated rules on hygiene took effect on January 1, 2006, with the introduction of Regulation (EC) No. 852/2004 on the hygiene of foodstuffs as well as other specific rules.

The European Food Safety Authority (EFSA), headquartered in Parma, Italy, plays a central role in providing scientific data for food safety. The EFSA can cover all stages of food production and supply, from primary production to the safety of animal feed, right through to the supply of food to consumers. It can also look into the properties of nonfood and feed GMOs and nutrition issues.

The EFSA provides the European Commission with independent, scientific advice that is also made public to enable it to be fully open to scrutiny. It provides input when legislation is being drafted and advice when policymakers are dealing with a food scare.

The Commission enforces EU feed and food law, seeing that EU legislation has been incorporated into Member State law. The Commission also checks compliance through reports from Member States and other countries and through on-the-spot inspections in the EU and abroad. The Commission's Food & Veterinary Office (FVO) based at Grange in Ireland conducts inspections. The FVO can check individual food production plants, but its main task is to check that EU governments and those of other countries have the necessary machinery for verifying their own food producers.

The Agreement on the Application of Sanitary and Phytosanitary (SAPS) Measures, established by the World Trade Organization in 1995, developed from the 1986–1993 Uruguay Round of Multilateral Trade Negotiations. The Agreement had a major focus on food-safety regulation and could be used in settling disputes and restrictions on trade. WTO members recognize the principles of transparency, equivalence, science-based measures, and harmonization (Buzby and Unnevehr, 2003). Codex Alimentarius Commission (Codex) provides standards for human health measures. The International Office of Epizootics (OIE) provides international standards for animal and human health measures, and the International Plant Protection Convention (IPPC), measures for plant health.

1.6 SUMMARY

Increase in global trade and organic farming expose consumers to produce with limited information on microbial safety. For example, limited data exist on the use of microbes for biocontrol of pests. In addition, chemical decontaminants and their impact on microbial food safety is not well known. There are no clear trends in consumption of specific categories of whole fruits and vegetables. However, convenience food items like cut and ready-to-eat fresh fruit and vegetables are increasing. This increase in handling could also increase the potential for product contamination. Water quality is a major factor affecting product contamination in the farm, packhouse, processing facility, and in the home.

Sprouted seeds and fruit juices have been associated with foodborne outbreaks and special legislation provides extra control measures to protect these products. Good Agricultural Practices and Good Handling Practices are thought to be the basis for safe production of fruits and vegetables. The industry relies on proactive systems to reduce risks during production and handling, because decontamination steps have limited effects on safety. There are insufficient scientific data to quantify actual risks at each stage in the fresh produce chain, but with increased research, industry can gain insight into developing adequate preventive measures.

Foodborne illnesses associated with fresh produce are low compared to products of animal origin. Detailed traceability programs can improve the epidemiological investigation and identification of factors causing foodborne outbreaks. These investigations would then lead to the control of the causative agent. Because only a few developed nations have surveillance programs on foodborne pathogens in place, it is not easy to compare outbreaks among many countries. An international effort should take place to introduce surveillance systems in densely populated nations. In addition, scientists should collect data on the potential for organic farming methods to contaminate fresh fruit and vegetables.

Governments could play a role in creating national regulations that influence the safe production of fruits and vegetables. The globalization of the food supply requires a concerted effort from all stakeholders, with preventive measures in place from farm to fork to protect the consumer. This publication focuses on the U.S. fresh produce industry and provides a summary of microbial hazards from production to consumption. Through identification of the hazards and review of mitigation measures, fresh produce suppliers can continue to learn and provide consumers everywhere with safe and healthy fresh fruit and vegetables.

REFERENCES

B.A. Annous, G.M. Sapers, A.M. Mattrazzo, and D.C.R. Riordan, Efficacy of washing with a commercial flat-bed brush washer, using conventional and experimental washing agents, in reducing populations of *Escherichia coli* on artificially inoculated apples. *J. Food Prot.*, 64 (2), 159–163 (2001).

L.R. Beuchat, Pathogenic microorganisms associated with fresh produce. *J. Food Prot.*, 59, 204–206 (1996).

L.R. Beuchat, *Surface Decontamination of Fruits and Vegetables Eaten Raw: A Review.* Food Safety Unit, World Health Organization: Geneva, Switzerland, WHO/FSF/FOS/98.2, 1998.

L.R. Beuchat and J.-H. Ryu, Produce handling and processing practices. *Emerg. Infect. Dis.*, 3 (4), 1–12 (1997). Available at http://www.cdc.gov.gov/ncidod/eid/vol3no4, accessed October 2005.

K.A. Blaine and D.A. Powell, Microbial food safety considerations for organic produce production: An analysis of Canadian organic production standards compared with US FDA guidelines for microbial food safety. *Food Protec. Trends.*, 24 (4), 246–252 (2004).

M. Breitenberg, *Questions and Answers on Quality, the ISO 9000 Standard Series, Quality System Registration, and Related Issues.* National Institute of Standards and Technology, Revised July 1992. Available at http://ts.nist.gov/ts/htdocs/210/gsig/ir4721. htmGaithersburg, MD 20899, accessed October 2005.

R.L. Buchanan, Acquisition of microbiological data to enhance food safety. *J. Food Prot.*, 63 (6), 832–838 (2000).

F. Busta, T. Suslow, M. Parish, L. Beuchat, J. Farber, E. Garrett, and L. Harris, Use of indicators and surrogate microorganisms for the evaluation of pathogens in fresh and fresh-cut produce. *Comp. Rev. Food Sci. Food Safety*, 2 (S1), 179–185 (2003).

J.C. Buzby and L. Unnevehr, "Introduction and Overview," in J.C. Buzby, Ed., *International Trade and Food Safety*, Chapter 1. Economic Research Service/USDA: Washington, D.C., 2003, pp. 1–9.

Cable News Network (CNN), *E. coli* poisoning leads to Odwalla juice recall. November 1, 1996. Web posted at 6:25 p.m. est. From Correspondent Eugenia Halsey. Available at http://www.cnn.com/HEALTH/9611/01/e.coli.poisoning/, accessed October 2005.

Cable News Network (CNN), Hepatitis outbreak spreads fear. Saturday, November 15, 2003, Posted 6:40 p.m. est. (23:40 GMT). Available at http://www.cnn.com/2003/HEALTH/conditions/11/15/hepatitis.outbreak.ap/, accessed October 2005.

L. Calvin, "Produce, Food Safety, and International Trade." Response to U.S. Foodborne Illness Outbreaks Associated With Imported Produce," in J.C. Buzby, Ed., *International Trade and Food Safety*, Chapter 5. Economic Research Service/USDA: Washington, D.C., 2003, pp. 74–96.

L. Calvin, W. Foster, L. Solorzano, D. Mooney, et al., "Response to a Food Safety Problem in Produce: A Case Study of a *Cyclosporiasis* Outbreak," in B. Krissoff, M. Bohman and

J. Caswell, Eds., *Global Food Trade and Consumer Demand for Quality*. Kluwer Academic/Plenum Publishers, New York, 2002.

D. Carrington and S. Cianci, "FDA's Food Program Priorities: A Look Back at 2004 Accomplishments," Regulatory Report, February/March, 11 (1), 13–17 (2005).

J.A. Caswell and S.J. Hensen, "Interaction of Private and Public Food Quality Control Systems in Global Markets," in R.J. Loader, S.J. Henson, and W.B. Traill, Eds., *Proceedings of Globalization of the Food Industry: Policy Implications*, 1997, pp. 217–234.

Centers for Disease Control and Prevention (CDC), Outbreaks of *Shigella sonnei* infection associated with eating fresh parsley – United States and Canada, July–August 1998. *MMWR*, 48 (14), 285–289 (1999).

Centers for Disease Control and Prevention (CDC), Preliminary FoodNet data on the incidence of infection with pathogens transmitted commonly through food – selected sites, United States, 2003. *MMWR*, 53 (16), 338–343 (2004).

CIES, The Business Forum (CIES), Global Food Safety Initiative (GFSI), 2005. Available at http://www.ciesnet.com/programmes/foodsafety/global_food/main.htm, accessed October 2005.

F.V. Corte, S.V. De Fabrizio, D.M. Salvatori, and S.M. Alzamora, Survival of *Listeria innocua* in apple juice as affected by vanillin or potassium sorbate. *J. Food Safety*, 24 (1), 1–16 (2004).

W.D. Dingman, Growth of *Escherichia coli* O157:H7 in bruised apple (*Malus domesticus*) as influenced by cultivar, date of harvest and source. *Appl. Environ. Microbiol.*, 66, 1077–1083 (2000).

Electronic Data Interchange (EDI), The Global Language of Business, 2005. Available at http://www.ean-int.org/locations.html. http://www.adams1.com/pub/russadam/ucc128.html, accessed October 2005.

European Union Commission, "Risk Profile on the Microbiological Contamination of Fruits and Vegetables Eaten Raw," Report of the Scientific Committee on Food. Health and Consumer Protection Directorate-General, Brussels, Belgium, 2002.

Food and Drug Administration (FDA), Guide to Minimize Microbial Food Safety Hazards for Fresh Fruits and Vegetables, 1998a. Available at http://vm.cfsan.fda.gov/~dms/prodguid.html, accessed October 2005.

Food and Drug Administration (FDA), Guide to minimize microbial food safety considerations for organic produce production: An analysis of Canadian organic production standards compared with US FDA Guidelines for Microbial Food Safety. *Food Protec. Trends*, 24 (4), 246–252 (1998b).

Food and Drug Administration (FDA), Hazard Analysis and Critical Control Point (HACCP); Procedures for the Safe and Sanitary Processing and Importing of Juice; Final Rule. *Fed. Regist.*, 66, 6138–6202 (2001).

Food and Drug Administation (FDA), The Bioterrorism Act of 2002, 2002. Available at http://www.fda.gov/oc/bioterrorism/bioact.html, accessed October 2005.

Food and Drug Administration (FDA), Code of Federal Regulations, Title 21 Part 110. Current Good Manufacturing Practice in Manufacturing, Packing or Holding Human Food, 2003. Available at http://www.cfsan.fda.gov//~lrd/cfr110.html., accessed February 2006.

Food and Drug Administration (FDA), Produce Safety. From Production to Consumption: 2004 Action Plan to Minimize Foodborne Illness Associated with Fresh Produce

Consumption, October 2004. Center for Food Safety and Applied Nutrition/ Office of Plant and Dairy Foods, Maryland. Available at http://www.cfsan.fda.gov/~dms/prod-pla2.html, accessed October 2005.

FDA/CFSAN, Analysis and Evaluation of Preventative Control Measures for the Control and Reduction/Elimination of Microbial Hazards on Fresh and Fresh-Cut Produce, 2001a. Available at http://www.cfsan.fda.gov/~comm/ift3-7.html, accessed January 2003.

FDA/CFSAN, The Use of Indicators and Surrogate Microoganisms for the Evaluation of Pathogens in Fresh and Fresh-Cut Produce, in Analysis and Evaluation of Preventive Control Measures for the Control and Reduction/Elimination of Microbial Hazards on Fresh and Fresh-Cut Produce, 2001b. Available at http://www.cfsan.fda.gov/~comm/ift3-7.html, accessed October 2005.

Food and Drug Administration, Center for Food Safety and Applied Nutrition, Office of Plant and Dairy Foods and Beverages (FDA/CFSAN), FDA Survey of Imported Fresh Produce, FY 1999 Field Assignment, 2001c. Available at http://www.cfsan.fda.gov/~dms/prodsur6.html, accessed April 20, 2004.

FDA/CFSAN, FDA Survey of Domestic Fresh Produce, 2003. Available at http://vm.cfsan.fda.gov/~dms/prodsur10.html, accessed April 20, 2004.

Food Safety Magazine, Global food safety testing market to reach $415.6 million. News Bites. *Food Safety Mag.*, 11 (1), 8 (2005).

J.G. Gorny, A Fresh Look at Produce Safety. *Food Safety Mag.*, 11 (1), 70–76 (2005).

N.H. Hooker and J.A. Caswell, A framework for evaluating non-tariff barriers to trade related to sanitary and phytosanitary regulation. *J. Agric. Econ.*, 50, 2 (1999).

C.B. Jaquette, L.R. Beuchat, and B.E. Mahon, Efficacy of chlorine and heat treatment in killing *Salmonella stanley* inoculated onto alfalfa seeds and growth and survival of the pathogen during sprouting and storage. *Appl. Environ. Microbiol.*, 62, 2212–2215 (1996).

J.M. Jay, "Indicators of Food Microbial Quality and Safety," in *Modern Food Microbiology*, 5th edn. Aspen P., Gaithersburg (MD); 1996, pp. 387–407.

J.M. Jay, "Indicators of Food Microbial Quality and Safety. Indicators of Food Microbial Quality and Safety." in *Modern Food Microbiology*, 6th edn. Aspen. P., Gaithersburg (MD); 2000, pp. 387–406.

A.E. Li-Cohen and C.M. Bruhn, Safety of consumer handling of fresh produce from the time of purchase to the plate: A comprehensive consumer survey. *J. Food Prot.*, 65 (8), 1287–1296 (2002).

B.-H. Lin, J. Guthress, and E. Frazão, "Nutrient Contribution of Food Away from Home," in *America's Eating Habits: Changes and Consequences*, Chapter 12. U.S. Dept. Agr. Econ. Res. Serv. AIB – 750, Apr. 1999. Available at www.ers.usda.gov/publications/aib750, accessed October 2005.

R. Linton, Food Safety Hazards in Foodservice and Food Retail Establishments. Available at http://www.foodsci.opurdue.edu/Publications/FoodSafety/food_safety-2.html, accessed October 2005.

S. Martinez and A. Reed, From Farmers to Consumers: Vertical Coordination in the Food Industry, 1996. U.S. Dept. Agr., Econ. Res. Serv., AIB-720, Washington, D.C., June 1996.

P. Millner, Composting: Improving On a Time-Tested Technique. *Agric. Res. Mag.*, 51, 8 (2003).

Ministry of Health and Welfare of Japan, National Institute of infectious Diseases and Infectious Disease Control Division. Verocytotoxin Producing *Escherichia coli* (Entero-Haemorrhagic *E. coli*) in Infection, Japan, 1996–June 1997. Infectious Agents Surveillance Report 18; 153–154 (1997).

J.F. Moncrief and P.R. Bloom, Generic Environmental Impact Statement for Animal Agriculture in Minnesota. Soils and Manure Issues. Available at www.eqb.state.mn.us/geis/TWP_Soil.pdf, accessed October 2005.

D.A.A. Mossel, J.E.L. Corry, C.B. Struijk, and R.M. Baird, *Essentials of the Microbiology of Foods: A Textbook for Advanced Studies.* John Wiley & Sons, Chichester, UK, 1995, pp. 287–289.

C. Nguyen-the and F. Carlin, The microbiology of minimally processed fresh fruits and vegetables. *Crit. Rev. Food Sci. Nutr.*, 34, 371–401 (1994).

C. Nguyen-the and F. Carlin, "Fresh and Processed Vegetables," in B.M. Lund, T.C. Baid-Parker, and G.W. Gould, Eds., *The Microbiological Safety and Quality of Foods.* Aspen Publication, Gaithersburg, 2000, pp. 620–684.

Public Health Agency Canada, Waterborne outbreak of gastroenteritis associated with a contaminated municipal water supply, Walkerton, Ontario, May–June, 2000, Canada Communicable Disease Report, Vol. 26; 20. Available at http://www.phac-aspc.gc.ca/publicat/ccdr-rmtc/00vol26/dr2620eb.html, accessed October 2005.

B. Ray, *Injured Index and Pathogenic Bacteria: Occurrence and Detection in Foods, Water and Feeds.* CRC Press, Boca Raton, FL, 1989.

R. Redemann, Basic elements of effective food plant cleaning and sanitizing. *Food Safety Mag.*, 11 (2), 22–28 (2005).

D.C.R. Riordan, G.M. Sapers, and B.A. Annous, The survival of *E. coli* O157:H7 in the presence of *Penicillium expansum* and *Glomerella cingulata* in wounds on apple surfaces. *J. Food Prot.*, 63 (12), 1637–1642 (2000).

G.M. Sapers, R.L. Miller, M. Jantschke, and A.M. Mattrazzo, Factors limiting the efficacy of hydrogen peroxide washes for decontamination of apples containing *Escherichia coli. J. Food Sci.* 65 (3), 529–532 (2000).

G.M. Sapers, R.L. Miller, V. Pilizota, and A.M. Mattrazzo, Anti-microbial treatments for minimally processed cantaloupe melon. *J. Food Sci.*, 66 (2), 345–349 (2001).

S. Sivapalasingam, E. Barrett, A. Kimura, S. Van Duyne, W. De Witt, M. Ying, A. Frisch, Q. Phan, E. Gould, P. Shillam, V. Reddy, T. Cooper, M. Hoekstra, C. Higgins, J.P. Sanders, R.V. Tauxe, and L. Slutsker, Multistate outbreak of *Salmonella enterica* serotype *Newport* infection linked to mango consumption: impact of water-dip disinfestation technology. *Clin. Infect. Dis.*, 37 (12), 1585–1590 (2003).

L.M. Smoot and M.D. Pierson, "Indicator microorganisms and microbiological criteria," in M.P. Doyle, L.R. Beuchat, T.J. Montville, Eds., *Food Microbiology: Fundamentals and Frontiers.* American Society for Microbiology, Washington; 1997, pp. 66–80.

R. Stier, Ten reasons why you should be using chlorine dioxide. Food Safety Insider: Sanitation Systems and Solutions. *Food Safety Mag.*, 11 (1), 50 (2005).

R.V. Tauxe, Emerging foodborne diseases: an evolving public health challenge. *Emerg. Infect. Dis.*, 3, 425–434 (1997).

United States Department of Agriculture/Agricultural Marketing Service (USDA/AMS), Microbiological Data Program (MDP). Science and Technology Programs, 2003. Available at http://www.ams.usda.gov/science/MPO/Mdp.htm, accessed October 2005.

WHO, *WHO Global Strategy for Food Safety.* World Health Organization, Geneva, 2002. Available at http://www.who.int/foodsafety/publications/general/en/strategy_en.pdf, accessed October 2005.

R.W. Worobo, *Efficacy of the CiderSure 3500 Ultraviolet Light Unit in Apple Cider.* Department of Food Science & Technology, Cornell University, 1999. Available at http://vm.cfsan.fda.gov/~comm/cidwworo.html, accessed October 2005.

S. Zhang and J.M. Farber, The effects of various disinfectants against *Listeria monocytogenes* on fresh-cut vegetables. *Food Microbiol.*, 13, 311–321 (1996).

CHAPTER 2

The Epidemiology of Produce-Associated Outbreaks of Foodborne Disease

LYNETTE M. JOHNSTON

Department of Food Science, North Carolina State University, Raleigh, NC, USA

CHRISTINE L. MOE

Department of International Health, Rollins School of Public Health, Emory University, Atlanta, GA, USA

DEBORAH MOLL

Health Studies Branch, Division of Environmental Hazards and Health Effects, National Center for Environmental Health, Centers for Disease Control and Prevention, Chamblee, GA, USA

LEE-ANN JAYKUS

Department of Food Science, North Carolina State University, Raleigh, NC, USA

Microbial Hazard Identification in Fresh Fruit and Vegetables, Edited by Jennylynd James

2.1 INTRODUCTION

The food supply in the United States is considered one of the safest in the world. In fact, recent reports indicate that from 1996 to 2003 the estimated incidence of several important foodborne diseases in the United States, including those caused by *Escherichia coli* O157:H7, *Salmonella*, *Campylobacter*, and *Cryptosporidium*, have declined (CDC/Foss, 2004). Regardless, in the United States alone, it is estimated that 76 million persons contract foodborne illness each year, with an associated 325,000 hospitalizations and 5000 deaths (Mead et al., 1999).

The epidemiology, or the occurrence and distribution, of foodborne disease in a population is the result of complex interactions among environmental, cultural, and socioeconomic factors (Potter et al., 1996). As new foods and those from alternative sources become available, new opportunities for transmission of foodborne disease often follow. Other factors that contribute to the dynamics of foodborne disease include changes in human demographics and behavior, as well as new food production and processing technologies. Some infectious agents have been either newly described or newly associated with foodborne transmission routes (Hedberg et al., 1994; Altekruse et al., 1997; Tauxe, 1997). Examples such as *Campylobacter jejuni*, *Escherichia coli* O157:H7, *Listeria monocytogenes*, and *Cyclospora cayetanensis* are illustrative of this phenomenon.

Fresh produce can be a vehicle for the transmission of bacterial, parasitic, and viral pathogens capable of causing human illness. Although low, the proportion of U.S. foodborne illness associated with both domestic and imported fresh fruits and vegetables has increased over the last several decades. In fact, the median number of reported produce-associated outbreaks increased from two outbreaks per year in the 1970s, to 16 per year in the 1990s (Sivapalasingam et al., 2004). In the decade between 1970 and 1980, 0.7 percent of reported outbreaks were associated with produce; however, in the 1990s, fresh produce was associated with over 5 percent of reported foodborne outbreaks (Olsen et al., 2000; Sivapalasingam et al., 2004). The majority of these were caused by pathogens transmitted by fecal–oral routes (Table 2.1). Produce outbreaks also accounted for an increased proportion of foodborne illnesses among all outbreaks, increasing from 1 percent (708 of 68,712 cases) in the 1970s to 12 percent (8808 of 74,592 cases) in the 1990s (Sivapalasingam et al., 2004). This increase is illustrated in Figure 2.1 by comparing the relative rates of yearly total foodborne cases and produce-associated cases from 1989 through 1997 to those occurring in 1988. Specifically, between

TABLE 2.1 Examples of Fecally Associated Pathogens

Bacteria	Viruses	Parasites
	Animal Source	
Aeromonas spp.	Coronavirus	*Cryptosporidium parvum*
Bacillus anthracis	Norovirus[a]	*Giardia lamblia*
Enterohemorrhagic *Escherichia coli*	Rotavirus	
Listeria monocytogenes		
Salmonella spp.		
Yersinia enterocolitica		
Campylobacter		
	Human Source	
Shigella	Hepatitis A	*Cryptosporidium parvum*
Salmonella spp.	Norovirus	*Giardia lamblia*
Campylobacter	Rotavirus	*Cyclospora cayetanensis*
Enterohemorrhagic *Escherichia coli*	Astrovirus	*Entamoeba histolytica*

[a]Ability of norovirus to cross species barrier is still unknown.

1992 and 1996, a substantial increase in the number of produce-associated cases occurred.

2.2 FACTORS AFFECTING THE EPIDEMIOLOGY OF PRODUCE-ASSOCIATED OUTBREAKS

Over the last 15 years, our knowledge of foodborne disease epidemiology has evolved at the same time the fresh fruit and vegetable industry has undergone

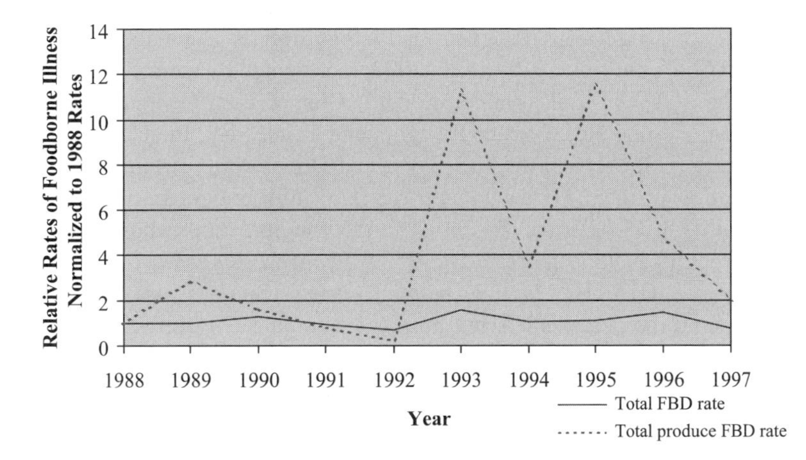

Figure 2.1 Relative rates of total foodborne disease (FBD) cases and produce-associated cases of FBD compared with 1988 based on CDC surveillance data (CDC, 2000b).

notable changes. Modifications in agronomic practices, processing and packaging technologies, along with global marketing strategies have allowed the fresh produce industry to supply consumers with a wide variety of products year round. Some of these same technologies and practices have also introduced an increased risk for human illness associated with pathogenic microorganisms. Furthermore, changes in population demographics, food consumption patterns, and increased awareness due to stepped-up epidemiologic surveillance programs and increased media attention have contributed to better documentation of produce-associated foodborne disease.

The proportion of the population in industrialized countries with heightened susceptibility to severe foodborne infections has increased as a result of demographic changes. In the United States, a growing segment of the population is immunocompromised because of such factors as infection with human immunodeficiency virus (HIV) or underlying chronic disease. For example, reported rates of salmonellosis, campylobacteriosis, and listeriosis were higher among HIV-infected persons than among those not infected with HIV (Altekruse et al., 1994). *Salmonella* (and possibly *Campylobacter*) infections are more likely to be severe, recurrent, or persistent in this population (Altekruse et al., 1994). The elderly are also at an increased risk of foodborne disease (Altekruse et al., 1997). According to the U.S. Census Bureau (1998), over the course of the next 25 years, the age structure of the world's population will continue to shift, with older age groups making up an increasingly larger share of the total. For example, during the 1998–2025 period, the world's elderly population (age 65 and above) will more than double, whereas the world's youth (population under age 15) will grow by 6 percent, and the number of children under age 5 will increase by less than 5 percent. As a result, world population will become progressively older during the coming decades.

Notable changes have occurred in the surveillance of foodborne disease. The Foodborne Diseases Active Surveillance Network (FoodNet) is the principal foodborne disease component of the U.S. Centers for Disease Control and Prevention (CDC) Emerging Infections Program (EIP). FoodNet is a collaborative project between the CDC, multiple EIP sites (in 2003, represented by various health departments in the states of California, Colorado, Connecticut, Georgia, New York, Maryland, Minnesota, New Mexico, Oregon, and Tennessee), the U.S. Department of Agriculture (USDA), and the U.S. Food and Drug Administration (FDA). The total population of the 2003 catchment was 37.6 million people, or 13.8 percent of the total U.S. population. Potential foodborne infections monitored in the FoodNet program include those caused by *Salmonella*, *Shigella*, *Campylobacter*, *Escherichia coli* O157:H7, *Listeria monocytogenes*, *Yersinia enterocolitica*, and *Vibrio* spp., and the parasites *Cryptosporidium* spp. and *Cyclospora cayetanensis*. It must be noted that FoodNet is a disease surveillance program, and except for a few case–control studies, the program cannot conclusively link these infections to foodborne transmission, nor can it routinely identify the food vehicle associated with the diseases under surveillance. Increased surveillance through programs such as FoodNet, along with the National Molecular Subtyping Network for Foodborne Disease Surveillance (PulseNet), a network of laboratories in state health departments, CDC, and food regulatory agencies, has contributed to an overall increased awareness of foodborne illness of bacterial etiology and its effect on public health (CDC, 1999a).

Because of the health benefits of fresh produce, public health officials have recently begun to recommend increased consumption of fruits and vegetables. For example, the USDA's *Food Guide Pyramid* has increased the recommended servings of produce to five to nine per day, as part of a healthier diet in order to potentially reduce the risks of heart disease and certain cancers (DHHS, 2000). At the same time, the World Health Organization (WHO) has developed a global initiative advising sufficient consumption of fruits and vegetables as part of a regular diet, stating that up to 2.7 million lives could be saved annually with adequate fresh produce consumption. According to this WHO report, inadequate fruit and vegetable intake is estimated to cause about 31 percent of heart disease and 11 percent of strokes worldwide (WHO, 2003).

Along with the significant rise in consumption of fresh produce, major shifts in consumer consumption trends have forced changes in the marketing of these commodities. For instance, as a consequence of the demand for variety, the number of different produce items offered by retailers doubled from 173 to 345 during the period from 1987 to 1997 (Supermarket Business, 1999; Dimitri et al., 2003). One explanation for the diversity of product selection is the growing consumer preference for foods that are convenient, minimally processed, and contain fewer preservatives (De Roever, 1999). For example, a sharp increase in convenience items such as precut, or minimally processed lettuce, has risen from 1 to 15 percent of total sales within the past 15 years (Handy et al., 2000). There has also been a growing trend toward food consumption outside the home and an increase in the popularity of salad bars. Larger quantities of intact or minimally processed fruits and vegetables are being shipped from central locations and distributed over vast geographical areas to diverse markets.

Growing consumer demand for organic produce, once thought of only as a niche market, has also contributed to market transformation. In 2000, more organic food was purchased from conventional markets than from any other venue (Dimitri and Greene, 2002). In the United States, freshness, taste, and quality rank among the top reasons for organic produce purchases. In fact, from 1997 to 2001, the number of U.S. acres dedicated to the production of organic fruits and vegetables increased by over 30 percent. It has been suggested that because of the agronomic practices used in organic farming, conventionally produced fruits and vegetables may be microbiologically safer (Mukherjee et al., 2004). However, there is little comprehensive research in this area. A recent study in Minnesota analyzed samples of fresh fruit and vegetables from 32 organic farms and 8 conventional farms to examine differences in microbiological indicator levels as well as the prevalence of select pathogens, such as *E. coli*, *Salmonella* spp., and *E. coli* O157:H7. These investigators found that organic produce was more susceptible to fecal contamination. For example, organic produce samples from farms that used manure or compost aged less than 12 months had an *E. coli* prevalence 19 times higher than product from farms that used manure aged more than 12 months. Nonetheless, pathogens were rarely found in either organic or conventionally produced items (Mukherjee et al., 2004).

U.S. consumers are currently able to purchase a vast array of produce items year round, many of which were once thought of only as seasonal. Between 1980 and 2001, fresh vegetable imports increased by over 250 percent, while fresh fruit imports increased by 155 percent (Clemens, 2004). Along with a rise in the sales of imported produce, an increasing number of foodborne disease outbreaks have been linked to imported products. In addition to recent outbreaks associated with imported cantaloupe and green onions (CDC, 2002b, 2003), several other instances of foodborne illness have been associated with produce originating from other countries. For example, in 1996, a total of 1465 cases of cyclosporiasis were reported from 20 states, the District of Columbia, and two Canadian provinces. Epidemiologic investigations found that the illnesses were caused by the consumption of raspberries imported from Guatemala (Herwaldt and Ackers, 1997). In March 1997, a total of 153 cases of hepatitis A were reported in Calhoun County, Michigan; subsequent epidemiological investigation implicated strawberries from Mexico as the source of the illnesses (CDC, 1997a). Furthermore, a 1999 multistate outbreak of *Salmonella* Newport infection was associated with the consumption of mangoes imported from a single farm in Brazil (Sivapalasingam et al., 2003). Despite such reports, little is known about the relative importance of imported product as compared to domestic produce when considering the overall burden of produce-associated foodborne disease.

2.3 PATHOGENS ASSOCIATED WITH FRESH PRODUCE

Epidemiologic evidence demonstrates a relationship between some foodborne pathogens and specific commodities. For example, the predominant reservoir for *Salmonella enterica* serovar enteritidis is contaminated raw eggs, whereas *E. coli* O157:H7 infection is often associated with the consumption of contaminated, inadequately cooked ground beef. Although fresh produce is not the most common vehicle for the transmission of the majority of foodborne pathogens, fruits and vegetables have gained notoriety as occasional vehicles of a wide array of pathogens, including bacterial, viral, and parasitic agents.

Numerous pathogens have been isolated from a wide variety of fresh fruits and vegetables. Table 2.2 lists various pathogens found in produce from research studies performed in several countries, together with their detection rates. Detection rates ranged from 0 percent to well over 50 percent. It is important to note that the number of samples in each study varied substantially. Although not all of the pathogens have been associated with produce-related foodborne disease outbreaks, they are all capable of causing illness. A wide variety of these are of considerable public health significance, including *Shigella* spp., enterotoxigenic and enterohemorrhagic *E. coli*, *Campylobacter* spp., *L. monocytogenes*, *Y. enterocolitica*, *Bacillus cereus*, *Clostridium botulinum*, enteric viruses, and parasitic protozoa such as *Cyclospora cayetanensis*, *Giardia lamblia*, and *Cryptosporidium parvum*.

TABLE 2.2 Detection Rate of Various Foodborne Pathogens in Representative Produce Items

Pathogen	Reservoir	Produce	Detection Rate (%)	Reference
Aeromonas hydrophila, caviae	Fresh, stagnant, estuarine, or brackish water	Chicory salads, lettuce, salad mix, watercress, endive, carrots	48–100	Marchetti et al., 1992; Nguyen-the and Carlin, 2000
Bacillus cereus	Frequently isolated from soil, growing plants	Alfalfa, mung bean, and wheat sprouts, broccoli	12–92.9	Harmon et al., 1987; Splittstoesser et al., 1983; Thunberg et al., 2002
Campylobacter	Zoonotic; isolation includes poultry, cattle, sheep, rodents, horses, pigs	Lettuce, parsley, mushrooms, green onion, potatoes, spinach, radish	0.6–1.5	Park and Sanders, 1992; Little et al., 1999; Doyle and Schoeni, 1986
Clostridium botulinum	Soil, sediments of streams, lakes, and coastal waters, intestinal tracts of fish and mammals, in the gills and viscera of crabs and other shellfish	Salad mix, shredded cabbage, chopped green pepper	0.3–0.7	Lilly et al., 1996; Notermans et al., 1989
Clostridium perfringens	Soil, dust, intestinal tract of humans and animals	Mixed raw vegetables	34	Nguyen-the and Carlin, 2000
Cryptosporidium	Human and animal intestines, surface waters	Lettuce, cilantro, carrots, cucumber, radish, tomato	1.3–8.7	Monge and Chinchilla, 1996
Escherichia coli O157:H7	Intestinal tract of cattle and other domestic animals and wildlife	Mixed raw vegetables, cilantro, corriander, celery	19–20	Nguyen-the and Carlin, 2000; Beuchat, 1996
Giardia spp.	Surface waters	Mixed raw vegetables	13	Nguyen-the and Carlin, 2000

(continued)

TABLE 2.2 *Continued*

Pathogen	Reservoir	Produce	Detection Rate (%)	Reference
Listeria monocytogenes	Decaying vegetation, surface waters, soil, feces of animals	Mung bean sprouts, lettuce, prepacked salads, salad mix, mixed vegetables, broccoli, cabbage, coleslaw, cucumber, green pepper, leafy vegetables, mushrooms, potatoes, radish, tomatoes, field cress	1.1–86	Nguyen-the and Carlin, 2000; Arumugaswamy et al., 1994; Odumeru et al., 1997; Francis et al., 1999; Szabo et al., 2000; Harvey and Gilmour, 1993; Sizmur and Walker, 1988; Beuchat, 1996a; Lin et al., 1996; de Simon et al., 1992
Salmonella	Poultry, eggs, swine, water, soil, insects	Cantaloupe, strawberries, mung bean sprouts, alfalfa seeds, lettuce, salad greens, mixed raw vegetables, chili, cilantro, culantro, parsley, artichoke, cabbage, cardoon, cauliflower, celery, eggplant, endive, fennel, green onions, spinach, beet leaves	0.4–72	FDA, 2001a; Madden, 1992; Splittstoesser et al., 1983; Beuchat 1996b; Jerngklinchan and Saitanu, 1993; Ercolani, 1976; Tamminga et al., 1978; Saddik et al., 1985; Lin et al., 1996; Garcia-Villanova et al., 1987; Rude et al., 1984; Al-Hindawi and Rished, 1979; Tauxe et al. 1997; Mukherjee et al., 2004
Shigella	Humans	Cantaloupe, lettuce, salad greens, salad mix, parsley, celery, green onion	1.1–2.4	FDA, 2001a; Saddik et al., 1985
Staphylococcus aureus	Humans and animals	Alfalfa, mixed, and onion sprouts, parsley, carrots, radish, lettuce	5.1–28.0	Prokopowich and Blank, 1991; Beuchat 1996b; Thunberg et al., 2002
Yersinia enterocolitica	Intestinal tract of mammalian species, birds, frogs, fish, flies, soil, surface waters	Lettuce, prepacked salad, mixed raw vegetables, watercress	1–100	Szabo et al., 2000; Brocklehurst et al., 1987; Darbas et al., 1985; Beuchat 1996b; dos Reis Tassinari et al., 1994

Figure 2.2 illustrates reported produce-associated outbreaks in the United States of known etiology by produce group and etiologic agent from 1990 to 2002. A specific etiologic agent was identified for 187 produce-associated outbreaks during this 13-year span. Among these outbreaks, 102 (55 percent) were caused by bacteria, 68 (36 percent) were caused by viruses, and 17 (9 percent) were caused by parasites. Among the bacterial agents, *Salmonella* accounted for 60 percent of outbreaks, and pathogenic *E. coli* was responsible for 25 percent of bacterial outbreaks. Norovirus caused a majority of viral outbreaks, accounting for over 80 percent. It must be noted that through improved surveillance and detection methods, the apparent prevalence of norovirus has increased. *Cyclospora* caused the majority (65 percent) of parasitic produce-associated outbreaks. Over 40 percent of the outbreaks were caused by salads (including lettuce and tomatoes), while outbreaks caused by fruit and fruit salads comprised 13 percent of the

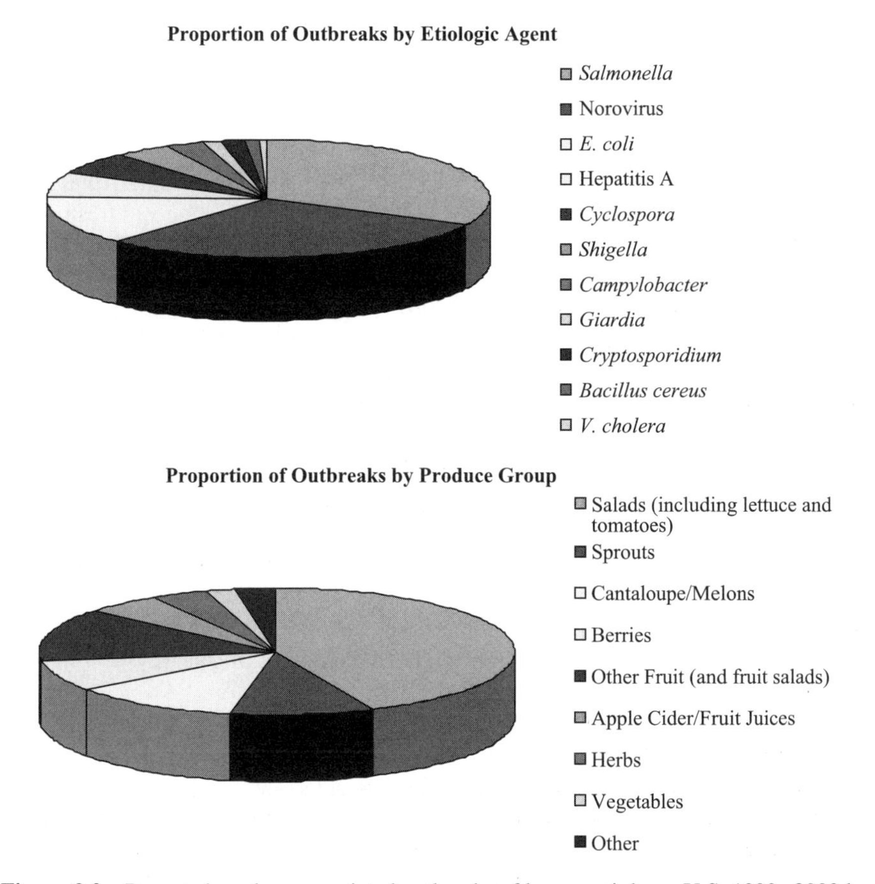

Figure 2.2 Reported produce-associated outbreaks of known etiology, U.S. 1990–2002 by etiologic agent and by produce group. (Data Source: CDC Foodborne Outbreak Surveillance System, www.cdc.gov)

outbreaks. Melons, including cantaloupe, honeydew, and watermelon, also represented 12 percent of produce-associated outbreaks, while sprouts comprised 10 percent of the outbreaks. Table 2.3 lists examples of produce-associated outbreaks from 1996 to 2004.

2.3.1 *Salmonella*

Salmonella is the most common cause of bacterial foodborne disease in the United States. Although the number of reported *Salmonella* cases actually increased between 2000 and 2002, recent preliminary reports have suggested an overall decrease in their number since then (CDC, 2004b.). Birds, reptiles, amphibians, and mammals are natural reservoirs of *Salmonella*, and several surveys have reported the presence of various *Salmonella* serotypes on certain produce items. A study carried out in Italy reported *Salmonella* to be present on 72 percent (64/89) of fennel samples and 68 percent (82/120) of lettuce samples (Ercolani, 1976). In an FDA survey of over 1000 imported produce samples originating from 21 countries, 35 samples were confirmed positive for *Salmonella* (FDA, 2001a). The contaminated items included cantaloupe (8/151), celery (1/84), cilantro (16/171), culantro (6/12), lettuce (1/116), parsley (2/84), scallions (1/180), and strawberries (1/143). The organism has also been detected on beet leaves, cardoon, cabbage, cauliflower, eggplant, peppers, endive, and spinach (Beuchat, 1996b).

Several outbreaks of salmonellosis in the United States have involved fresh fruits, particularly imported cantaloupe. In 1990, a large outbreak of salmonellosis, affecting at least 245 people in 30 states, was associated with the consumption of cantaloupe served at salad bars (Ries et al., 1990). Two deaths were reported and it was estimated that 25,000 individuals were eventually infected (CDC, 1991). During June and July of 1991, more than 400 laboratory-confirmed infections of *Salmonella enterica* serovar Poona reported in 23 states and Canada were linked to the consumption of contaminated cantaloupe. Thereafter, three high-profile multistate outbreaks of serovar Poona associated with eating cantaloupe imported from Mexico occurred during the spring of consecutive years between 2000 and 2002. Outbreaks were first identified by the California Department of Health Services (2000 and 2001) and the Washington State Department of Health (2002) and involved residents of 12 states and Canada (CDC, 2002b). Following these outbreaks, the FDA conducted farm investigations in Mexico, issued press releases to warn consumers, placed implicated farms on detention, and conducted sampling surveys of imported cantaloupe.

Acidic foods, such as orange juice, which were once thought to be an unlikely vehicle for *Salmonella*, have recently been associated with outbreaks. For instance, 62 visitors to a large theme park in Orlando contracted salmonellosis following the consumption of unpasteurized orange juice (Cook et al., 1998). Unpasteurized apple cider and apple juice have also been associated with outbreaks of *Salmonella typhimurium* (CDC, 1975). Laboratory-based studies have shown that some *Salmonella* strains can survive in orange juice at detectable levels for 27 days (pH 3.5 product) and 73 days (pH 4.4 product) (Parish et al., 1997).

TABLE 2.3 Examples of Recent Outbreaks of Foodborne Disease in the United States and Canada Associated with Fruits and Vegetables

Pathogen	Year	Food Source	No. of States	No. of Cases	Food Origin	Comments	Reference
				Bacterial			
Escherichia coli O157:H7	1996	Unpasteurized apple juice	8, Canada	70	U.S.	Dropped apples used, apple orchard near deer and cattle; incorrect use of disinfectant during wash	CDC, 1996a; Cody et al., 1999
Escherichia coli O157:H8	1996	Mesclun	2	49	U.S.	Contaminated lettuce was traced to a single grower; cattle found near lettuce fields	Hilborn et al., 1999
Escherichia coli O157:H9	1998	Alfalfa sprouts	1	8	California and Nevada	Contaminated seeds; sprouts were traced to single sprouter	Taormina et al., 1999; Harris, 2001
Salmonella Enteriditis	2000	Citrus juices	Multistate	14	California	Gallon-sized containers of domestic citrus juices were implicated in the outbreak	Harris, 2001
Salmonella Kottbus	2001	Alfalfa sprouts	4	24	California	Traceback investigation identified a single sprout producer as the source of the contaminated sprouts; sample of seed lot yielded S. Kottbus; seeds were imported from Australia	CDC, 2002c
Salmonella Poona	2002	Cantaloupe	10, Canada	58	Mexico	Possible sources of contamination included irrigation water contaminated with sewage, cleaning and cooling produce with *Salmonella*-contaminated water, poor hygiene of harvesters and processors, pests in packing facilities, and inadequate cleaning and sanitizing of equipment packing facilities, and inadequate cleaning and sanitizing of equipment	CDC, 2002b

(continued)

TABLE 2.3 *Continued*

Pathogen	Year	Food Source	No. of States	No. of Cases	Food Origin	Comments	Reference
Salmonella Newport	2002	Tomatoes	22	404	Mid-Atlantic region, U.S.	All tracebacks implicated the same tomato packing shed	Kretsinger et al., 2003
Shigella flexneri 6A	1994	Green onions	2	72	Mexico	Possible contamination during harvest or packaging	Tauxe, 1997
Shigella sonnei	1998	Parsley	4	310	Mexico	Poor water quality in packing shed and poor worker hygiene; in restaurant, product was chopped and left at room temperature	CDC, 1999b
				Viruses			
Hepatitis A	2003	Green onions	4	>900	Mexico	Raw green onions from three firms in Mexico have been implicated in the Tennessee and Georgia outbreaks	CDC, 2003
Hepatitis A	1997	Strawberries	4	242 (+14 suspect)	Mexico	Possible contamination during harvesting	CDC, 1997a
				Parasites			
Cryptosporidium parvum	1996	Apple juice	1	20 (+11 suspect)	New York	Dairy farm nearby; *E. coli* was detected in well water, indicating fecal contamination	CDC, 1997c
Cyclospora cayetanesis	2004	Snow peas	Pennsylvania	96	Guatemala	Source of contamination unknown	CDC, 2004a

Raw seed sprouts have also been recognized as significant vehicles for the transmission of foodborne illness within the last ten years. Seed sprouts are considered a high-risk produce item because seeds are sometimes contaminated with enteric pathogens and the high temperature and humidity of the sprout germination process is conducive to the proliferation of these organisms. Since 1996, the U.S. Food and Drug Administration (FDA) has responded to 27 outbreaks, accounting for over 1500 cases of salmonellosis and *Escherichia coli* O157:H7 infection, all of which were associated with raw or lightly cooked sprouts (FDA, 2004). For instance, alfalfa sprouts produced at a single facility were implicated as the cause of a *Salmonella enterica* serovar Kottbus outbreak, affecting 32 patients in four states (Arizona, California, Colorado, and New Mexico) (CDC, 2002c).

2.3.2 *Escherichia coli* O157:H7

Enterohemorrhagic *E. coli* was first recognized as a human pathogen in 1982 when it was identified as the cause of two outbreaks of hemorrhagic colitis (Wells et al., 1983). Since then, sporadic infections and outbreaks have been reported from many parts of the world, including North America, Australia, Asia, Western Europe, and Africa. Undercooked or raw hamburger (ground beef) has been implicated in most of the documented outbreaks; however, *E. coli* O157:H7 has also been found on other products, including contaminated produce items. For example, Zepeda-Lopez et al. (1995) detected *E. coli* O157:H7 on 25, 19.5, and 20 percent of cabbage, cilantro, and coriander, respectively. On the other hand, several studies have reported the absence of *E. coli* O157:H7 on produce (Lin et al., 1996; Little et al., 1999; FDA, 2001a, 2003). It should be noted that in those studies in which *E. coli* O157:H7 contamination was absent, sample sizes ranged from 51 to over 1000 produce items.

Along with *Salmonella*, *E. coli* O157:H7 has also been linked to outbreaks associated with raw seed sprouts. The first reported outbreak of *E. coli* O157:H7 infection associated with eating alfalfa sprouts occurred in the summer of 1997. In this outbreak, a total of 60 cases from 16 countries were reported to the Michigan Department of Community Health; simultaneously, the Virginia Department of Health received reports of 48 cases. Isolates from both the Michigan and Virginia outbreaks were compared using a molecular subtyping method (pulsed field gel electrophoresis, PFGE) and the analysis revealed that the isolates were identical and both outbreaks were associated with alfalfa seeds originating from the same supplier and lot (CDC, 1997d). The world's largest reported *E. coli* O157:H7 outbreak occurred in Japan during 1996 and was linked to the consumption of white radish sprouts. Approximately 6000 schoolchildren were infected and 17 people died. Interestingly, during May through August of that same year, approximately 10,000 cases of *E. coli* O157:H7 infection occurring in at least 14 separate disease clusters were reported in Japan (Watanabe et al., 1999). In the following year, white radish sprouts were once again implicated in a Japanese outbreak of *E. coli* O157:H7 that affected a total of 126 people (Gutierrez, 1997). Although no decontamination steps exist that completely eliminate foodborne pathogens from seed sprouts, the FDA recommends the application of a seed disinfection treatment of 20,000 ppm calcium hypochlorite prior to sprout production (FDA, 1999).

Along with *Salmonella* spp., *E. coli* O157:H7 has been implicated in outbreaks involving the consumption of unpasteurized fruit juices. In 1991, the first confirmed outbreak of *E. coli* O157:H7 associated with apple cider occurred in Massachusetts, affecting 23 people (Meng et al., 2001). In 1996, three outbreaks of *E. coli* O157:H7 were associated with unpasteurized apple juice, the largest of which occurred in the western United States and Canada, with 71 confirmed cases and one death (CDC, 1996a). The source of the pathogen was suspected to be animal manure, because "drops" or apples that had fallen on the ground during season were used to produce the cider. Although the definitive source of contamination is still not clear, further investigation of the 1991 outbreak revealed that cattle grazed in a location adjacent to the orchard and could have served as the source of fecal contamination (Besser et al., 1993).

Researchers have demonstrated that *E. coli* O157:H7, as well as some *Salmonella* strains, can withstand the low pH (3.3–4.2) characteristic of many juices (Dingman, 2000; Koodie and Dhople, 2001; Janes et al., 2002). As a result, the FDA mandated the use of Hazard Analysis and Critical Control Point (HACCP) principles for juice processing (FDA, 2001c). Manufacturers are also required to use processes that achieve a 5-log reduction in the numbers of the most resistant pathogen in their finished products, effectively resulting in pasteurization (FDA, 2001b).

2.3.3 *Shigella*

In 1999, the WHO estimated that *Shigella* spp. were responsible for over 166 million illnesses annually (Kotloff et al., 1999). This organism is also the third leading cause of bacterial foodborne outbreaks in the United States, infecting 450,000 people each year (Mead et al., 1999) with no significant change in the occurrence of reported cases over the last six years (CDC, 2004b). This pathogen is not indigenous to foods, but rather introduced through the fecal–oral route by contact with human excreta; fecally contaminated water and poor hygiene of infected food handlers are the most common causes of contamination of fresh produce.

Shigella spp. have been isolated from several produce items. In FDA 1000-sample surveys of imported and domestic produce, *Shigella* was isolated from less than 2.5 and 0.5 percent of the samples, respectively; commodities that were occasionally contaminated included cantaloupe, celery, lettuce, parsley, and scallions (FDA, 2001a, 2003). In 1985, Saddick and colleagues found *Shigella* in one of 57 salad green samples obtained from Egyptian retail outlets (Saddik et al., 1985). Although only reported occasionally, several foodborne outbreaks of shigellosis have been associated with raw produce, including green onions (Cook et al., 1995), iceberg lettuce (Frost et al., 1996), and uncooked baby maize (Eurosurveillance Weekly, 1998). In the summer of 1998, eight outbreaks of *Shigella sonnei*, occurring in Minnesota, California, Massachusetts, and Florida, as well as in Ontario and Alberta, Canada, were linked to chopped, uncooked, curly parsley. In this case, molecular subtyping revealed that strains from the seven outbreaks for which isolates were available shared the same PFGE pattern, indicating a common source (CDC, 1999b).

2.3.4 *Listeria monocytogenes*

Although the causative agent of listeriosis was discovered more than 70 years ago, the significance of the foodborne route of transmission has only been recognized within the last 20 years. In fact, *L. monocytogenes* came to the forefront as an emerging pathogen after a 1981 outbreak of listeriosis in Nova Scotia, Canada, which was traced back to the consumption of contaminated coleslaw (Schlech III et al., 1983). Thiry-four perinatal and seven adult listeriosis cases initiated the epidemiological investigation. Coleslaw obtained from a patient's refrigerator was positive for the epidemic strain, *L. monocytogenes* serotype 4b, which was the same strain isolated from the patient's blood. Further investigation led to the identification of a cabbage grower whose farming practices provided sufficient opportunities for the introduction of *L. monocytogenes* in the food chain. In this case, the cabbage was grown in fields fertilized with raw sheep manure, and interestingly, two of the farmer's sheep had previously died of listeriosis.

Listeria monocytogenes is a major public health concern because of its severe disease manifestations (meningitis, septicemia, and abortion), high case-fatality rate (approximately 20–30 percent), long incubation period (one to six weeks for severe cases), and predilection for immunocompromised individuals. In 2000, the CDC included listeriosis among its list of notifiable diseases (CDC, 2002d). The estimated incidence of *Listeria* infections shows considerable variation and did not change significantly between 1998 and 2003, with the number of cases ranging from 94 to 114 during this period (CDC, 2004b).

Listeria monocytogenes is quite different from other foodborne pathogens for several reasons. First, it can persist under diverse environmental conditions, including reduced pH and relatively high salt concentrations (Lou and Yousef, 1999). The pathogen is also microaerobic and is able to grow at very low temperatures (2 to 4°C) (Swaminathan, 2001). *L. monocytogenes* is found in soil and water and is widely distributed among animals, humans, and, particularly, plant vegetation (Beuchat, 1996a).

Although *L. monocytogenes* has been isolated from a wide variety of raw and ready-to-eat meat products, it can also be associated with produce items. In 1994, Arumugaswamy and coworkers surveyed various food products from a Malaysian market and found that 6 of 7 samples of bean sprouts and 5 of 22 leafy vegetable samples were contaminated with *L. monocytogenes* (Arumugaswamy et al., 1994). The prevalence of *L. monocytogenes* during production and postharvest processing of cabbage was recently examined in farms and packing sheds in south Texas (Prazak et al., 2002). The pathogen was isolated from 26 of 855 (3 percent) of the total samples. Twenty isolates originated from cabbage samples that were obtained from farms and packing sheds, while three isolates were from water samples. Additionally, three isolates were obtained from environmental sponge samples taken from packing shed surfaces. *L. monocytogenes* can grow on fresh produce stored at refrigeration temperatures, as has been demonstrated for asparagus, cauliflower, and broccoli held at 4°C (Berrang et al., 1989). The pathogen has also been reported to grow on lettuce at 5°C (Steinbrugge et al., 1988; Berrang et al., 1989). Growth of *L. monocytogenes* can occur on the surface of tomatoes held at 21°C, but not at 10°C (Beuchat and Brackett, 1991). Interestingly, carrot juice has been reported to have an inhibitory effect on *L. monocytogenes* growth (Beuchat and Brackett, 1990; Beuchat et al., 1994; Beuchat and Doyle, 1995).

2.3.5 *Campylobacter*

Before 1972, when methods were first reported for its isolation from feces, *Campylobacter* was believed to be primarily an animal pathogen causing abortion and enteritis in sheep and cattle. Since then, epidemiological studies have shown that *Campylobacter jejuni* and *C. coli* together are the leading cause of bacterial diarrheal illness in the United States. Raw milk and undercooked foods of animal origin are recognized as the primary vehicles for infection, although the potential for cross-contamination of fresh produce during the preparation of poultry and other meats is significant (Nachamkin, 2001).

Park and Sanders studied the occurrence of thermotolerant campylobacters in fresh vegetables sold at supermarkets and farmers' markets in Canada (Park and Sanders, 1992). Of a total of 1564 samples of 10 vegetable types, *Campylobacter* was detected on spinach (3.3 percent), lettuce (3.1 percent), radish (2.7 percent), green onion (2.5 percent), parsley (2.4 percent), and potatoes (1.6 percent). Interestingly, *Campylobacter* was found among only those samples collected from outdoor markets with no microbial decontamination step. Produce samples collected from supermarkets and outdoor markets, which had been thoroughly washed with chlorinated water, were all negative for *Campylobacter*.

Produce is an infrequent vehicle of *Campylobacter* enteritis and very few incidents have occurred in which vegetables have been involved. Only one outbreak of *Campylobacter* enteritis associated with the contamination of fruits or vegetables was reported for the period 1973–1987, although another *Campylobacter* outbreak due to fresh produce was reported during 1989 (Bean et al., 1996; Bean and Griffin, 1990). An outbreak of campylobacteriosis caused by the ingestion of a cabbage–beef stew by schoolchildren was reported in Germany. Because the stew was cooked, cross-contamination with *C. jejuni* was a likely contributing factor (Steffen et al., 1986). Similarly, cross-contamination of salad lettuce by raw chicken led to a small outbreak of campylobacteriosis in an Oklahoma restaurant in 1996 (CDC, 1998b). Harris et al. (1986) analyzed the dietary histories of individuals who had campylobacteriosis and determined that out of the several vegetable items surveyed, only mushrooms were significantly associated with *C. jejuni* enteritis cases.

2.3.6 Potential Produce-Associated Pathogens

Aeromonas strains can be present in drinking, fresh, saline, and brackish waters, as well as in sewage. The organism is considered ubiquitous, having been found in a wide range of seafoods, meats, and poultry. Several studies have isolated *Aeromonas* spp. from produce (Callister and Agger, 1989). For example, Szabo et al. (2000) found *Aeromonas* spp. in over 50 percent (66/120) of lettuce samples. Although no produce-associated outbreaks with this organism have been reported, several characteristics of the *Aeromonas* genus suggest its potential as an infectious agent (Llopis et al., 2004; Martins et al., 2002). *Aeromonas* spp. can grow rapidly on raw vegetables and seed sprouts at refrigeration temperatures (Harris et al., 2003).

Yersinia enterocolitica has also been found in a wide variety of environments, including the intestinal tract of various mammalian species, as well as in birds, frogs, fleas, oysters, flies, and fish (Cover and Aber, 1989) Like *L. monocytogenes*, *Y. enterocolitica* can grow at refrigeration temperatures, making it a potential concern for food manufacturers. *Y. enterocolitica* has been isolated from many foods, including fresh produce. Szabo et al. (2000) detected *Y. enterocolitica* in 71 of 120 (59 percent) packaged lettuce samples; however, the strains isolated were not pathogenic. Another study reported a 4 percent prevalence (27/673) of *Yersinia* spp. on ready-to-eat vegetables; of the 27 strains isolated in this study, 18 were *Y. enterocolitica* (Lee et al., 2004).

2.3.7 Human Enteric Viruses

Hepatitis A virus (HAV) and the noroviruses are the most commonly documented viral agents to contaminate food. Because of their low infective dose, many foodborne viral infections are transmitted via infected food handlers. HAV is primarily transmitted by the fecal–oral route, either by person-to-person contact or by ingestion of food or water contaminated with human feces (Fiore, 2004). Between 1980 and 2001, an average of 25,000 cases of HAV were reported annually to the CDC; however, when corrected for asymptomatic infections and underreporting, this number probably exceeds 250,000 (CDC, 2002a). Fortunately, foodborne transmission accounts for only about 5% of HAV cases of known etiology (Mead et al., 1999).

The source of most reported foodborne hepatitis A outbreaks has been infected food handlers who contaminate the product at the point of sale or those who prepare food for social events. A single HAV-infected food handler can transmit the virus to dozens or even hundreds of persons and cause a substantial public health problem. A common theme for such outbreaks includes the presence of an HAV-infected food handler who worked while viremic (two weeks before to one week after the onset of symptoms) and who had contact with ready-to-eat foods; subsequently, the appearance of secondary cases occurs among other food handlers and patrons who ate product contaminated by the index case (Fiore, 2004).

Hepatitis A outbreaks have also been associated with the consumption of fresh produce contaminated during cultivation, harvesting, processing, or distribution. Outbreaks involving a food item that was contaminated before distribution are particularly difficult to identify as the product might be widely distributed geographically before recognition of the first cases. Low attack rates are common probably because contamination is only found in a small proportion of the distributed food.

Several recent clusters of hepatitis A infection occurred in the fall of 2003 in four states, Tennessee, Georgia, North Carolina, and Pennsylvania. Raw or undercooked green onions served in restaurants were the implicated source, with disease manifested in over 900 people, three of whom died. The nucleic acid sequences of the Pennsylvania outbreak strains were very similar to sequences obtained from persons involved in hepatitis A outbreaks in the other three states. Raw green onions from three farms in Mexico have since been implicated in the Tennessee and Georgia outbreaks (CDC, 2003).

Noroviruses (formerly known as the Norwalk-like viruses) are regarded as the most common of the foodborne viruses, and are the most significant cause of acute nonbacterial gastroenteritis in both children and adults. Transmitted by the fecal–oral route, the two most likely ways by which to become exposed to noroviruses are through person-to-person contact or the consumption of contaminated food or water. Food items implicated in norovirus outbreaks include molluscan shellfish and ready-to-eat foods that become contaminated by human handling, such as fruit salad, raspberries, cake icing, and deli meat. Noroviruses have also been associated with produce outbreaks. For instance, an outbreak caused by norovirus occurred in December 1979 at a luncheon banquet (Griffin et al., 1982). Among all the foods served, consumption of green salad was epidemiologically associated with the disease, which affected 63 persons. Contamination by infected food handlers was thought to be the likely source of the virus. In April 1998, an outbreak of viral gastroenteritis linked to the consumption of imported frozen raspberries occurred in Helsinki, Finland (Ponka et al., 1999). It was suspected that the raspberries, imported from Eastern European countries, were contaminated through the use of fecally impacted waters, either during irrigation or by postharvest sprays that were applied immediately before freezing.

2.3.8 Parasitic Protozoa

In 1999, Mead et al. (1999) estimated that foodborne transmission of parasitic agents accounted for over 350,000 cases of illness annually in the United States alone. From 1993 to 1997, 19 foodborne outbreaks of parasitic etiology were reported to the CDC, resulting in a total of 2325 cases (CDC, 2000b). The environmental routes of transmission for the parasitic protozoa include water, soil, and food that become contaminated by contact with animal or human fecal matter (Slifko et al., 2000).

Cyclospora cayetanensis, *Giardia lamblia*, and *Cryptosporidium parvum* are the most common human enteric protozoan infections. Outbreaks of cyclosporiasis have occurred in the last decade and have been associated with the consumption of fresh raspberries, basil, mesclun lettuce, and snow peas (CDC, 2004a; Herwaldt, 2000). In the mid- to late 1990s, several multistate outbreaks of cyclosporiasis were associated with the consumption of raspberries imported from Guatemala. An outbreak in the summer of 1997, with 57 reported clusters of cyclosporiasis and 341 cases in Northern Virginia, Washington DC, and Baltimore, Maryland, was associated with the consumption of basil (CDC, 1997b; Herwaldt, 2000). Nonhuman animal species serve as reservoir hosts of *C. cayetanensis*. In contrast to other foodborne enteric pathogens, *Cyclospora* oocysts are not immediately infective after excretion but require a period of days to weeks in favorable environmental conditions in order to sporulate and become infectious. Therefore, person-to-person transmission or transmission via infected foodhandlers is considered unlikely (Herwaldt, 2000). Anecdotal evidence from outbreak investigations suggests that the infective dose for *Cyclospora* is low (Herwaldt, 2000).

Outbreaks of giardiasis have been associated with the consumption of fruit salad (two outbreaks), iceberg lettuce (one outbreak), and salad (one outbreak) (Juranak, 2005, personal communication) An outbreak caused by *Giardia lamblia* occurred in a cafeteria at a corporate office building for which an asymptomatic food handler was the probable source of contamination to raw sliced vegetables (Mintz et al., 1993). In 1997, an outbreak of cryptosporidiosis in Washington, which caused 54 laboratory-confirmed illnesses, was linked to the consumption of green onions. In this instance, the onions reportedly had not been washed by either the supplier or the restaurant where they were served (CDC, 1998a). Two outbreaks of cryptosporidiosis were also associated with the consumption of apple cider (Juranek, 2005, personal communication).

2.3.9 Spore-Forming Bacteria

The contamination of fruits and vegetables with spores of *Bacillus cereus*, *Clostridium botulinum*, or *Clostridium perfringens* is not uncommon. Harmon et al. (1987) isolated *Bacillus cereus* from 83 percent (33/40) of mung bean sprout samples collected from a health food store and *B. cereus* and *C. perfringens* were isolated from over one-third (34/100) of assorted vegetables collected from retail outlets in the United Kingdom (Roberts et al., 1982). However, spore-forming bacteria become a public health threat only when produce is handled in a way that will support the germination of spores and growth of vegetative cells, which is not a common practice.

Nevertheless, there have been disease outbreaks in fresh produce associated with contamination by these organisms. In 1993, *C. perfringens* caused an outbreak affecting 48 people in Ontario, Canada, for which the probable vehicle was salad. In 1985, an outbreak of botulism in British Columbia, Canada, was caused by the consumption of chopped garlic in oil (St. Louis, 1988). A subsequent outbreak of botulism occurred in 1989 in New York, and was traced back to the same processor implicated in the 1985 outbreak (Morse et al., 1990). The product implicated in the 1989 outbreak had been made between 1985 and 1987, contained no preservatives, and was kept at room temperature for approximately three months after purchase prior to opening. In response to the 1989 outbreak, the FDA took steps to prevent a recurrence by requiring the use of microbial inhibitors or acidifying agents such as phosphoric or citric acid in vegetable tubers or roots cooked or coated in oil (Morse et al., 1990).

2.4 FACTORS CONTRIBUTING TO PRODUCE CONTAMINATION

Fresh produce may become contaminated with pathogens at any point during cultivation, harvesting, processing, distribution, or preparation. In 1998, the FDA, USDA, and CDC published voluntary guidelines to address produce safety issues, entitled *The Guide to Minimize Microbial Food Safety Hazards for Fresh Fruits and Vegetables* (FDA et al., 1998). The purpose of the Guide was to provide a framework for the identification and implementation of practices likely to decrease the risk of pathogen contamination in fresh produce from production, packaging, and transport based on Good Agricultural Practices (GAPs) and Good Manufacturing Practices (GMPs). Table 2.4 presents the major sources of pathogen contamination that can occur during both preharvest and postharvest phases. Although there are numerous potential sources of contamination along the "farm-to-fork" continuum, several central themes emerge from the document and these will be discussed in the following sections.

2.4.1 Use of Animal Manure and Biosolids

Animal wastes are commonly recycled to agricultural land and provide an economical and environmentally sustainable means of disposal. Although animal manures have beneficial fertilizer value for field crops, the use of animal manure has resulted in the contamination of produce with pathogenic bacteria, viruses, and parasites. Animal manure may harbor a plethora of pathogens, including, but not limited to, *Salmonella* spp., *Campylobacter* spp., *E. coli* O157:H7, *L. monocytogenes*, *Giardia* spp., and *C. parvum* (Mawdsley et al., 1995). Although frequently used as a crop fertilizer in the developing world, in the United States, excess animal manure may be spread on land only in the vicinity of animal or produce farms, but not directly on land intended for fruit and vegetable production. According to a USDA survey, organic sources of fertilizers are not commonly used by conventional fruit and vegetable growers (Suslow, 2001).

TABLE 2.4 Sources of Pathogenic Microorganisms on Fresh Fruits and Vegetables

Preharvest	Postharvest
Feces	Feces
Soil	Human handling
Irrigation water	Harvesting equipment
Water used to apply fungicides, insecticides	Transport containers
Green or inadequately composted manure	Wild and domestic animals
Dust	Insects
Wild and domestic animals	Dust
Insects	Wash and rinse water
Human handling	Processing equipment; ice; transport vehicles; improper storage

Several studies have reported extended survival of pathogens in animal manure (Himathognkham et al., 1999; Lung et al., 2001; Jiang et al., 2002; Islam et al., 2004). Kudva et al. (1998) reported that *E. coli* O157:H7 survived in ovine manure for as long as 21 months. Furthermore, an ongoing study in the United Kingdom has found that *E. coli* O157, *Salmonella*, and *Campylobacter* survived in stored manure slurries for up to three months, with *L. monocytogenes* surviving for up to six months. Following land application of inoculated manure, *E. coli* O157:H7, *Salmonella*, and *Campylobacter* survived in sandy and clay loam soil for up to one month, and *L. monocytogenes* survived for more than one month (Nicholson et al., 2004). Although survival of pathogens has been documented in manure and soil environments, significant natural die-off often occurs, with the result that relatively few organisms may survive at the end of the reported time period. Inactivation of pathogens can be further augmented by manipulation of the physical conditions of the environment. For example, these same investigators found that pathogens could not be detected after one week in solid manure heaps where temperatures greater than 55°C were obtained, supporting the use of high-temperature composting as an effective measure to reduce the microbial load in solid manure (Nicholson et al., 2004).

The listeriosis outbreak in the Maritime Provinces of Canada is a significant example of the proper use of fertilizer. A review of agronomic practices revealed that the implicated product was produced by a farmer who also maintained a flock of sheep. Both composted and raw sheep manure had been applied to fields in which the cabbage was grown. It has been reported that sheep manure from flocks with known cases of "circling disease" (listeriosis) may contain viable *L. monocytogenes* cells (Bojsen-Moller, 1972). In this particular case, ovine manure may have been the source of the pathogen on the cabbage crop. Furthermore, cold storage of the cabbage from the last harvest in October through the winter and early spring provided conditions under which *L. monocytogenes* could thrive. An *E. coli* O157:H7 outbreak in Montana affecting 70 people was caused by the consumption of lettuce, possibly contaminated from irrigation run-off or compost used for fertilization. In this case, it was found that cattle had access to the source of the irrigation water that may have contaminated the water supply (Ackers et al., 1998). Islam et al. (2004) studied the fate of *Salmonella enterica* serovar Typhimurium on carrots and radishes grown in fields treated with contaminated irrigation water and manure composts and found that the pathogen could survive in the soil for over 200 days, regardless of contamination mode. Furthermore, *Salmonella* persisted for 84 and 203 days on radishes and carrots, respectively.

2.4.2 Water Quality

During production, irrigation and surface run-off waters can be sources of pathogenic microorganisms that contaminate fruits and vegetables in the field. The use of water during production includes irrigation and the application of pesticides and fertilizer. Typical sources of agricultural water are surface water from rivers, streams, irrigation ditches, and open canals, impounded water such as ponds, reservoirs, and lakes, locally collected water such as cisterns and rain barrels, groundwater from wells, and municipal supplies. It was speculated that the Guatemalan raspberries implicated in the 1996 cyclosporiasis outbreak became contaminated by water used in insecticide and fungicide sprayers (Herwaldt and Ackers, 1997; Herwaldt, 2000). In addition, although the source of irrigation water may influence the safety and quality of fresh produce, the type of irrigation method is also important. For example, overhead irrigation has been shown to have a higher probability of contaminating produce than drip or furrow irrigation methods (Suslow et al., 2001).

Most fruits and vegetables receive some sort of processing before being distributed to commercial locations. Within the packing house, water quality is one of the key issues in maintaining a safe product. In the 1998 outbreaks of *Shigella sonnei* associated with Mexican parsley, field investigations revealed that the packing shed used unchlorinated municipal water in a hydrocooler to chill the product immediately after harvest. Furthermore, the water was recirculated, and this same water was used to produce the ice for transport packaging (CDC 1999b). Taken together, these practices facilitated the survival of the organism and promoted widespread cross-contamination of parsley.

2.4.3 Worker Hygiene

During harvest, one of the most important factors in preventing the contamination of fresh produce is worker hygiene. The potential for transferring fecal contamination to product is increased in the absence of sanitary handwashing facilities in the production area. The FDA suggests that packing sheds educate workers on the importance of good hygiene, as promoted through the use of GAPs and GMPs. Several HAV outbreaks have been associated with poor worker hygiene. For instance, a 1983 outbreak of hepatitis A associated with contaminated strawberries was apparently caused by infected food handlers picking the berries (Reid and Robinson, 1987). Over 100 people became ill in Florida after consuming lettuce that had been handled by an HAV-infected food handler who practiced poor hygiene (Lowry et al., 1989).

2.4.4 Environmental Sources

The production environment itself may also be the source of pathogens (Luechtefeld et al., 1980; Lee et al., 1982; Fenlow, 1985; Quessy and Messier, 1992). For instance, both insects and wild birds are important vectors in the transfer of fecal material (Geldreich, 1964). Allowing domestic animals access to field crops and orchards may result in contamination of produce and subsequent human infection with enteric pathogens. In the 1998 *E. coli* O157:H7 outbreak linked to unpasteurized apple cider, the juice was traced back to an orchard in which cattle were kept prior to harvest and from which "drops" or fallen apples were used to produce the cider (Tamblyn et al., 1999). An inspection following a *Salmonella* Hartford outbreak associated with the consumption of unpasteurized orange juice at a popular theme park revealed signs of environmental contamination, including a poorly sealed processing room, which may have facilitated unrestricted access by feral rodents, birds, and amphibians (Cook et al., 1998).

Some of the pathogens that can contaminate vegetables are natural inhabitants of the growing environment. Soil that has not been contaminated with feces is generally not a source of enteric microorganisms. However, spores of *Clostridium botulinum*, *C. perfringens*, and *Bacillus cereus* have been isolated from soils free of fecal contamination. *L. monocytogenes*, an environmentally ubiquitous organism, has been found in soil and decaying vegetation (Beuchat, 1996a). Twenty-seven strains of *L. monocytogenes* were isolated from soil and decaying corn, soybean plants, and wild grasses taken from 19 sites in the Netherlands (Welshimer and Donker-Voet, 1971; Beuchat and Ryu, 1997). The survival of enteric organisms in soil is dependent on factors such as soil type, moisture retention, pH, microbial activity, nutrient availability, and inoculum level. A French study reported that *L. monocytogenes* was able to survive in chalky and peaty soil for well over one year when incubated at 20°C, whereas at 4°C, the pathogen survived up to 1500 days (Picard-Bonnaud et al., 1989).

2.5 RECENT PRODUCE-ASSOCIATED OUTBREAKS: LESSONS LEARNED

2.5.1 Cyclosporiasis and Guatemalan Raspberries

Prior to 1996, *Cyclospora* caused sporadic cases of traveller's diarrhea in developed countries and its disease was considered one of several occurring among the immunocompromised and children in the developing world (Shields and Olson, 2003). In fact, the majority of documented cases of cyclosporiasis in North America were in overseas travelers, and only four clusters of cases, affecting 91 people, had been reported before 1996 (Huang et al., 1995; Carter et al., 1996; Koumans et al., 1996; Herwaldt, 2000). However, in the spring of 1996, an increase in cyclosporiasis cases was reported to the CDC and Health Canada, which subsequently focused attention on this little-known pathogen (CDC, 1996b,c, 1997e; Chew et al., 1996; Colley, 1996; Fleming et al., 1998; Herwaldt, 2000; Katz et al., 1996; Neamatullah et al., 1996). Over 1400 cases of cyclosporiasis were reported, which were epidemiologically linked to the consumption of Guatemalan raspberries.

In the late 1980s, raspberries were introduced into the Guatemalan market as an export crop (Herwaldt, 2000). By the mid-1990s, exports of Guatemalan raspberries were increasing rapidly, escalating 113 percent from 1995 to 1996. The Guatemalan spring growing season was essentially over by the time raspberries were recognized as the vehicle of the outbreak in 1996, and no regulatory action was taken. After the 1996 outbreak, the CDC and the FDA investigated Guatemalan farming practices; however, the mode of contamination was never definitively established. The investigators' leading hypothesis was that the raspberries became contaminated by exposure to agricultural water, specifically when the berries were sprayed with insecticides, fungicides, and fertilizers that had been mixed with water (Hoge et al., 1993; Rabold et al., 1994; Huang et al., 1995; Herwaldt, 2000).

In response to this outbreak, the FDA provided technical assistance and suggested the application of GAPs, GMPs, and Standard Sanitation Operating Procedures. In addition, the Guatemalan Berry Commission (GBC) was developed and voluntarily implemented control measures on farms that focused on improving employee hygiene, water quality, and sanitation. In the spring of 1997, the commission allowed only farms that were classified as low risk to export raspberries to the United States (CDC, 1997e). However, despite these measures, another multistate outbreak of cyclosporiasis, linked to Guatemalan raspberries, occurred in the United States and Canada in 1997, with 1012 reported cases (Herwaldt and Beach, 1999; Herwaldt, 2000). After consultation with the FDA, the GBC voluntarily stopped exporting raspberries to the United States. Interestingly, importation to Canada continued, and in 1998 a multicluster outbreak of cyclosporiasis linked to the consumption of Guatemalan raspberries occurred in Ontario, Canada, affecting over 300 people (Herwaldt, 2000).

This series of outbreaks was unique because it was caused by an obscure pathogen, involved prolonged illness, and had widespread exposure with many cases of illness. These large outbreaks alerted public health officials to the hazards associated with a pathogen once thought to be problematic only in the developing world. They also illustrated how such a problem can be transferred to developed countries through food importation. Indeed, it has become increasingly apparent that global intervention strategies are needed to assure produce safety. Unfortunately, the impact of these outbreaks was not limited to public health; Guatemalan farmers suffered substantial economic losses as well. It was estimated that ceasing raspberry exports to the United States, midseason, resulted in a loss of $10 million in income for the Guatemalan industry (Powell, 1998).

2.5.2 Hepatitis A and Mexican Strawberries

Many hepatitis A outbreaks occur as the result of a single infected food handler at a single food establishment. Sporadically, widespread foodborne outbreaks of HAV are associated with the contamination of product before distribution; furthermore, the source of infection cannot be identified in approximately 50 percent of reported cases of hepatitis A infection (CDC, 1999c, 2000a; Fiore, 2004). In 1997, the epidemiologic investigation of an outbreak and several sporadic cases of HAV in multiple states revealed a single, predistribution contamination source. This particular investigation was facilitated by new molecular biology detection and typing methods (CDC, 1997a).

Between January and May 1997, a total of 258 cases of hepatitis A were reported in Michigan and Maine. Of these, 242 (94%) occurred among school employees and students in 23 Michigan schools, with 39 additional cases reported from 13 schools in Maine (Shields and Olson, 2003). The incidence peaked during March and then declined rapidly, suggesting a common source. Case–control studies in both states indicated that the frequency of eating school lunch was associated with an increased likelihood of illness, and food items containing frozen strawberries were the likely vehicles of infection.

The strawberries implicated in this outbreak were grown in Mexico, processed and frozen at a California facility, and distributed commercially through the USDA school lunch programs. Traceback information indicated that 22 lots of frozen strawberries processed in April and May 1996 were sent to Michigan and Maine school lunch programs. Regulatory agency inspection of both processing and school facilities did not reveal a likely source of contamination. Investigation of three Mexican growing sites showed inadequate hygiene and toilet facilities. The only handwashing facilities were on trucks that circulated through the fields, and the pickers did not wear gloves when removing the strawberry stems with their fingernails. Additionally, only a few slit latrines were available for field workers' use.

Other states also received strawberries from the same lot and reported sporadic cases of hepatitis A infection. These states included Tennessee (two cases), Arizona (nine cases), Wisconsin (five cases), and Louisiana (four cases). In order to determine if the sporadic cases were related to the Michigan and Maine outbreaks, viral nucleic acid obtained from clinical specimens was sequenced. Amplified regions of the viral RNA obtained from all the Michigan patients tested had identical sequences; the sequences of 8 of 10 of the Maine patients matched those of the Michigan patients. Sequence analysis further determined that the five cases in Wisconsin, two of the Louisiana cases, seven from Arizona, and one of the two patients in Tennessee were identical to the Michigan strain. In this case, sequencing a part of the viral genome detected in outbreak cases in Michigan and Maine, along with sporadic cases in other states, confirmed the epidemiologic evidence indicating that the outbreak originated from a common source exposure.

2.5.3 *Escherichia coli* O157:H7 in Apple Juice

Produce-associated outbreaks have been responsible for important new food-safety regulations within the past 10 years. *E. coli* O157:H7, once considered an obscure pathogen, has challenged clinicians, alarmed food producers, and transformed public perception about food safety. Outbreaks of *E. coli* O157:H7 and *Salmonella* associated with the consumption of contaminated juices have dramatically altered the production and processing of these food products. Since 1990, at least six reported outbreaks of *E. coli* O157:H7 infection in North America were associated with unpasteurized apple cider (Health, 1999). Five of these outbreaks occurred in the United States over a three-year period (Besser et al., 1993; CDC, 1997c; Luedtke and Powell, 2002).

In the past, it was assumed that unpasteurized fruit juices were unlikely to harbor or support the growth of pathogens, primarily because of the low pH of the product. However, research has confirmed that *E. coli* O157:H7 can survive for a relatively long time in fruit juices. In response to these outbreaks, the FDA proposed two regulations to improve the safety of fresh and processed juices. In 1998, it began mandating the use of warning labels on all unpasteurized juices or juices not otherwise treated to control disease-causing pathogens. In January 2001, the FDA published final regulations requiring all domestic and foreign fruit and vegetable processors to use HACCP approaches to prevent, reduce, or eliminate microbiological hazards in juices. Large companies had one year after publication of the regulation to implement HACCP; small companies were required to comply in two years and very small companies within three years. Specifically, the rule stipulated that juice manufacturers in the United States implement a process able to achieve a 5-$\log_{10}$ reduction of the most resistant pathogen(s) of concern in their product (FDA, 2001b).

2.6 CONCLUSIONS

Advanced production practices, new packaging technologies, global marketing, and ever-changing consumer lifestyles have forced public health officials to respond to new and increasing microbial threats from the consumption of fresh produce. Production and processing technologies in the produce industry have evolved to meet consumer demands at the same time that epidemiologic surveillance tools for foodborne disease have also advanced. In 1993, CDC began using molecular analyses of *E. coli* O157:H7 to augment their epidemiologic investigations involving that pathogen (Tauxe, 2005, personal communication). Since then, the use of molecular analytical techniques in outbreak investigations of foodborne disease has progressed to include a number of bacterial, protozoan, and viral pathogens. The 1998 hepatitis A outbreak associated with the consumption of strawberries represented the first instance in which molecular analyses were used to supplement an ongoing epidemiologic investigation involving that pathogen. Indeed, federal public health systems have initiated significant active surveillance programs and engaged multiple relevant state and local health constituencies, as illustrated in programs such as PulseNet and FoodNet. Voluntary preventive and control measures such as HACCP are being implemented not only at a national level, but also internationally. The 1996 outbreak of cyclosporiasis associated with the consumption of Guatemalan raspberries not only added to concern about newly characterized and emerging pathogens, but was also an example of how the food safety sector must be prepared to control and react globally in order to keep the U.S. food supply safe. However, at the same time, health officials and regulators are reminded that the epidemiology of foodborne disease, particularly infections associated with fresh produce, is constantly changing and there are new challenges to face and lessons to learn. Research continues to examine the ecology, growth, and survival of pathogens on fresh produce. The food safety sector has demonstrated its ability to keep up with changes in the produce industry and will continue to respond with new technologies and regulations in order to maintain a safe and wholesome food supply.

REFERENCES

M. Ackers, B.E. Mahon, E. Leahy, B. Goode, T. Damrow, P.S. Hayes, W.F. Bibb, D.H. Rice, T.J. Barrett, L. Hutwagner, P.M. Griffin, and L. Slutsker, An outbreak of *Escherichia coli* O157:H7 infections associated with leaf lettuce consumption. *J. Infect Dis.*, 177, 1588–1593 (1998).

N. Al-Hindawi and R. Rished, Presence and distribution of *Salmonella* species in some local foods from Baghdad City, Iraq. *J. Food Prot.*, 42 (11), 877–880 (1979).

S.F. Altekruse, M.L. Cohen, and D.L. Swerdlow, Foodborne bacterial infections in individuals with the human immunodeficiency virus. *South Med. J.*, 87, 169–173 (1994).

S.F. Altekruse, M.L. Cohen, and D.L. Swerdlow, Emerging foodborne diseases. *Emerg. Infect. Dis.*, 3 (3), 285–293 (1997).

R.K. Arumugaswamy, G. Rusul Rahamat Ali, and S. Nadzriah Bte Abd. Hamid, Prevalence of *Listeria monocytogenes* in foods in Malaysia. *Int. J. Food Microbiol.*, 23, 117–121 (1994).

N.H. Bean, J.S. Goulding, C.L. Frederick, and J. Angulo, Surveillance for foodborne-disease outbreaks – United States, 1988–1992, CDC surveillance summaries (October). *MMWR*, 45 (no. SS-5) (1996).

N.H. Bean and P.M. Griffin, Foodborne disease outbreaks in the United States, 1973–1987: pathogens, vehicles, and trends. *J. Food Prot.*, 53 (9), 804–817 (1990).

M.E. Berrang, R.E. Brackett, and L.R. Beuchat, Growth of *Listeria monocytogenes* on fresh vegetables stored under controlled atmosphere. *J. Food Prot.*, 52 (10), 702–705 (1989).

R.E. Besser, S.M. Lett, J.T. Weber, M.P. Doyle, T.J. Barret, J.G. Wells, and P.M. Griffin, An outbreak of diarrhea and hemolytic uremic syndrome from *Escherichia coli* O157:H7 in fresh-pressed apple cider. *JAMA*, 269 (17), 2217–2220 (1993).

L.R. Beuchat, *Listeria monocytogenes:* incidence on vegetables. *Food Control*, 4/5, 223–228 (1996a).

L.R. Beuchat, Pathogenic microorganisms associated with fresh produce. *J. Food Prot.*, 59 (2), 204–216 (1996b).

L.R. Beuchat and R.E. Brackett, Inhibitory effects of raw carrots on *Listeria monocytogenes*. *J. Food Prot.*, 56 (6), 1734–1742 (1990).

L.R. Beuchat and R.E. Brackett, Behavior of *Listeria monocytogenes* inoculated into raw tomatoes and processed tomato products. *Appl. Environ. Microbiol.*, 57, 1367–1371 (1991).

L.R. Beuchat, R.E. Brackett, and M.P. Doyle, Lethality of carrot juice to *Listeria monocytogenes* as affected by pH, sodium chloride and temperature. *J. Food Prot.*, 57, 470–474 (1994).

L.R. Beuchat and M.P. Doyle, Survival and growth of *Listeria monocytogenes* in foods treated or supplemented with carrot juice. *Food Microbiol.*, 12, 73–80 (1995).

L.R. Beuchat and J.H. Ryu, Produce handling and processing practices. *Emerg. Infect. Dis.*, 3 (4), 459–465 (1997).

J. Bojsen-Moller, Human listeriosis: diagnostic, epidemiological and clinical studies. *Acta Pathol. Microbiol. Scand.*, 229 (Suppl.), 1–155 (1972).

C.M. Brocklehurst, C.M. Zaman-Wong, and B.M. Lund, A note on the microbiology of retail packs of prepared salad vegetables. *J. Appl. Bacteriol.*, 63, 409–415 (1987).

S.M. Callister and W.A. Agger, Enumeration and characterization of *Aeromonas hydrophila* and *Aeromonas caviae* isolated from grocery store produce. *Appl. Environ. Microbiol.*, 53, 249–253 (1989).

R. Carter, F. Guido, G. Jacquette, and M. Rapoport, Outbreak of cyclosporiasis associated with drinking water. Program of the 30th Interscience Conference on Antimicrobial Agents and Chemotherapy, New Orleans, LA, American Society for Microbiology, 1996.

Centers for Disease Control and Prevention (CDC), *Salmonella typhimurium* outbreak traced to commercial apple cider. *MMWR*, 24, 87–88 (1975).

Centers for Disease Control and Prevention (CDC), Epidemiologic notes and reports multi-state outbreak of *Salmonella Poona* infections – United States and Canada, 1991. *MMWR*, 40 (32), 549–552 (1991).

Centers for Disease Control and Prevention (CDC), Outbreak of *Escherichia coli* O157:H7 infections associated with drinking unpasteurized commercial apple juice – British Columbia, California, Colorado, and Washington, October 1996. *MMWR*, 45 (44), 975 (1996a).

Centers for Disease Control and Prevention (CDC), Outbreaks of *Cyclospora cayetanensis* infection – United States. *MMWR*, 45, 549–551 (1996b).

Centers for Disease Control and Prevention (CDC), Update: outbreaks of *Cyclospora cayetanensis* infection – United States and Canada. *MMWR*, 45, 611–612 (1996c).

Centers for Disease Control and Prevention, Hepatitis A associated with consumption of frozen strawberries – Michigan, March 1997. *MMWR*, 46 (13), 288 (1997a).

Centers for Disease Control and Prevention, Outbreak of Cyclosporiasis – Nothern Virginia – Washington, D.C. – Baltimore, Maryland, Metropolitan Area. *MMWR*, 46 (30), 689–691 (1997b).

Centers for Disease Control and Prevention, Outbreaks of *Escherichia coli* O157:H7 infection and cryptosporidiosis associated with drinking unpasteurized apple cider – Connecticut and New York, October 1996. *MMWR*, 46 (1), 4–8 (1997c).

Centers for Disease Control and Prevention (CDC), Outbreaks of *Escherichia coli* O157:H7 infection associated with eating alfalfa sprouts – Michigan and Virginia, June–July 1997. *MMWR*, 46 (32), 741–744 (1997d).

Centers for Disease Control and Prevention, Update: outbreaks of cyclosporiasis – United States and Canada. *MMWR*, 46, 521–530 (1997c).

Centers for Disease Control and Prevention, Foodborne outbreak of cryptosporidiosis – Spokane, Washington, 1997. *MMWR*, 47, 565–567 (1998a).

Centers for Disease Control and Prevention (CDC), Outbreak of *Campylobacter* enteritis associated with cross-contamination of food – Oklahoma, 1996. *MMWR*, 47, 129–131 (1998b).

Centers for Disease Control and Prevention (CDC), Achievements in public health, 1900–1999: safer and healthier foods. *MMWR*, 48 (40), 905–913 (1999a).

Centers for Disease Control and Prevention (CDC), Outbreaks of *Shigella sonnei* infection associated with eating fresh parsley – United States and Canada, July–August 1998. *MMWR*, 48 (14), 285–289 (1999b).

Centers for Disease Control and Prevention (CDC), Prevention of hepatitis A through active or passive immunization: recommendations of the Advisory Committee on Immunization Practices (ACIP). *MMWR*, Recomm. Rep. 48 (RR-12), 1–37 (1999c).

Centers for Disease Control and Prevention (CDC), *Hepatitis Surveillance Report no. 57.* Centers for Disease Control and Prevention, Atlanta, 2000a.

Centers for Disease Control and Prevention (CDC), Surveillance for foodborne disease outbreaks – United States, 1993–1997. *MMWR*, 49 (SS01), 1–51 (2000b).

Centers for Disease Control and Prevention (CDC), Disease burden from hepatitis A, B, and C in the United States, 2002a. Available at: http://www.hepatitisresources-calif.org/pdf/disease_burden2002.pdf, accessed June 2004.

Centers for Disease Control and Prevention (CDC), Multistate outbreaks of *Salmonella* serotype Poona infections associated with eating cantaloupe from Mexico – United States and Canada, 2000–2002. *MMWR*, 51 (46), 1044–1047 (2002b).

Centers for Disease Control and Prevention (CDC), Outbreak of *Salmonella* serotype Kottbus infections with eating alfalfa sprouts – Arizona, California, Colorado, and New Mexico, February–April, 2001. *MMWR*, 51 (10), 7–9 (2002c).

Centers for Disease Control and Prevention (CDC), Summary of notifiable diseases – United States. *MMWR*, 51 (53), 1–84 (2002d).

Centers for Disease Control and Prevention (CDC), Hepatitis A outbreak associated with green onions at a restaurant – Monaca, Pennsylvania. *MMWR*, 52 (47), 1155–1157 (2003).

Centers for Disease Control and Prevention (CDC), Outbreak of cyclosporiasis associated with snow peas – Pennsylvania. *MMWR*, 53 (37), 876–878 (2004a).

Centers for Disease Control and Prevention (CDC), Preliminary FoodNet data on the incidence of infection with pathogens transmitted commonly through food – selected sites, United States, 2003. *MMWR*, 53 (16), 338–343 (2004b).

Centers for Disease Control and Prevention, Foodborne Outbreak Surveillance System (CDC/FOSS), 2004. Available at www.cdc.gov/foodborneoutbreaks/, accessed November 2004.

D. Chew, R. Caraballo, J. Hofmann, Z. Liu, B. Herwaldt, and L. Finelli, *Cyclospora cayetanensis* infection associated with consumption of raspberries, 1996. Program of the 36th Interscience Conference on Antimicrobial Agents and Chemotherapy, New Orleans, LA, American Society for Microbiology.

R. Clemens, The expanding U.S. market for fresh produce, review paper. Center for Agriculture and Rural Development. Available at http://www.agmrc.org/markets/info/expandingusmarketproduce.pdf, accessed April, 2004.

S.H. Cody, M.K. Glynn, J.A. Farrar, K.L. Cairns, P.M. Griffin, J. Kobayashi, M. Fyfe, R. Hoffman, A.S. King, J.H. Lewis, B. Swaminathan, R.G. Bryant, and D.J. Vugia, An outbreak of *Escherichia coli* O157:H7 infection from unpasteurized commercial apple juice. *Ann. Intern. Med.*, 130, 202–209 (1999).

D.G. Colley, Widespread foodborne cyclosporiasis outbreaks present major challenges. *Emerg. Infect. Dis.*, 2, 354–356 (1996).

K. Cook, T. Boyce, C. Langkop, et al., A multistate outbreak of *Shigella flexneri* 6 traced to imported green onions. Presented at the 35th Interscience Conference on Antimicrobial Agents and Chemotherapy, San Francisco, CA, 1995.

K.A. Cook, T.E. Dobbs, W.G. Hlady, J.G. Wells, T.J. Barrett, N.D. Puhr, G.A. Lancette, D.W. Bodager, B.L. Toth, C.A. Genese, A.K. Highsmith, K.E. Pilot, L. Finelli, and D.L. Swerdlow, Outbreak of *Salmonella* serotype Hartford infections associated with unpasteurized orange juice. *JAMA*, 280 (17), 1504–1509 (1998).

T.L. Cover and R.C. Aber, *Yersinia enterocolitica*. *New Engl. J. Med.*, 321, 16–24 (1989).

H. Darbas, M. Riviere, and J. Oberti, *Yersinia* in refrigerated vegetables. *Sci. Aliments*, 5, 81–84 (1985).

C. De Roever, Microbiological safety evaluations and recommendations on fresh produce. *Food Control*, 10, 117–143 (1999).

M. de Simón, C. Tarragó, and M.D. Ferrer, Incidence of *Listeria monocytogenes* in fresh foods in Barcelona [Spain]. *Int. J. Food Microbiol.*, 16, 153–156 (1992).

Department of Health and Human Servies, U.S. Department of Agriculture (DHHS), Nutrition and Your Health: Dietary Guidelines for Americans, 2000. Available at http://www.health.gov/dietaryguidelines/, accessed May 2004.

C. Dimitri and C. Greene, Recent growth patterns in the U.S. organic foods market. Economic Research Service, U.S. Department of Agriculture, 2002. Available at http://www.ers.usda.gov/publications/aib777/, accessed May, 2004.

C. Dimitri, A. Tegene, and P.R. Kaufman, U.S. Fresh Produce Markets: marketing channels, trade practices, and retail pricing behavior. Economic Research Service, U.S. Department of Agriculture, 2003. Available at http://www.ers.usda.gov/publications/aer825/aer825fm.pdf, accessed April 2004.

D.W. Dingman, Growth of *Escherichia coli* O157:H7 in bruised apple (*Malus domestica*) tissue as influenced by cultivar, date of harvest, and source. *Appl. Environ. Microbiol.*, 66 (3), 1077–1083 (2000).

A. dos Reis Tassinari, B. Dora Gombossy de Melo Franco, and M. Landgraf, Incidence of *Yersinia* spp. in foods in Sao Paulo, Brazil. *Int. J. Food Microbiol.* 21 (3), 263–270 (1994).

M.P. Doyle and J.L. Schoeni, Isolation of *Campylobacter jejuni* from retail mushrooms. *Appl. Environ. Microbiol.*, 51 (2), 449–450 (1986).

G.L. Ercolani, Bacteriological quality assessment of fresh marketed lettuce and fennel. *Appl. Environ. Microbiol.*, 31 (6), 847–852 (1976).

Food and Drug Administration (FDA), U.S. Department of Agriculture, Centers for Disease Control and Prevention, Guide to minimize microbial food safety hazards for fresh fruits and vegetables, 1998. Available at http://vm.cfsan.fda.gov/~dms/prodguid.html, accessed November 2000.

Food and Drug Administration, Guidance for Industry: Reducing microbial food safety hazards for sprouted seeds, 1999. Available at http://vm.cfsan.fda.gov/~dms/sprougd1.html, accessed December 2004.

Food and Drug Administration, Center for Food Safety and Applied Nutrition, Office of Plant and Dairy Foods and Beverages, FDA survey of imported fresh produce. FY 1999 Field Assignment, 2001a. Available at http://www.cfsan.fda.gov/~dms/prodsur6.html, accessed April 2004.

Food and Drug Administration, Department of Health and Human Services, FDA publishes final rule to increase safety of fruit and vegetable juices, 2001b. Available at http://vm.cfsan.fda.gov/~lrd/hhsjuic4.html, accessed June 2004.

Food and Drug Administration, Hazard analysis and critical control point (HACCP); procedures for the safe and sanitary processing and importing of juice; final rule. *Fed. Regist.*, 66, 6138–6202 (2001c).

Food and Drug Administration, Center for Food Safety and Applied, Office of Plant and Dairy Foods and Beverages, FDA survey of domestic fresh produce, 2003. Available at http://vm.cfsan.fda.gov/~dms/prodsu10.html, accessed April, 2004.

Food and Drug Administration, FDA issues alert on foodborne illness associated with certain basil and mesculin/spring mix salad products. FDA News, May 21, 2004. Available at http://www.fda.gov/bbs/topics/news/2004/NEW01071.html, accessed November 2004.

D.R. Fenlow, Wild birds and silage as reservoirs of *Listeria* in the agricultural environment, *J. Appl. Bacteriol.*, 59, 537–544 (1985).

A.E. Fiore, Hepatitis A transmitted by food. *Clin. Infect. Dis.*, 38, 705–715 (2004).

C.A. Fleming, D. Caron, J.E. Gunn, and M.A. Barry, A foodborne outbreak of *Cyclospora cayetanensis* at a wedding: clinical features and risk factors for illness. *Arch. Intern. Med.*, 158, 1121–1125 (1998).

G.A. Francis, C. Thomas, and D. Ó Beirne, The microbiological safety of minimally processed vegetables [review article]. *Int. J. Food Sci. Technol.*, 34, 1–22 (1999).

J.A. Frost, M.B. McEvoy, C.A. Bentley, Y. Andersson, and B. Rowe, An outbreak of *Shigella sonnei* infection associated with consumption of iceberg lettuce. *Epidemiol. Infect. Dis.*, 1 (1), 1–4 (1996).

B. Garcia-Villanova Ruiz, R. Galvez Vargas, and R. Garcia-Villanova, Contamination on fresh vegetables during cultivation and marketing. *Int. J. Food Microbiol.*, 4, 285–291 (1987).

E.E. Geldreich, The occurrence of coliforms, fecal coliforms, and streptococci on vegetation and insects. *Appl. Microbiol.*, 12, 63–69 (1964).

M.R. Griffin, J.J. Surowiec, D.I. McCloskey, B. Capuano, B. Pierzynski, M. Quinn, R. Wojnarski, W.E. Parkin, H. Greenberg, and G.W. Gary, Foodborne Norwalk virus. *Am. J. Epidemiol.*, 115, 178–184 (1982).

E. Gutierrez, Japan prepares as O157 strikes again. *Lancet*, 349, 1156 (1997).

C.R. Handy, K. Park, and G.M. Green, Evolving market channels reveal dynamic U.S. produce industry. *Food Rev.*, 23 (2), 14–20 (2000).

S.M. Harmon, D.A. Kautter, and H.M. Solomon, *Bacillus cereus* contamination of seeds and vegetable sprouts grown in a home sprouting kit. *J. Food Prot.*, 50, 62–65 (1987).

L. Harris, J.N. Farber, L.R. Beuchat, M.E. Parish, T.V. Suslow, E.H. Garrett, and F.F. Busta, Outbreaks associated with fresh produce: incidence, growth, and survival of pathogens in fresh and fresh-cut produce, in Analysis and Evaluation of Preventive Control Measures for the Control and Reduction/Elimination of Microbial Hazards on Fresh and Fresh-Cut Produce. *Comprehensive Reviews in Food Science and Food Safety*, 2, 78S–141S (2001).

N.V. Harris, T. Kimball, N.S. Weiss, and C. Nolan, Dairy products, produce and other non-meat foods as possible sources of *Campylobacter jejuni* enteritis. *J. Food Prot.*, 49, 347–351 (1986).

J. Harvey and A. Gilmour, Occurrence and characteristics of *Listeria* in foods produced in Northern Ireland. *Int. J. Food Microbiol.*, 19, 193–205 (1993).

C. Health, An outbreak of *Escherichia coli* O157:H7 infection associated with unpasteurized non-commercial, custom-pressed apple cider – Ontario, 1998. *Can. Communicable Dis. Rep.*, 25 (13), 113–120 (1999).

C.W. Hedberg, K.L. MacDonald, and C. Shapiro, Changing epidemiology of food-borne disease: a Minnesota perspective. *Clin. Infect. Dis.* 18 (5), 671S–680S (1994).

B.L. Herwaldt and M.L. Ackers, An outbreak in 1996 of the cyclosporiasis associated with imported raspberries. *New Engl. J. Med.*, 336, 1548–1556 (1997).

B.L. Herwaldt and M.J. Beach, The return of *Cyclospora* in 1997: another outbreak of cyclosporiasis in North America associated with imported raspberries. *Ann. Int. Med.*, 130 (3), 210–220 (1999).

L.B. Herwaldt, *Cyclospora cayetanenesis*: a review, focusing on the outbreaks of cyclosporiasis in the 1990s. *Clin. Infect. Dis.*, 31 (4), 1040–1057 (2000).

E.D. Hillborn, J.H. Mermin, P.A. Mshar, J.L. Hadler, A. Voetsch, C. Wojtkunski, M. Swartz, R. Mshar, M.A. Lambert-Fair, J.A. Farrar, et al., A multistate outbreak of *Escherichia coli* O157:H7 infections associated with consumption of mesclun lettuce. *J. Am. Med. Assoc.*, 159, 1758–1764 (1999).

S. Himathognkham, H. Riemann, and D. Cliver, Survival of *Escherichia coli* O157:H7 and *Salmonella typhimurium* in cow manure and cow manure slurry. *FEMS Microbiol. Lett.*, 178, 251–257 (1999).

C.W. Hoge, D.R. Shlim, R. Rajah, J. Triplett, M. Shear, J.G. Rabold, and P. Echeverria, Epidemiology of diarrhoeal illness associated with coccidian-like organism among travellers and foreign residents in Nepal. *Lancet*, 341, 1175–1179 (1993).

P. Huang, J.T. Weber, D.M. Sosin, P.M. Griffin, E.G. Long, J.J. Murphy, F. Kocka, C. Peters, and C. Kallick, The first reported outbreak of diarrheal illness associated with *Cyclospora* in the United States. *Ann. Intern. Med.*, 123, 409–414 (1995).

M. Islam, J. Morgan, M.P. Doyle, S.C. Phatak, P. Millner, and X. Jiang, Fate of *Salmonella enterica* serovar Typhimurium on carrots and radishes grown in field treated with contaminated manure composts or irrigation water. *Appl. Environ. Microbiol.*, 70 (4), 2497–2502 (2004).

M.E. Janes, T. Cobbs, S. Kooshesh, and M.G. Johnson, Survival difference of *Escherichia coli* O157:H7 strains in apples of three varieties stored at various temperatures. *J. Food Protect.*, 65 (7), 1075–1080 (2002).

J. Jerngklinchan and K. Saitanu, The occurrence of *Salmonellae* in bean sprouts in Thailand. *Southeast Asean. J. Trop. Med. Public Health*, 24 (1), 114–118 (1993).

X. Jiang, J. Morgan, and M.P. Doyle, Fate of *Escherichia coli* O157:H7 in manure-amended soil. *Appl. Environ. Microbiol.*, 68, 2605–2609 (2002).

D. Katz, S. Kumar, J. Malecki, M. Lowdermilk, E.H. Koumans, and R. Hopkins, Cyclosporiasis and consumption of Guatemalan raspberries – Florida. Presented at the 124th Annual Meeting and Exposition of the American Public Health Association, 1996.

L. Koodie and A.M. Dhople, Acid tolerance of *Escherichia coli* O157:H7 and its survival in apple juice. *Microbios.*, 104 (409), 167–175 (2001).

K.L. Kotloff, J.P. Winickoff, B. Ivanoff, J.D. Clemens, D.L. Swerdlow, P.J. Sansonetti, G.K. Adak, and M.M. Levine, Global burden of *Shigella* infections: implications for vaccine development and implementation of control strategies. *Bull. World Health Organization*, 77, 651–666 (1999).

E.H. Koumans, D. Katz, J. Malecki, S. Wahlquist, S. Kumar, A. Hightower, et al., Novel parasite and mode of transmission: *Cyclospora* infection – Florida. 45th Annual Epidemic Intelligence Service (EIS) Conference, Hyattsville, MD, Public Health Service, 1996.

K. Kretsinger, S. Noviello, M. Moll, et al., Tomatoes sicken hundreds: multistate outbreak of *Salmonella* Newport. Presented at the 52nd Annual Epidemic Intelligence Service Conference, Atlanta, GA, April 2003.

I.T. Kudva, K. Blanch and C.J. Hovde, Analysis of *Escherichia coli* O157:H7 survival in orine or borine manure and manure slurry. *Appl. Environ. Microbiol.*, 64 (9), 3166–3174 (1998).

J.V. Lee, D. Bashford, T. Donovan, A. Furniss, and D. West, The incidence of *Vibrio cholerae* in water, animals and birds in Kent, England. *J. Appl. Bacteriol.*, 52, 281–291 (1982).

T.-S. Lee, S.-W. Lee, W.-S. Seok, M.-Y. Yoo, J.-W. Yoon, B.-K. Park, K.-D. Moon, and D.-H. Oh, Prevalence, antibiotic susceptibility, and virulence factors of *Yersinia enterocolitica* and related species from ready-to-eat vegetables available in Korea. *J. Food Prot.*, 67 (6), 1123–1127 (2004).

T. Lilly, H.M. Solomon, and E.J. Rhodehamel, Incidence of *Clostridium botulinum* in vegetables packaged under vacuum or modified atmosphere. *J. Food Prot.*, 59 (1); 59–61 (1996).

C.-M. Lin, S.Y. Fernando, and C.-I. Wei, Occurrence of *Listeria monocytogenes, Salmonella* spp., *Escherichia coli*, and *E. coli* O157:H7 in vegetable salads. *Food Control*, 7, 135–140 (1996).

C.L. Little, D. Roberts, E. Youngs, and J. de Louvois, Microbiological quality of retail imported unprepared whole lettuces: a PHLS Food Working Group Study. *J. Food Prot.*, 62 (4), 325–328 (1999).

F. Llopis, I. Grau, F. Tubau, M. Cisnal, and R. Pallares, Epidemiological and clinical characteristics of bacteraemia caused by *Aeromonas* spp. as compared with *Escherichia coli* and *Pseudomonas aeruginosa*. *Scand. J. Infect. Dis.*, 35 (5), 335–341 (2004).

Y. Lou and A.E. Yousef, "Characteristics of *Listeria monocytogenes* important to the food processor", in E.T. Ryser and E.H. Marth, Eds., *Listeria, Listeriosis, and Food Safety*. Marcel Dekker, New York, 1999, pp. 131–224.

P.W. Lowry, R. Levine, D.F. Stroup, R.A. Gunn, M.H. Wilder, and C. Konisberg, Hepatitis A outbreak on a floating restaurant in Florida. *Am. J. Epidemiol.*, 129, 155–164 (1989).

N. Luechtefeld, M. Blaser, L. Reller, and W. Wang, Isolation of *Campylobacter* fetus subsp. *jejuni* from migratory wildfowl. *J. Clin. Microbiol.* 12, 406–408 (1980).

A. Luedtke and D.N. Powell, A review of North American apple cider–associated *E. coli* O157:H7 outbreaks, media coverage and a comparative analysis of Ontario apple cider producers' information sources and production practices. *Dairy Food Environ. Sanit.*, 22, 590–598 (2002).

A.J. Lung, C.-M. Lin, J.M. Kim, M.R. Marshall, R. Nordstedt, N.P. Thompson, and C.-I. Wei, Destruction of *Escherichia coli* O157:H7 and *Salmonella enteritidis* in cow manure composting. *J. Food Prot.*, 64 (9), 1309–1314 (2001).

J.M. Madden, Microbial pathogens in fresh produce – the regulatory perspective. *J. Food Prot.* 55 (10), 821–823 (1992).

R. Marchetti, M.A. Casadei, and M.E. Guerzoni, Microbial population dynamics in ready-to-use vegetable salads. *Ital. J. Food Sci.*, 2, 97–108 (1992).

L.M. Martins, R.F. Marquez, and T. Yano, Incidence of toxic *Aeromonas* isolated from food and human infection. *Immunol. Med. Microbiol.*, 32, 237–242 (2002).

J.L. Mawdsley, R.D. Bardgett, R.J. Merry, B.F. Pain, and M.K. Theodorou, Pathogens in livestock waste their potential for movement through soil and environmental pollution. *Appl. Soil Ecology*, 2, 1–15 (1995).

P.S. Mead, L. Slutsker, V. Dietz, L.F. McCaig, J.S. Bresee, C. Shapiro, P.M. Griffin, and R.V. Tauxe, Food-related illness and death in the United States. *Emerg. Infect. Dis.*, 5, 607–625 (1999).

J. Meng, M.P. Doyle, T. Zhao, and S. Zhao, "Enterohemorrhagic *Escherichia coli*," in M.P. Doyle, L.R. Beuchat, and T.J. Montville, Eds., *Food Microbiology, Fundamentals and Frontiers*. ASM Press, Washington, DC, 2001, pp. 193–213.

E.D. Mintz, M. Hudson-Wragg, P. Mshar, M.L. Cartter, and J.L. Hadler, Foodborne giardiasis in a corporate office setting. *J. Infect. Dis.*, 167, 250–253 (1993).

R. Monge and M. Chinchilla, Presence of *Cryptosporidium* oocysts in fresh vegetables. *J. Food Prot.*, 59 (2), 202–203 (1996).

D.L. Morse, L.K. Pickard, J.J. Guzewich, B.D. Devine, and M. Shayegani, Garlic-in-oil associated botulism: episode leads to product modification. *Am. J. Pub. Health*, 80 (11), 1372–1373 (1990).

A. Mukherjee, D. Speh, E. Dyck, and F. Diez-Gonzalez, Preharvest evaluation of coliforms, *Escherichia coli, Salmonella*, and *Escherichia coli* O157:H7 in organic and conventional produce grown by Minnesota farmers. *J. Food Prot.*, 67 (5), 894–900 (2004).

I. Nachamkin, "*Campylobacter jejuni*," in M.P. Doyle, L.R. Beuchat, and T.J. Montville, Eds., *Food Microbiology, Fundamentals and Frontiers.* ASM Press, Washington, DC, 2001, pp. 179–192.

S. Neamatullah, D. Manuel, and D. Werker, Investigation of Cyclospora outbreak associated with consumption of fresh berries. 64th Conjoint Meeting on Infectious Diseases, Edmonton, Alta. Canadian Association for Clinical Microbiology and Infectious Diseases, 1996.

F.A. Nicholson, S.J. Groves, and B.J. Chambers, Pathogen survival during livestock manure storage and following land application. *Bioresource Technology* **96**: 135–143 (2005).

C. Nguyen-the and F. Carlin, "Fresh and Processed Vegetables," in B.M. Lund, T.C. Baird-Parker, and G.W. Gould, Eds., *The Microbiological Safety and Quality of Food.* Aspen publishers, Gaithersburg, MD, 2000, pp. 620–684.

S. Notermans, J. Dufrenne, and J.P.G. Gerrits, Natural occurrence of *Clostridium botulinum* on fresh mushrooms (*Agaricus bisporus*). *J. Food Prot.*, 52 (10), 733–736 (1989).

J.A. Odumeru, S.J. Mitchell, D.M. Alves, J.A. Lynch, A.J. Yee, S.L. Wang, S. Styliadis, and J.M. Farber, Assessment of the microbiological quality of ready-to-use vegetables for health-care food services. *J. Food Prot.*, 60 (8), 954–960 (1997).

S.J. Olsen, L.C. Mackinon, J.S. Goulding, N.H. Bean, and L. Slutsker, Surveillance for food-borne disease outbreaks – United States, 1993–1997. *MMWR*, 49, 1–62 (2000).

Outbreak in Denmark of *Shigella sonnei* infections related to uncooked "baby maize" imported from Thailand. Eurosurveillance Weekly, 1998. Available at http://www.euro surveillance.org/ew/1998/980813.asp#1, accessed June 2004.

M.E. Parish, J.A. Narciso, and L.M. Friedrich, Survival of *Salmonellae* in orange juice. *J. Food Safety*, 17, 273–281 (1997).

C.E. Park and G.W. Sanders, Occurrence of thermotolerant campylobacters in fresh vegetables sold at farmers' outdoor markets and supermarkets. *Can. J. Microbiol.*, 38, 313–316 (1992).

F. Picard-Bonnaud, J. Cottin, and B. Carbonnelle, Persistence of *Listeria monocytogenes* in three sorts of soil. *Acta Microbiol. Hung.*, 36 (2–3), 263–270 (1989).

A. Ponka, L. Maunula, C.-H. von Bonsdorff, and O. Lyytikäinen, An outbreak of calicivirus associated with consumption of frozen raspberries. *Epidemiol. Infect.*, 123, 469–474 (1999).

M.E. Potter, S.G. Ayala, and N. Silarug, "Epidemiology of Foodborne Diseases," in M.P. Doyle, L.R. Beuchat, and T.J. Montville, Eds., *Food Microbiology: Fundamentals and Frontiers.* American Society for Microbiology, Washington, DC, 1996, pp. 376–390.

D. Powell, Risk-Based Regulatory Responses in Global Food Trade: Guatemalan Raspberry Imports into the U.S. and Canada, 1996–1998. Paper presented at the Conference on

Science, Government, and Global Markets: the State of Canada's Science-Based Regulatory Institutions, Carleton Research Unit of Innovation, Science and Environment, Ottawa, October 1–2, 1998.

A.M. Prazak, E.A. Murano, I. Mercado, and G.R. Acuff, Prevalence of *Listeria monocytogenes* during production and postharvest processing of cabbage. *J. Food Prot.*, 65 (11), 1728–1734 (2002).

D. Prokopowich and G. Blank, Microbiological evaluation of vegetable sprouts and seeds. *J. Food Prot.*, 54 (7), 560–562 (1991).

S. Quessy and S. Messier, *Campylobacter* spp. and *Listeria* spp. in ring-billed gulls (*Larus delawarensis*). *J. Wildl. Dis.*, 28, 526–531 (1992).

J.G. Rabold, C.W. Hoge, D.R. Shlim, C. Kefford, R. Rajah, and P. Echeverria, *Cyclospora* outbreak associated with chlorinated drinking water. *Lancet*, 344, 1360–1361 (1994).

T.M.S. Reid and H.G. Robinson, Frozen raspberries and hepatitis A. *Epidemiol. Infect.*, 98, 109–112 (1987).

A.A. Ries, S. Zaza, C. Langkop, R.V. Tauxe, and P.A. Blake, A multistate outbreak of *Salmonella chester* linked to imported cantaloupe. Interscience Conf. Antimicrob. Agents Chemother., American Society of Microbiology, Washington, DC, 1990.

D. Roberts, G.N. Watson, and R.J. Gilbert, "Contamination of Food Plants and Plant Products with Bacteria of Public Health Significance," in R. Rhodes-Roberts and F.A. Skinner, Eds., *Bacteria and Plants*. Academic Press, London, 1982, pp. 169–195.

R.A. Rude, G.J. Jackson, J.W. Bier, T.K. Sawyer, and N.G. Risty, Survey of fresh vegetables for nematodes, amoebae, and *Salmonella*. *J. AOAC Int.* 67, 613–615 (1984).

M.F. Saddik, M.R. El-Sherbeeny, and F.L. Bryan, Microbiological profiles of Egyptian raw vegetables and salads. *J. Food Prot.* 48 (10), 883–886 (1985).

W.F. Schlech III, P.M. Lavigne, R.A. Bortolussi, A.C. Allen, E.V. Haldane, A.J. Wort, A.W. Hightower, S.E. Johnson, S.H. King, E.S. Nicholls, and C.V. Broome, Epidemic Listeriosis – Evidence for transmission by food. *N. Engl. J. Med.*, 308 (4), 203–206 (1983).

J.M. Shields and B.H. Olson, *Cyclospora cayetanensis*: a review of an emerging parasitic coccidian. *Int. J. Parasitol.*, 33, 371–391 (2003).

S. Sivapalasingam, C.R. Friedman, L. Cohen, and R.V. Tauxe, Fresh produce: a growing cause of outbreaks of foodborne illness in the United States, 1973 through 1997. *J. Food Prot.*, 67 (10), 2342–2353 (2004).

S. Sivapalasingam, E. Barrett, A. Kimura, S. Van Duyne, W. De Witt, M. Ying, A. Frisch, Q. Phan, E. Gould, P. Shillam, V. Reddy, T. Cooper, M. Hoekstra, C. Higgins, J.P. Sanders, R.V. Tauxe, and L. Slutsker, A multistate outbreak of *Salmonella enterica* serotype Newport infections linked to mango consumption: impact of water dip disinfestation technology. *Clin. Infect. Dis.*, 37, 1585–1590 (2003).

K. Sizmur and C.W. Walker, *Listeria* in prepacked salads. *Lancet*, 1167 (1988).

T.R. Slifko, et al., Emerging parasite zoonoses associated with water and food. *Int. J. Parasitol.*, 30, 1379–1393 (2000).

D.F. Splittstoesser, D.T. Queale, and B.W. Andaloro, The microbiology of vegetable sprouts during commercial production. *J. Food Safety*, 5, 79–86 (1983).

W. Steffen, H. Mochmann, I. Kontny, U. Richter, U. Werner, and O. el Naeem, A food-borne infection caused by *Campylobacter jejuni* serotype Lauwers 19. *Zentralbl. Bakteriol. Mikrobiol. Hyg. [B]*, 183, 28–35 (Abstract) (1986).

E.D. Steinbrugge, R.B. Maxcy, and M.B. Liewen, Fate of *Listeria monocytogenes* on ready to serve lettuce. *J. Food Prot.*, 51, 596–599 (1988).

Supermarket Business, Annual Produce Department Review, October 1999.

T.V. Suslow, M.P. Oria, L.R. Beuchat, E.H. Garrett, M.E. Parish, L.J. Harris, J.N. Farber, and F.F. Busta, "Production Practices as Risk Factors in Microbial Food Safety of Fresh and Fresh-Cut Produce," in *Analysis and Evaluation of Preventive Control Measures for the Control and Reduction/Elimination of Microbial Hazards on Fresh and Fresh-Cut Produce. Comprehensive Reviews in Food Science and Food Safety*, 2, 38S–77S (2001).

B. Swaminathan. "*Listeria monocytogenes*," in M.P. Doyle, L.R. Beuchat, and T.J. Montville, Eds., *Food Microbiology, Fundamentals and Frontiers*, ASM Press; Washington, DC, 2001, pp. 383–409.

E.A. Szabo, K.J. Scurrah, and J.M. Burrows, Survey for psychrotrophic bacterial pathogens in minimally processed lettuce. *Lett. Appl. Microbiol.*, 30 (6), 456–460 (2000).

S. Tamblyn, S. Tamblyn, J. deGrosbois, D. Taylor, and J. Stratton, An outbreak of *Escherichia coli* O157:H7 infection associated with unpasteurized non-commercial, custom-pressed apple cider – Ontario. *Can. Commun. Dis. Rep.*, 25, 113–117 (1999).

S.K. Tamminga, R.R. Beumer, and E.H. Kampelmacher, The hygienic quality of vegetables grown in or imported into the Netherlands: a tentative study. *J. Hyg. Camb.*, 80, 143–154, (1978).

P.J. Taormina, L.R. Beuchat, and L. Slutsker, Infections associated with eating seed sprouts: an international concern. *Emerg. Infect. Dis.*, 62, 626–634 (1999).

R.V. Tauxe, Emerging foodborne diseases: an evolving public health challenge. *Emerg. Infect. Dis.*, 3 (4), 425–434 (1997).

R.L. Thunberg, T.T. Tran, R.W. Bennett, R.N. Matthews, and N. Belay, Microbial evaluation of selected fresh produce obtained at retail markets. *J. Food Prot.*, 65 (4), 677–682 (2002).

U.S. Bureau of the Census, World Population Profile: 1998, pp. 1–2. Available at http://www.census.gov/ipc/www/wp98001.html, accessed October 2004.

United States Department of Agriculture (USDA), Economic Research Service. Food Review: Consumer-driven agriculture 2002, R. Mentzer Morrison, Ed., 25 (1).

Y. Watanabe, K. Ozasa, J.H. Mermin, P.M. Griffin, K. Masuda, S. Imashuku, and T. Sawada, Factory outbreak of *Escherichia coli* O157:H7 infection in Japan. *Emerg. Infect. Dis.*, 5 (3), 424–428 (1999).

J.G. Wells, B.R. Davis, I.K. Wachsmuth, L.W. Riley, R.S. Remis, R. Sokolow, and G.K. Morris, Laboratory investigation of hemorrhagic colitis outbreaks associated with a rare *Escherichia coli* serotype. *J. Clin. Microbiol.*, 18, 512–520 (1983).

H.J. Welshimer and J. Donker-Voet, *Listeria monocytogenes* in nature. *Appl. Environ. Microbiol.* 21, 516–519 (1971).

World Health Organization and Food and Agriculture Organization (WHO), WHO and FAO announce global initiative to promote consumption of fruit and vegetables. Media centre, 2003. Available at http://www.who.int/mediacentre/releases/2003/pr84/en/, accessed June 2004.

H. Zepeda-Lopez, M. Ortega-Rodriquez, E.I. Quinonez-Ramirez, and C. Vazguez-Salinas, Isolation of *Escherichia coli* O157:H7 from vegetables. Annual Mtg Am. Soc. Microbiol (abstracts), Washington, DC, 1995.

CHAPTER 3

Microbiological Risk in Produce from the Field to Packing

ELMÉ COETZER

EUREPGAP, c/o FoodPLUS GmbH, Köln, Germany

3.1 INTRODUCTION

It is well known that the incorporation of fresh fruit and vegetables into the diet of modern man has various health benefits. However, because these products are

Microbial Hazard Identification in Fresh Fruit and Vegetables, Edited by Jennylynd James

consumed raw or minimally processed, there are several serious heath risks involved. Fruit and vegetables, as a result of production methods, and because they are often eaten without washing, can be carriers of pathogens, known as foodborne pathogens. Outbreaks associated with fruit and vegetables include bacteria such as *Escherichia coli* O157:H7, *Samonella* spp., *Shigella* spp., *Listeria monocytogenes*, *Campylobacter*, and *Bacillus cereus*, of which *E. coli* and *Salmonella* are the most important. Parasites such as *Cryptosporidium parvum* and *Cyclospora cayetanensis*, and viruses like hepatitis A and Norwalk have also been indicated as causative agents in several produce-associated outbreaks (McCabe-Sellers and Beattie, 2004; Sperber, 2003; De Roever, 1999). Fruit and vegetables that have been associated with these disease-causing microorganisms include lettuce, green onions, tomatoes, sprouts, cantaloupe, carrots, and raspberries, as well as unpasteurized apple cider and orange juice. For this reason, fresh produce has to be added to the traditional list of foods requiring careful selection and handling to prevent foodborne disease.

This chapter will focus on the microbiological risks associated with production of fruit and vegetables (production processes, harvesting, and low-risk, postharvest handling) as well as control measures to minimize risks. It is generally not believed that all potential food-safety hazards associated with fresh produce that are eaten raw can be eliminated by current technologies. Therefore, this chapter will emphasize risk reduction. This chapter will not specifically address other concerns for the food supply or the environment such as crop protection product residues or chemical contamination. However, possible environmental issues resulting from some practices in crop production causing transfer of pathogens to food will be addressed.

Prevention of foodborne disease in fresh produce must start at the farm level. According to Sperber (2005), the supply chain consists of seven links: crop production, harvest, raw product production, processed product production, food service or retail operations, and, finally, consumption (Fig. 3.1). It is relatively easy to implement effective controls, such as cooking and drying during the processing step. It is, however, not so easy to implement effective control measures at the crop production, harvest, or raw product production stages. Hazard Analysis Critical Control Points (HACCP), a preventive system in which food safety can be designed into the product and the process by which it is produced, is a useful system for processing of food where the procedures include steps in which the potential risk can be measured and action be taken to negate it. When considering applying HACCP principles to a farm operation, one can immediately see the difficulty in controlling naturally occurring hazards. For example, bird or other wild animal droppings in an orchard may potentially represent a hazard from the spread of *E. coli* O157:H7 or *Salmonella* spp. According to the principles of HACCP, this should be a critical control point (CCP), but in reality, it may not be one, because there is no way to prevent this hazard by controlling a production process. Furthermore, there is no way to quantify and measure the droppings to know if they are within critical limits. This would also be true of *Clostridium botulinum* spores in soil. Although they may represent a potential hazard, it would not be appropriate to establish the

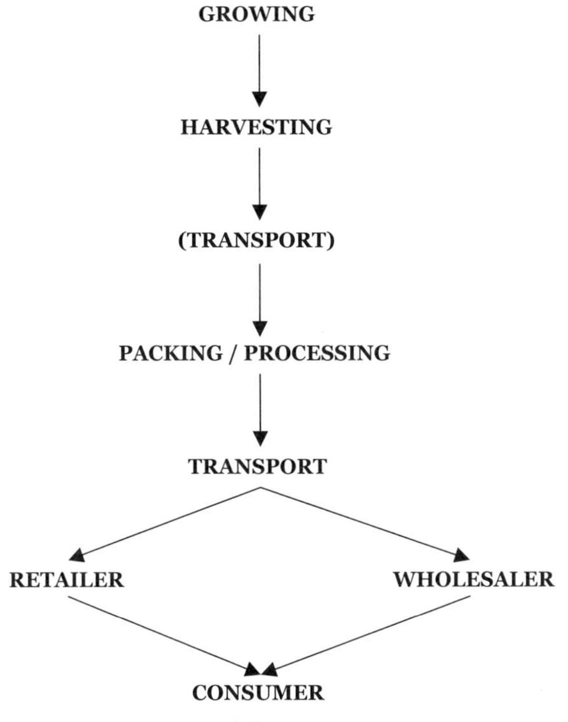

Figure 3.1 Supply chain.

microbiological safety of soil as a CCP because it is not practical to measure the spores in soil or to control them through any known process.

In fact, most agricultural hazards cannot, and should not, be prevented through a HACCP program. Instead, the use of Good Agricultural Practices (GAPs) has been identified by the produce industry as a more appropriate way to address these hazards. Prospective interventions, as part of the prerequisite program, include cleaning and sanitation, worker hygiene and training, pest control, and GAPs (clean water for irrigation and postharvest washing, and at handwashing facilities for workers; chemical control, and so on). Various field operations during crop production can increase the risk of microbiological contamination of the produce. The potential control points that can reduce microbiological hazards in fresh produce during preharvest field operations as well as during harvest and postharvest are set in out in Table 3.1.

Fresh produce can become microbiologically contaminated at any point along the farm-to-table chain. The major source of microbial contamination in fresh produce is associated with human or animal feces. A major hazard is the use of contaminated water that comes in contact with produce. Its source and quality dictate the potential for contamination. Because animal manure or municipal biosolid wastes are

TABLE 3.1 Potential Control Points Pre- and Postharvest

Stage	Operation	Potential Risk-Reducing Control Point
Production	Crop–site selection	Prior use, soil, water, wild and domestic animals, run-off, drift
	Irrigation	Water quality and method
	Preplant and in-season fertilization	Biosolids/animal manure, Organic fertilizers, foliar fertilizers (water quality)
	Pest control	Vermin, wild and domestic animals
	Other (dust control with water)	Water quality
Harvest	Harvesting	Worker hygiene, equipment, water quality, harvesting containers, packaging material
	Cooling	Water quality, air quality, facility
Postharvest	Sorting/washing	Worker hygiene, water quality, equipment, facility, wild and domestic animals
	Packing	Worker hygiene, facility, packaging material
	Storage	Facility, temperature

potential sources of pathogens, practices using it should be managed closely to minimize the potential microbial contamination of fresh produce. Worker hygiene and sanitation practices during production, harvesting, sorting, packing, and transport can also play a critical role in minimizing the potential for microbial contamination.

The discussion in this chapter will therefore specifically focus at pathogen contamination during (1) production, (2) harvesting, and (3) postharvest handling, and on ways to minimize these risks in each situation.

Several food-safety inspection systems have been developed to ensure that GAPs are implemented on farms. In response to consumer concerns, retailers have started to require an independent inspection for their suppliers. The benefits of auditing programs for growing and packing operations as well as retailers and consumers are discussed.

3.2 PRODUCTION

3.2.1 Land History and Adjacent Land Use

The main concern with site selection is the possibility that crops can be contaminated by growing in soils that are microbiologically tainted. The contaminants could have been introduced if the land was previously used for animal husbandry practices such as a feedlot or chicken hatchery, industrial dumping, or if biosolids/sludge or animal waste have been applied.

To ensure that land is suitable for the intended use when purchasing or leasing new ground with the intent to grow produce for human consumption, a risk assessment should be considered. One should be able to use land that grew produce for human consumption previously, without incidence. Fields must be upstream and upwind from any animal containment (Brackett, 1999).

Adjacent land use can also cause microbial contamination of produce. When adjacent grounds or nearby farms are used for animal husbandry or rendering (e.g., grazing, housing, feeding, or slaughtering), microbial contamination (*E. coli* O157:H7 in cattle and dairy cows, *Salmonella* in fowls) resulting from fecal matter, and sick or dead animals can occur. The contaminants can be introduced by

1. Movement of animals from pasture, feedlot, and so on;
2. Movement of animal waste;
3. Ground sloping toward the crop, contaminated by run-off from rain;
4. Application of manure on adjacent growing areas; and
5. Contamination from decaying carcasses.

Animal movement should be restricted with proper fencing or other suitable physical barriers. Rain induced run-off of animal waste should be diverted by trenching or similar land preparation. The use of electrical charge on fencing to scare off wild animals and to keep, for example, cattle out of a production area is an option. The value of this option will depend on the severity of the identified risk. If there is not a high risk of wild or stray animals contaminating the production area, this is probably not the best method from a financial viewpoint. Overall, the decision to fence a farm will depend on many factors, such as size of the farm, legislation, production of crops and livestock, and so on. Contamination caused by diseased or dead animals can be avoided with monitoring and a procedure for quick removal and disposal of the carcasses.

3.2.2 Irrigation

Water can be a vector for many microorganisms, including pathogenic strains of *E. coli*, *Salmonella* spp., *Vibrio cholerae*, *Shigella* spp., *Cryptosporidium parvum*, *Giardia lamblia*, *Cyclospora cayetanesis*, *Toxoplasma gondii*, and the Norwalk and hepatitis A viruses (WHO, 1989, 2004). Even small amounts of contamination with some of these organisms can result in foodborne illness. Irrigation water and water used to mix and apply crop protection products, if contaminated with harmful microorganisms, can spread pathogens to the crops. Considering surface water is commonly used for irrigation on fruit and vegetable farms, care must be taken that upstream neighbors keep animals out of the waterways and prevent feedlot run-off from entering streams. Fencing around water storage would keep animals out (Fig. 3.2).

Water quality is more important when water comes in direct contact with the edible parts of the plant, especially close to harvest. Irrigation methods that will

Figure 3.2 Fenced water storage to keep animals from contaminating surface water.

reduce the risk of spreading pathogens to fresh produce need to be selected. This will depend on the crop as well as the water quality. Water samples must be sent to a reputable laboratory for analysis of fecal coliforms at a frequency based on a risk assessment.

Drinking water must be free of waterborne pathogens. *E. coli* is the most numerous and specific bacterial indicator of fecal pollution from humans and animals. Thus, it must not be present in 100 ml samples of any water intended for drinking (WHO, 2004). According to the Engelberg Guidelines for the microbiological quality of treated wastewater intended for crop irrigation, <1000 fecal coliform bacteria per 100 ml for unrestricted irrigation are allowed. Included in unrestricted irrigation is the irrigation of industrial crops, fruit trees, and pastures, and restricted irrigation of edible crops (WHO, 1989).

3.2.3 Fertilization

Fertilizer use increased steadily between 1950 and 1988, but has stabilized since then (Lægreid et al., 1999). The use of organic fertilizers has increased over the last few years, as the demand for "organic production" has increased. Manure can be used successfully to replace or supplement fertilizers depending on the amount of nutrients available. There are, however, risks involved when utilizing the benefits of organic fertilizers. The use of synthetic fertilizers does not contribute to microbiological risk as is associated with the use organic fertilizers such as manure, guano, and biosolids. Proper treatment is critical to ensure safety as the use of untreated or improperly treated manure or biosolids for fertilizer or soil enhancement may cause run-off to surface or groundwaters. Run-off may contain pathogens that can contaminate produce, such as *E. coli*, *Salmonella*, and *Cryptosporidium* harbored in human and animal feces.

Manure. It has been shown that *E. coli* O157:H7, *Salmonella*, *Listeria*, and *Campylobacter* survive in stored slurries and dirty water for up to three months. In contrast, all these pathogens survive for less than one month in solid manure heaps where temperatures greater than 55°C are obtained. Following manure spreading to land, *E. coli* O157:H7, *Salmonella*, and *Campylobacter* generally survived in the soil for up to one month after application to both the sandy arable and clay loam grassland soils, whereas *Listeria* commonly survived for more than one month (Nicholson et al., 2005). There are conflicting opinions on the effect of bovine manure as fertilizer. In a trial where manure was used in production of lettuce (Johannessen et al., 2004), *E. coli* O157:H7 was isolated in the fertilized soil, but none was recovered from the lettuce. This suggested that the use of manure (as compost, firm manure, or slurry) does not influence the bacteriological quality of organic lettuce considerably under Norwegian growing conditions. However, as humans may ingest soil adhering to crops, there should be sufficient time interval between manure application and the harvest or planting of crops to allow any pathogens to die off. This is especially important for those crops that are likely to be consumed raw.

Farmers handling and applying animal manure should follow GAPs to reduce the introduction of microbial hazards to the produce. Composting is a method that can be used successfully to reduce possible levels of pathogens in manure. Another practice to decrease the risk of contamination of produce is to avoid direct contact of manure with the produce close to harvest.

Pathogens in animal manure can be reduced through composting, an active and controlled process. The methodology relies on high temperatures (reached during the appropriate conditions) to kill most pathogens. Although the appropriate conditions are also reached over a period of time, composting must be differentiated from other methods where manure is aged over a longer period of time without exposure to high temperatures or high pH. Detailed records of the composting process such as temperature logs, turn frequency, and microbial test results must be kept. If manure is bought, it must be sourced from a reputable supplier that can supply the grower with this information.

Care should be taken when storing organic fertilizers and biosolids on the farm to avoid contamination of surface water or well/borehole water as well as fresh produce production and handling sites. Slope of the land, likelihood of rainfall, and quantity of manure are but a few examples of factors that should be taken into account when laying out an area specific for manure storage. It might also be necessary to consider physical barriers such as concrete blocks or lagoons. Stored animal waste or manure should be covered with plastic tarps or comparable material to minimize the possibility of movement as well as recontamination of treated compost by domestic animals or birds.

Other good practices that should be considered to minimize the potential of contamination include the following:

1. Manure applications should be timed to avoid conflicts with growing scheduled in adjacent fields. This reduces possible contamination of these fields and/or crops.
2. Crop rotation should be planned so manure applications are made to fields planted with crops that are to be cooked or heat-processed prior to consumption.
3. No root or leafy crops must be grown in the year that manure is applied to the field.
4. Growers should consider incorporating raw manure into the soil prior to planting, because competition with soil microorganisms might reduce pathogens.
5. Equipment such as tractors that were in contact with untreated manure and that are also used in fields with produce must be cleaned to prevent spread of pathogens to the produce.
6. The time between application of raw manure and harvest must be maximized and applications should not be made during the growing season prior to harvest (Rangarajan et al., 2000; FDA, 1998).

Sewage Sludge. *Sewage sludge* and *biosolids* are two technical terms that are often used when talking about organic soil conditioners and fertilizers and is what is left behind after water is cleaned in waste treatment works. The term biosolids is now widely used instead of sewage sludge. Biosolids specifically refers to sewage sludge that has undergone treatment and meets set standards for beneficial use. It is high in organic content and plant nutrients and, in theory, makes good fertilizer. However, it may also contain pathogenic bacteria, viruses, protozoa, parasites, and other microorganisms that are disease-causing agents. Most developed countries regulate its use because of this and also because it can contain a multitude of metals and organic pollutants (EPA, 1993).

In the United Kingdom, biosolid use ceased at the end of 1999 as part of the "Safe Sludge Matrix" agreement between the UK Water Industry and the British Retail Consortium (ADAS, 2001). The matrix provided clear guidance on the minimum acceptable level of treatment of any sewage sludge (biosolids) based product that may be applied to crops. The main impact of this agreement was the phasing out of untreated sewage sludge use on agricultural land for food production. The use of untreated sludge on nonfood crops (such as willow and poplar for coppicing, hemp for fiber, and so on) was acceptable until 31 December 2005. Stringent requirements were set for biosolids applied to land growing vegetable crops, particularly those crops usually consumed raw. The matrix differentiated between conventionally treated and enhanced treated sludges. Conventionally treated sludge was subjected to treatment processes that ensure at least 99.99 percent of the pathogens were destroyed. There are different processes that can be used to reduce the

TABLE 3.2 Sewage Sludge Matrix (Adapted from ADAS, 2001)

Crop Group	Untreated Sludge	Conventionally Treated Sludge	Harvest Interval (Months)	Enhanced Treated Sludge	Harvest Interval
Salads	×	×	30	✓	10
Vegetables	×	×	12	✓	10
Fruits	×	×	—	✓	10
Horticulture	×	×	—	✓	10

×, applications are not allowed, except where stated harvest interval conditions apply.
✓, applications are allowed with 10 months harvest interval.
Salads = ready to eat crops, e.g. lettuce, cabbage, carrots, spinach.
Vegetables = potatoes, marrows, sweetcorn, etc.
Fruits = top fruits, stone fruit, soft fruit, vines, hops, nuts.
Horticulture = Crops grown under protection (e.g. tomatoes, cucumbers), mushrooms, seed potatoes.

fermentation and these rely on biological, chemical, or heat treatments. Enhanced treated sludge refers to sludge that has undergone treatment, which is capable of virtually eliminating the pathogens present. It is stated that enhanced treated sludge will be free from *Salmonella* and that there will be a 6-log reduction of other pathogens. Table 3.2 is an extract of the Safe Sludge Matrix showing the restrictions for untreated sludge and treated sludge.

In the European Union, Council Directive 86/278/EEC (EEC, 1986) on the "protection of environment, and in particular of the soil, when sewage sludge is used in agriculture," Article 7(b) prohibits the use of sludge on fruit and vegetable crops during the growing season, except for tree fruit crops. Article 7(c) states that sludge application to ground intended for the cultivation of fruit and vegetables that are normally in direct contact with the soil and eaten raw must be avoided for a period of 10 months preceding the harvest of the crops and application must be prohibited during harvest itself. According to Langenkamp and Part (2001), organic contaminants in sludge are not expected to pose major health problems to the human population when sludge is reused for agricultural production. Metal contamination of sludge on the other hand is much more important with respect to human health.

In the United States, the use of treated sewage sludge is regulated under 40 CFR (Code of Federal Regulations) Part 503, 1993. Subpart D of the 503 Regulation also makes provision for protection of public health and the environment through requirements designed to reduce the potential for contact with the disease-bearing pathogens in sewage sludge applied to land. These requirements are divided into requirements designed to control and reduce pathogens in biosolids and requirements designed to reduce the ability of the biosolids to attract vectors. Biosolids are classified either as Class A or as Class B with respect to pathogens. If pathogens, including *Salmonella* spp., bacteria, enteric viruses, and viable helminth ova, are below detectable levels, the biosolids meet Class A designation. Class B biosolids are those in which pathogens are detectable but have been reduced to levels that do not pose a threat to public health and the environment, as long as actions are taken to prevent exposure after their use. Certain restrictions, similar to those in

TABLE 3.3 Harvest Restrictions when Class B Biosolids are Used for Agricultural Production

Crop	Harvest Interval After Biosolids Application
Edible parts do not touch the soil surface	30 days
Edible parts touch biosolids/soil mixture, but totally above ground	14 months
Edible parts below ground, where biosolids remain on the soil surface for 4 months or longer prior to incorporation into the soil	20 months
Edible parts below ground, where biosolids remain on the soil surface for less than 4 months prior to incorporation into the soil	38 months

the UK, are imposed when Class B biosolids are applied to agricultural land for production of fresh produce (Table 3.3).

This chapter deals with the microbiological aspects of biosolids. However, there is evidence to suggest that possible contamination by heavy metals presents a more serious hazard than microorganisms when using biosolids as fertilizer in production of fresh produce (Harrison and Oakes, 2002; Harrison et al., 1999).

Other. Other organic manures that are less frequently used as fertilizers for production of fresh produce on a large scale include bat guano, fish emulsions, and chicken manure. As previously said, manure must be composted to lessen the chance of harmful pathogens contaminating the produce. Chicken manure decomposes the fastest of all manures. It must be composted to decrease *Salmonella* and other harmful bacteria. Bat guano (feces) harvested from caves is powdered. It can be applied directly to the soil, made into a tea and applied as a foliar spray, or injected into an irrigation system. Processing by beetles and decomposing microbes render bat guano free of toxins and pathogens dangerous to humans. Although relatively safe to use, bird and bat guano, as well as chicken manure from old poultry houses, can harbor *Histoplasm capsulatum*, a fungus, which causes histoplasmosis in humans. In severe cases this leads to symptoms similar to pneumonia and can even be fatal (Kuepper, 2003). The fungal spores are airborne, and it is advisable to wet down dried guano to reduce dust. Infamous for its foul smell, fish emulsions are soluble, liquid fertilizers that have been heat- and acid-processed from fish waste. Killing aerobic bacteria during composting will break down any disease-causing pathogens that may exist in decaying fish.

3.2.4 Other

Feces from other animals, such as birds and dogs, can also cause contamination of produce during its production period. Although it is not possible to exclude completely all animals from the production areas, growers have to assess the risk posed by animals entering the production area (birds, deer, chickens, dogs, and so on) and minimize access by fencing or redirecting wildlife to areas away from the crops for the fresh produce market (FDA, 1998).

Rodents and other animals can also be vectors by contaminating produce through transferring untreated manure from one place to the other. Rodent control and the use of bait stations, in open field production where possible as well as in greenhouse production, are advisable.

Field sanitation is important, because poor management of human and other waste in the production area can significantly increase the risk of contaminating produce. It is possible that cross-contamination and/or microbial contamination can occur if harvested produce comes in contact with soil, dirty containers, or workers' hands. Produce destined for human consumption must never be picked up from the ground. The effects of dirty harvesting containers and lack of personal hygiene will be discussed.

3.3 HARVEST

3.3.1 Harvesting

During harvesting, it happens that workers and equipment come in direct contact with the produce. This contact can often be the way harmful pathogens are transmitted to the produce. Measures must be in place to eliminate or diminish the possibility of contamination of produce. Failure of workers to implement hygienic practices and equipment sanitation may lead to contamination of food, resulting in adulteration and/or foodborne illnesses.

Personal Hygiene. All persons that come in direct contact with the produce, food contact surfaces, and/or packing materials must follow proper food handling techniques and food protection principles. Poor personal hygiene practices and bad work habits and conditions that can introduce contaminants include

1. Not washing hands after using toilet facilities;
2. Not washing hands properly;
3. Inadequate hand-washing and toilet facilities;
4. Employees with communicable diseases, open wounds, or infections.

All employees that handle produce must receive training according to their tasks and responsibilities. They must understand the importance of good hygiene and hand washing. Workers with any illness and/or diarrhea must notify their employers and not work with produce. Skin cuts or other lesions must be protected and workers must preferably work in an area where they do not have direct contact with the produce sorting or packing processes. Workers must not only use hand-washing facilities correctly, but must also understand the importance of using toilet facilities. This will reduce potential contamination of fields, produce, their

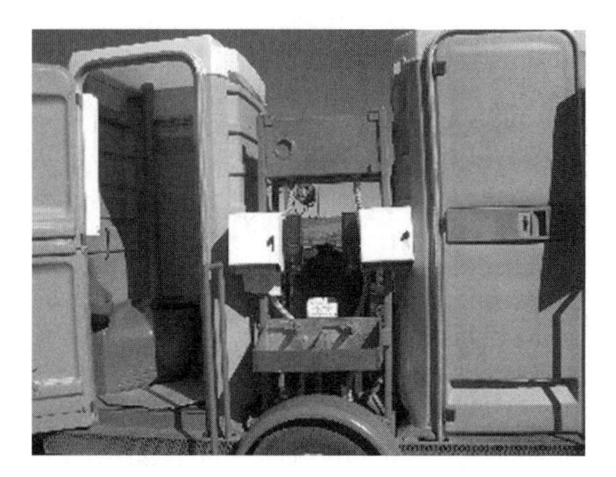

Figure 3.3 Portable toilets with hand-washing facilities.

hands, and even other workers. Potable water for hand washing must be available close to the toilets (Fig. 3.3). If running water is not available, the water must be regularly replaced and containers must be cleaned and sanitized. Care must be taken that wastewater from portable toilet facilities does not touch the produce or contaminate surface waters, and wastewater must be discarded appropriately (FDA, 1998). All visitors and subcontractors that come in contact with the produce must also be aware of personal hygiene rules and comply with them.

Equipment. All containers and equipment that come into contact with the produce must be washable and must be kept clean and sanitized. This includes harvest containers, harvesting machinery, knives, packaging materials, brushes, pallets, trailers in which produce is transported, and so on. Equipment and containers used for harvesting must be dedicated for this and not used to haul garbage, manure, or any other substance that can cause contamination. If harvest containers are stored outside, they should be cleaned and sanitized before use.

3.3.2 In-Field Packing

Packaging material and worker (picker) personal sanitation are critical where produce is packed in field, ready for use. Produce is not washed and any contamination that occurs during the harvesting/packing in the field is a potential problem. Crops packed in the field (berries, melons, and so on) will usually be consumed raw and possibly without the consumer washing them first (Fig. 3.4).

Packaging Material. Packaging material (and packed produce) must not be stored in the field overnight. Active rodent control is necessary where packaging is stored, and packaging should be covered to minimize contamination. Single-use packaging material must be used for produce and must be inspected for rodent droppings, insects, and any other form of contamination that might be a

Figure 3.4 Worker hygiene and clean packaging material are important when packing in the field.

possible source of microbial pathogens. Storage areas must be cleaned and sanitized when there are no perishable products in the areas.

3.4 POSTHARVEST HANDLING

Good Manufacturing Practices (GMPs) are every foodhandler's responsibility. Site sanitation and personal hygiene are an integral part of each employee's responsibility (Aramouni and Lindquist 2001; FDA, 2003). Good Manufacturing Practices are measures designed to ensure an effective overall approach to product quality control and risk management. They do so by setting appropriate standards and practices for product testing, manufacturing, storage, handling, and distribution.

3.4.1 Worker Hygiene and Health

As described in Section 3.3, good health and hygiene are important for food handlers. It must also be remembered that the more the produce is handled, the higher the likelihood of introducing foodborne pathogens. Handling of produce is, however, in most cases unavoidable, and personal hygiene is therefore extremely important. Leading hazards include *Salmonella typhi*, *Shigella* species, *E. coli* O157:H7, and Norwalk and hepatitis A viruses. Less common pathogens include *Staphylococcus aureus*, *Streptococcus pyrogenes*, *Cryptosporidium parvum*, and other strains of *Salmonella* and *E. coli* (FDA, 1998; Aramouni and Lindquist, 2001).

Basic hygiene principles should be instilled in all food handlers. Although management is ultimately responsible for the conduct and practices of employees, there are some responsibilities assigned to employees at the time employment begins:

1. Employees who are ill with vomiting or diarrhoea must be excluded from working in the establishment. Unhealthy conditions such as respiratory or gastrointestinal complications should be reported to the supervisor.
2. Injuries including cuts, burns, boils, and skin eruptions should be reported to the supervisor. Any cuts, wounds, or open sores on the hands and arms must be completely covered by a waterproof bandage. Single use gloves must be worn to cover bandages.
3. Because employees are the most important link in preventing foodborne illness, good personal hygiene and cleanliness, including proper and frequent handwashing is the best way to prevent foodborne illness.
4. Employees should be instructed to inform the supervisor when the toilet facilities' water, soap, or towels need to be replenished.
5. The mouth and nose should be covered when coughing or sneezing.
6. The hands should be washed after visiting the toilet, smoking, eating, handling soiled utensils, after coughing, and so on.
7. A notice board with personal hygiene rules will not only help with training of permanent staff, but will also inform visitors to the facility of the rules.

3.4.2 Washing Water

Water can be a carrier of many microorganisms, including pathogenic strains of *E. coli*, *Salmonella* spp., *Vibrio cholerae*, *Shigella* spp., *C. parvum*, *G. lamblia*, *C. cayetanesis*, *Toxoplasma gondii*, and the Norwalk and hepatitis A viruses (Gast and Holt, 2000; McCabe-Sellers and Beattie, 2004). Even small amounts of contamination with some of these organisms can result in foodborne illness. Great amounts of water are used postharvest and include processes such as produce rinsing, cooling, washing, waxing, and transport. Water of inadequate quality has the potential to be a direct source of contamination and a vehicle for spreading localized contamination in the field, facility, or transportation environments.

If potable water is used, it can reduce pathogens, but if reused or contaminated water is used, it can also increase contamination. Potable water of good quality must always be used for the final rinsing of produce. If water is reused, it must be filtered, treated, and tested to ensure that it is of good quality. Replacing washing water as frequently as possible is preferred to using antimicrobial chemicals, such as chlorine.

Washing of produce with water at the correct temperature is also important. There is great concern when washing tomatoes, peppers, apples, celery, and potatoes with water that is cooler than the produce pulp. These produce items can draw water into the stem area. If there is a pathogen on the surface of the produce or in the water it can be internalized (Gast and Holt 2000; Rangarajan et al., 2000). "Soft rot" bacteria strains of *Erwinia carotovora* are ubiquitous and have been found in surface water and, in some cases, in well water. The *Erwinia* spp. can grow on the surfaces of plants and cause soft rot of succulent plants such as tomatoes during wet conditions and can be spread by picking containers, harvest crews, and packinghouse equipment. A wax layer usually prevents bacteria from entering the fruit. But when, for example, the fruit surface is water congested because it is harvested from wet plants or washed with water that is cooler than the pulp, internalization takes place (Mahovic et al., 2004).

3.4.3 Cooling

Harvested produce must be cooled as quickly as possible to minimize the growth of pathogens and also to maintain good quality and ensure a longer shelf life. Temperatures must be maintained at levels suitable for the specific produce being cooled. Cold storage should not be overloaded to ensure even cooling throughout. High-quality products are more resistant to microbial contamination.

Different cooling methods are available, such as forced air cooling, hydrocooling, room/ambient cooling, vacuum cooling, and icing (not used much in practice). Forced air cooling is probably the safest means of cooling produce as there is no contact with water. Air quality must be checked and air intake areas must be clean from potential sources of contamination. The same is true when vacuum cooling is used. Room cooling can be used depending on the produce, the internal produce temperature, and the quantity of produce. This is a slow cooling process, especially if boxes of produce are tightly packed. When cooling produce in the field using ice or a hydrocooler, the water used must be potable, as the water can be a potential source of pathogenic contamination. Hydrocooling systems require daily cleaning and water needs to be changed as required to maintain quality. Water that is reused to cool continuous loads of produce is a potential source of cross-contamination as it can contaminate new incoming produce (Gast and Holt, 2000).

3.4.4 Facility

The grounds around the facility must be kept clean of litter and other waste and must not be a breeding place for rodents and birds. These animals can carry diseases that may be transmitted to humans. There must also be adequate drainage to eliminate seepage and pest breeding grounds. The facility itself must be designed and constructed in such a way that it can be cleaned and sanitized easily. Potential for contamination and cross-contamination can be reduced by considering the process flow. Produce processing areas must be separated from the rest of the facility.

Once the produce is cleaned (low risk), it should not come in contact with high-risk areas such as areas where product is received from the field. The floors must be sloped to allow proper drainage and avoid water accumulation. All equipment in the facility must also be made from materials that are easy to clean and nonporous. For example, no wooden surfaces are allowed (FDA, 2003).

Sanitation of the whole facility (washing, sorting, and packing lines) and all equipment must be carried out routinely according to a schedule. Product storage areas must also be cleaned on a regular basis and free-floating dust and other airborne contaminants must be minimized. Trash, fruit, and vegetable waste can be a source of microbiological contaminants. Decomposing organic matter can serve to spread microorganisms around the facility and attract insects and other pests bearing pathogenic organisms. Rejected material must be stored in a designated area and cleared on an ongoing basis (FDA, 1998).

3.4.5 Pest Control

Rodents, birds, and insects can be vectors of pathogens and carry bacteria over long distances. Rodents and insects carry pathogenic bacteria both internally and on their bodies. Birds may sometimes become a problem in food-processing areas and pose a potential health hazard. Pest-control programs for rodents, insects, and any other pests must be implemented at the perimeter of facilities as well as in the interior, if needed. Strategically placed bait stations, regular inspection and monitoring for activity and replacing of bait can be effective in decreasing rodent populations. Proper inspection of the perimeters around the facility will help to perform this task. Chemicals should be avoided as much as possible and integrated pest management (IPM) must be implemented. Cleaning and maintenance of the building and surroundings on a regular basis will reduce pest activity. The importance of sanitation must not be underestimated, because a clean environment without breeding places or free water for pests will reduce their presence significantly (Aramouni and Lindquist, 2001). Facility layout plays an important role in minimizing the ingress of pests. Packaging should be stored separately from produce and unprocessed or raw produce must also be separated from clean and packed produce, to prevent migration of flies or rodents. Domestic animals and birds must not be allowed in packing and processing areas.

3.5 INDEPENDENT VERIFICATION SYSTEMS

3.5.1 Food Safety Systems

The need for food safety regulations has increased because of global changes in eating habits, increased numbers of immunocompromised groups, elderly and young in the First World countries, and increased antibiotic resistance. In addition, trends include improved nutritional value of food commodities, availability of produce year round, and increased reporting and detection of foodborne pathogens.

Globalization of the food trade offers many benefits to consumers as well as suppliers. However, this challenges safe food production and distribution on a large scale. Information on the country of origin and assurance of production on the farm has become an increasingly important issue in recent years. Consumers and regulators fear that produce grown in Third World countries may pose a greater risk of being contaminated with foodborne pathogens. Much of this fear is based on assumptions that sanitary standards, particularly in developing countries, are inferior to those of developed countries (WHO, 2002).

As an effective means of reducing foodborne hazards, food-safety programs are increasingly focused on a farm-to-table approach. This holistic approach to the control of food-related risks involves consideration of every step in the chain, from raw material to food consumption. As has been shown throughout this chapter, hazards can enter the food chain on the farm and can continue to be introduced or exacerbated at any point in the chain until the food reaches the consumer.

Several standards exist globally, either on a regional, national, or even global scale. Food-safety standards/systems/programs focus most of the time on the food supply chain "prefarm gate" or "postfarm gate." "Prefarm gate" standards cover the production processes on the farm and do not include produce processing covered in "postfarm gate" standards. The main differences among food-assurance systems can be interpreted as policy-related issues, because all of them cover GAP, GMP, and HACCP programs. The differences then lie in whether the system

1. Attempts to cover all stages of the food chain or just a part of it;
2. Manages the supply chain information with central database systems to exercise control and support the interface to upstream and downstream systems as well as traceability;
3. Intends to build up a final consumer market or remain in the business-to-business communication in order not to create more labels in the market;
4. Is owned publicly or privately; or
5. Is driven by a supply chain partnership or by a single sector (Möller, 2005).

To harmonize standards and expand global trade, the United Nations Food and Agriculture Organization and the World Health Organization created the Codex Alimentarius Commission to help experts enhance domestic food safety, set benchmarks, and use best practices for hygiene, import, and export. Food safety is just one of many emerging standards for expanding market access.

Prefarm Gate Standards. Prefarm gate standards that have the minimization of food-safety hazards (including biological, physical, and chemical) as core, are all based on HACCP principles. Examples of global standards are the EurepGAP Fruit and Vegetable standard and Safe Quality Food 1000 Code. Standards on a national or regional level include the Assured Produce Scheme (UK), SwissGAP, US-GAP, and many others.

EurepGAP, the global partnership for safe and sustainable agriculture, is an agricultural standard that covers the principles of GAP. Established by a consortium of members of the agribusiness industry, EurepGAP requires producers to demonstrate their ability for sustainable agriculture by means of integrated pest and crop management, and continuous improvement of farming systems (www.eurep.org). EurepGAP provides the standard (General Regulations and Control Points and Compliance Criteria) and framework for recognized, independent, third-party certification of farm production processes, based on ISO Guide 65 (General Requirements for Bodies Operating Product Certification Systems) (EurepGAP, 2004). The standard covers the whole agricultural production process, from before the plant is in the ground, up to and including harvest. It does not include processing (changing the shape and structure of the produce). The scheme principles are based on

1. Food safety (control points are derived from the application of generic HACCP principles);
2. Environmental protection (based on GAP, which is designed to minimize the negative impact of agricultural production on the environment);
3. Occupational Health, Safety and Welfare (establishes awareness and responsibility regarding socially related issues).

There are more than 80 accredited and EurepGAP approved certification bodies (CBs) certifying against the standard in more than 60 countries (http://www.eurep.org/fruit/Languages/English/cbs.html).

Safe Quality Food (SQF) 1000 is another food-safety audit and certification program for agricultural producers. The Food Marketing Institute (FMI) currently owns the SQF Marketing scheme. The SQF Institute administers a rigorous, reliable, independent certification program, including standards for all food commodities from farm to retail, third-party auditing, and training (http://www.sqfi.com). Safe Quality Food is designed to provide benefits to suppliers, retailers, and consumers. The main feature of SQF 1000 is that it is based on a full HACCP application on the farm and relies on independent third-party assurance.

In the United States alone, foreign producers must meet standards for certification of agricultural and horticultural products. In the United States, GAPs standards are applied at all stages of production, processing, transportation, and storage. The U.S. Environmental Protection Agency sets the minimum residue level for pesticides, whereas the U.S. Food and Drug Administration and the Department of Health and Human Services determine risk prevention and risk analysis through HACCP, a system to prevent physical, chemical, or microbial contamination of food. The Food and Drug Administration (FDA) has designed a guide to provide voluntary guidance on GAPs to minimize microbial food-safety hazards common to the growing, harvesting, packing, and transport of most fruits and vegetables sold to consumers in an unprocessed or minimally processed (i.e., raw) form. US-GAP was developed and implemented in the United States. British standards, such as Assured Produce, combine those of the International Standards Organization

(ISO), HACCP, and GAP. Australia also has its own standards, which sometimes differ from one retail store to the other, but are also based on GAP.

Postfarm Gate Standard. Several postfarm gate standards exist, but only a few are recognized on a global scale. These include the International Food Standard (IFS), the British Retail Consortium (BRC) Global Standard for food, the Dutch HACCP Code, and Safe Quality Food 2000.

Both the IFS and BRC standards require

1. The adoption of HACCP and GMPs;
2. A documented quality management system; and
3. Control of facility environment standard, product, process, and personnel.

The IFS has been developed by German and French food retailers to create a consistent evaluation system for all companies supplying retailer branded food products (http://www.food-care.info/). The BRC is the lead trade association representing UK retailers (http://www.brc.org.uk/mission04.asp) and the majority of UK and Scandinavian retailers will only source from suppliers who have BRC certification.

Safe Quality Food 2000 is based on universally accepted CODEX Alimentarius HACCP guidelines, ISO, and Quality Management Systems, and offers the food sector a way to manage food safety and food quality, simultaneously (http://www.sqfi.com/).

The International Standard, ISO 22000 – Food Safety Management System (published on 1 September 2005), aims to harmonize the requirements for food-safety management in food and food-related business on a global level. It is particularly intended for application by organizations that seek a more focused, coherent, and integrated food-safety management system than is normally required by law. However, this International Standard is not intended for application as minimum requirements for regulatory purposes. While ISO 22000 can be implemented on its own, it is designed to be fully compatible with ISO 9001:2000 and companies already certified to ISO 9001 will find it easy to extend this certification to ISO 22000 (http://www.iso.org/iso/en/commcentre/pressreleases/2005/Ref966.html).

3.5.2 Benefits of Independent Verification Systems

Certification provides a grower with the means to demonstrate his produce has been grown in full compliance with the protocols and guidelines laid down by the standard requested by his customer (retailer). Adherence to the standards' requirements gives the suppliers (growers), exporter, retailer, and consumer the assurance that their produce consistently meets retailer specifications on food safety. Adherence to environmental and social issues is assured with some standards. When a grower complies with a recognized, accredited standard, it also means that the grower can be assured that the verification of the production processes has taken place according to a specific standard. All CBs that inspect against an accredited standard must employ inspectors who comply with strict criteria and are trained to ensure that farm visits are certified impartially and to a consistent standard.

One of the advantages of a globally accepted standard such as BRC or EurepGAP is that a grower who complies and receives certification decreases the number of inspections performed on the farm and in the packhouse. These standards are required by multiple retailers. This results in substantial savings in time and costs.

Implementation of a standard on the primary production level also leads to an increase in profitability. Processes become more efficient and streamlined, waste is reduced, and, in most cases, product quality increases. This happens because a system requiring documentation and traceability is put in place and all the standards require management to be involved. Beyond the requirement for compliance with GAPs, GMPs, HACCP, and so on, the standards enable growers to meet food-safety legislation in the country of production as well as the country of destination, underpinning due diligence responsibilities.

3.5.3 Concerns of Verification Systems

The existence of the different verification or food assurance systems can result in several visits to a producer, which can be very costly. Harmonization among standards is therefore critical. If a producer only needs to comply with one standard, and this standard is accepted by the markets supplied, verification costs per producer will decrease. The Global Food Safety Initiative (GFSI) has set out a guidance document for the benchmarking of food-safety management systems (http://www.ciesnet.com/programmes/foodsafety/global_food/main.htm), which is mainly applied to postfarm gate standards. EurepGAP has its own benchmarking program by which agricultural standards that cover more than food safety can achieve equivalence to a specific EurepGAP scope (http://www.eurep.org/fruit/Languages/English/approvals.html) by independent assessment of GAPs in detail.

Another possible concern of verification systems is differing interpretation among certifiers and consultants that can put a standard's integrity at risk. Inspector competency and training of consultants and everybody involved with the implementation of the standard are of major importance. Clear guidelines to explain the requirements for implementation and relevant standard rules can greatly reduce confusion. Close cooperation with accreditation bodies is important to manage certification bodies and to ensure a unified interpretation and level of competency.

When the use of consultants is necessary to implement and understand the standard, certification costs can be very high. A standard must have universal applicability and not be too rigid, because this can make the standard very complex.

3.6 CONCLUSIONS

Food safety has become an important focus of consumer concerns, policy responses, and strategic industry initiatives in many countries. The importance of bacterial pathogens in the transmission of foodborne illness has become apparent in recent years. Several large and well-publicized outbreaks of foodborne illness have been linked to cantaloupe, tomatoes, lettuce, and alfalfa sprouts. Cross-contamination with fecal matter, both domestic as well as from wild animals, has been suggested. Contact with contaminated water has also been identified as a possible source of contamination. The use of untreated manure or sewage, lack of field sanitation, poorly sanitized transportation vehicles, and contamination by handlers are also suggested as potential contributing factors. It is critical that control of foodborne pathogens in produce must begin before produce is even planted by avoiding fields that have been subjected to flooding or have otherwise been contaminated. Only clean potable water should be used for irrigation and postharvest washing of produce. All food handlers (harvesters and packers) should be trained in proper personal hygiene and provided with adequate sanitary and hand-washing facilities. To demonstrate implementation of Good Agricultural and Good Manufacturing Practices, producers can take part in third-party, independent certification systems and have their produce certified to a recognized standard.

REFERENCES

ADAS, *The Safe Sludge Matrix: Guidelines for the Application of Sewage Sludge to Agricultural Land*, 3rd edn. ADAS, Wolverhampton, 2001. Available at http://www.adas.co.uk/matrix.

F. Aramouni and T.B. Lindquist, *Current Good Manufacturing Practices (cGMPs) in Manufacturing, Packing, or Holding Human Foods*. Kansas State University, 2001. Available at http://www.oznet.ksu.edu/library/fntr2/mf2505.pdf.

R.F. Brackett, Incidence, contributing factors, and control of bacterial pathogens in produce. *Postharvest Biol. Technol.*, 15 (3), 305–311 (1999).

Council Directive 86.278/EEC, OLJ 181, 04/07/1986, 6–12, 1986. Available at http://europa.eu.int/smartapi/cgi/sga_doc?smartapi!celexapi!prod!CELEXnumdoc&lg=EN&numdoc=31986L0278&model=guichett.

C. De Roever, Microbiological safety evaluations and recommendations on fresh produce. *Food Contr.*, 10, 117–143 (1999).

Environmental Protection Agency, Code of Federal Regulations, Title 40 Part 503, 1993. Standards for the use or disposal of sewage sludge. Available at http://www.access.gpo.gov/nara/cfr/waisidx_02/40cfr503_02.html.

EurepGAP, General Regulations Fruit and Vegetables, V2.1–Jan04, 2004. Available at http://www.eurep.org/fruit/Languages/English/documents.html.

Food and Drug Administration, *Guide to Minimize Microbial Food Safety Hazards for Fresh Fruits and Vegetables*. FDA, Washington, DC, 1998. Available at http://www.vm.cfsan.fda.gov/~dms/prodguid.html.

Food and Drug Administration, Code of Federal Regulations, Title 21 Part 110. Current Good Manufacturing Practice in manufacturing, packing or holding human food. Available at http://www.cfsan.fda.gov/~lrd/cfr110.html.

K.L.B. Gast and K. Holt, *Minimizing Microbial Hazards for Fresh Fruit and Vegetables, Processing Water*. Kansas State University, 2000. Available at http://www.oznet.ksu.edu/library/hort2/mf2480.pdf.

E.Z. Harrison, M.B. McBride, and D.R. Bouldin, Land application of sewage sludges: an appraisal of the US regulations. *Int. J. Environ. Pollut.*, 11 (1), 1–36 (1999).

E.Z. Harrison and S.R. Oakes, Investigation of alleged health incidents associated with land application of sewage sludges. *New Solutions*, 12 (4), 387–408 (2002).

G.S. Johannessen, R.B. Froseth, L. Solemdal, J. Jarp, Y. Wasteson, and L. Rorvik, Influence of bovine manure as fertilizer on the bacteriological quality of organice Iceberg lettuce. *J. Appl. Microbiol.*, 96 (4), 787–794 (2004).

G. Kuepper, Manures for organic crop production. Appropriate Technology Transfer for Rural Areas, 2003. Available at http://www.attra.ncat.org/attra-pub/PDF/manures.pdf.

M. Lægreid, O.C. Bøckman, and O. Kaarstad, *Agriculture, Fertilizers and the Environment.* Norsk Hydro ASA, CABI Publishing, Norway, 1999.

H. Langenkamp and P. Part, Organic contaminants in sewage sludge for agricultural use. Available at http://europa.eu.int/comm/environment/waste/sludge/organics_in_sludge.pdf.

M. Mahovic, S.A. Sargent, and J.A. Bartz, *Identifying and Controlling Postharvest Tomato Diseases in Florida.* University of Florida, IFAS Extension HS866, 2004. Available at http://edis.ifas.ufl.edu/pdffiles/HS/HS13100.pdf.

B.J. McCabe-Sellers, and S.E. Beattie, Food Safety: Emerging trends in foodborne illness surveillance and prevention. *J. Am. Diet Assoc.*, 104, 1708–1717 (2004).

K. Möller, Economics of Standard Owners: Competition as Barrier to Global Harmonisation of Food Assurance Systems. 92nd Quality Management and Quality Assurance in Food Chains, EAAE Seminar, Göttingen, 2005.

F.A. Nicholson, S.J. Groves, and B.J. Chambers, Pathogen survival during livestock manure storage and following land application. *Biores. Technol.*, 96 (2), 135–143 (2005).

A. Rangarajan, E.A. Bihn, R.B. Gravani, D.L. Scott, and M.P. Pritts, *Safety Begins on the Farm. A Grower's Guide.* Cornell Good Agricultural Practices Program, Cornell University, USA, 2000.

W.H. Sperber, HACCP does not work from Farm to Table. *Food Contr.*, 16 (2005), 511–514 (2003).

WHO, *Health Guidelines for the Use of Wastewater in Agriculture and Aquaculture.* World Health Organization, Geneva, 1989. Available at http://whqlibdoc.who.int/trs/WHO_TRS_778.pdf.

WHO, *WHO Global Strategy for Food Safety.* World Health Organization, Geneva, 2002. Available at http://www.who.int/foodsafety/publications/general/en/strategy_en.pdf.

WHO, *The Guidelines: A Framework for Safe Drinking-water.* World Health Organization, Geneva, 2004. Available at http://www.who.int/water_sanitation_health/dwq/gdwq3_2.pdf.

CHAPTER 4

Produce Quality and Foodborne Disease: Assessing Water's Role

KRISTINA D. MENA

University of Texas Health Science Center at Houston School of Public Health, El Paso, TX, USA

4.1 WATER AND THE PRODUCE INDUSTRY

Approximately 76 million cases of foodborne illness occur in the United States every year, resulting in 325,000 hospitalizations and 5000 deaths (Mead et al., 1999). Although the incidence of such illnesses associated with fresh produce is relatively low, the reported number of outbreaks doubled between the documented

Microbial Hazard Identification in Fresh Fruit and Vegetables, Edited by Jennylynd James

periods of 1973–1987 and 1988–1992 (Mead et al., 1999; Bean et al., 1997). The trend toward increasing consumption of fruits and vegetables as well as a change in the geographical patterns of distribution may have contributed to this observed increase. In addition, outbreaks associated with fruits and vegetables lead to a greater number of reported illnesses than outbreaks associated with other food vehicles (CDC, 2000a). It is understood that most foodborne illnesses are not reported and those that are documented are those associated with outbreaks involving either many people and/or severe human health consequences. Moreover, even when foodborne outbreaks are observed and investigated, many are not fully characterized regarding etiological agent(s), vehicle(s) of transmission, and contributing factor(s) (inadequate cooking, improper holding temperature, poor food-handler hygiene, and so on). Foodborne illnesses generally result in diarrhea, nausea and vomiting, and other symptoms that can be related to a variety of causative agents and situations. Cases either do not seek medical attention or if medical treatment is sought, the attending physician does not routinely identify (and report) the etiological agent. Information regarding the incidence of foodborne illnesses and specific etiology in the United States has been estimated based on the data that have been reported and compiled (CDC, 2000a, 1996).

Produce can become contaminated with disease-causing microorganisms (pathogens) in a variety of ways, including through contact with contaminated water. Studies have documented the occurrence of pathogenic bacteria, viruses, and protozoan parasites on a variety of fresh fruits and vegetables (Beuchat, 1996) and contamination can occur during preharvest or postharvest (USFDA, 2001; Beuchat and Ryu, 1997). Water quality has traditionally been addressed separately in terms of the public health impact of waterborne pathogens and outbreaks due to contaminated water. Recently, however, the role of water as a contributing factor to foodborne disease has been explored, although minimally. Water has been considered the "forgotten food", particularly in terms of its impact as an ingredient of a manufactured food product, and now water source and quality are becoming more recognized as an important and potentially impacting component during food production and processing, particularly for minimally processed produce and that eaten raw.

4.2 WATER APPLICATIONS IN PRODUCE PRODUCTION AND PROCESSING

Water comes into contact with produce in a variety of ways during production and processing. During preharvest, water is used for irrigation and to apply insecticides and fungicides. In addition, surface run-off waters can contact produce. Postharvest avenues of water contact include wash and rinse water, ice, and harvesting equipment previously exposed to water. Workers during both preharvest and postharvest may provide produce contact with water after handwashing, for example. Food handlers may also contaminate produce prior to consumption either because of poor hygienic practices or through washing hands with low-quality water.

TABLE 4.1 Water Contact Routes in Produce Production and Processing

Irrigation water
Surface water run-off
Contact sprays
(fertility application; insecticide/fungicide application)
Harvesting equipment
Wash and rinse water
Ice
Hands of field workers and packers
Food handlers and packers

Table 4.1 provides a list of possible contact routes for water and produce. These various water contact routes yield opportunities for produce to become contaminated with disease-causing microorganisms.

Although it is difficult to definitively link an illness to a particular food item, it can be even more challenging to connect a foodborne illness to a specific production source or practice. There has been some published evidence implicating production practices, however, including the use of microbial contaminated water for pesticide application and irrigation, the use of manure, and poor personal hygiene of farm workers (Brackett, 1999). The importance of such studies is to identify risk factors for produce contamination that may result in consumer health consequences. The public health implications associated with microbial contaminated produce are discussed later in Sections 4.5 and 4.6.

4.3 WATER SOURCE AND QUALITY

In a survey conducted by the U.S. Department of Agriculture in 1999, deep ground wells were the most common source of water used for irrigation (51 percent of vegetable growers and 39 percent of fruit growers), followed by surface water (19 percent of vegetable growers and 38 percent of fruit growers), and municipal water (only 5 percent of growers). Irrigation with reclaimed water is also an option and, unlike the previously listed water sources, there are recommended guidelines for reclaimed water (USEPA, 1992; WHO, 1989). Studies have been conducted involving the microbial analysis of reclaimed water to be used for irrigation purposes and concluded that these waters pose negligible human health risks as long as the recommended disinfection guidelines are met (York et al., 2000; Sheikh et al., 1999; Shuval et al., 1997). Water quantity and quality are global concerns with water availability impacting water quality. In addition, other factors that may impair the integrity of water quality used in the agricultural arena include stormwater run-off, sewage overflows, discharge from confined animal-feeding operations and industry, and climatic conditions. Furthermore, topological factors, such as slopes, could enhance the conditions for contamination from run-off pollutants.

Whether microbial-contaminated water contacts fruits and vegetables during preharvest or postharvest depends on a variety of factors. Management choices on the farm setting as well as during packing operations may have the greatest impact. Water availability may not only affect produce quality by directly impairing source water quality, but it also drives costs, which may influence such management choices as source water selection for certain agricultural purposes (so as to minimize costs). If water availability is lacking, water of questionable quality may be applied for agricultural purposes to meet the needs of the grower. In addition, studies have been conducted regarding how different irrigation modes affect microbial quality of the produce. For example, drip irrigation methods are considered to present a lower risk of possible microbial contamination of produce via water compared to overhead irrigation (USDA, 2001; Bastos and Mara, 1995). However, the decision to use overhead irrigation may be an attractive option for the grower, because overhead irrigation offers the advantage of having better control over water usage. The type of irrigation method chosen by a grower depends on several issues, including water quantity and cost, depth of water table, soil type and slope, and crop rotations (USDA, 2001), and has yet to be considerably influenced by microbial contamination concerns. Most importantly, it is necessary to recognize that a variety of factors affect source water quality and a combination of management decisions may subsequently increase the probability of produce becoming contaminated with pathogens. It is not practical to suggest to growers which management choices would minimize this probability, because every grower routinely faces different and sometimes unique situations that may be impacted by outside influences (such as cost, water availability, changing climatic conditions, and so on).

4.4 WATER TREATMENT

Fruit and vegetable growers routinely confront water quality issues, although they are not given specific guidance on water source quality parameters and water use practices. Although an objective of the Good Agricultural Practices (GAP) is to reduce the potential for poor quality water to be used in the agricultural environment (USFDA, 2001), it is costly to monitor source water quality and specific criteria are not available to make this practice practical or economical. To date, little is known regarding source water treatment practices of produce growers; however, some studies have been conducted to evaluate treatment efficacy. Information on microbial inactivation following UV irradiation of agricultural water, for example, has been explored (Robinson and Adams, 1978). Various treatments of agricultural waters are applied in different situations, including chlorination, peroxyacetic acid, and ozone (USFDA, 2001). At this time, a preventive approach to controlling source water quality by growers may be more practical, including infrastructure maintenance and protection of wellheads. Guidelines for growers based on scientific studies need to be provided for routine microbial water monitoring and treatment to be both practical and meaningful.

TABLE 4.2 Waterborne Pathogens Affecting the Produce Industry

Bacteria
Campylobacter spp.
Escherichia coli (ETEC and EHEC)
Salmonella
Shigella spp.
Vibrio cholerae
Yersinia enterolitica
Enteric viruses
Hepatitis A Virus
Norovirus
Protozoa
Cyclospora
Cryptosporidium
Giardia

4.5 WATERBORNE PATHOGENS AFFECTING THE PRODUCE INDUSTRY

There are many microorganisms that have been detected on produce (some associated with foodborne outbreaks) that are also waterborne. Table 4.2 lists the bacteria, enteric viruses, and protozoan parasites that are known water contaminants that have also been detected on fruits and vegetables. Studies have documented the occurrence of these pathogens on various fruits and vegetables; however, caution should be exercised when interpreting the results of such studies, particularly in regard to determining incidence. Many studies report that a particular microorganism was detected on a type of fruit or vegetable, but do not report information regarding samples that are negative for the particular pathogen. This makes it difficult to identify risk factors for the produce contamination, such as source (was it waterborne?). Waterborne pathogens have been detected on fresh fruits and vegetables during different stages of production, but such studies have been conducted postharvest in particular (FDA, 2001; Nguyen-the and Carlin, 2000; Odumeru et al., 1997; Monge and Chinchilla, 1996; Madden, 1992; Park and Sanders, 1992; Garcia-Villanova Ruiz et al., 1987).

There are a variety of factors that influence whether microbes will contact produce via water and these factors are discussed in the previous section. In addition, there are variables that determine whether a pathogen can survive or even multiply on the produce. Survivability and/or multiplication depends on the microorganism itself, type of produce, and the environmental conditions during both pre- and postharvest (USFDA, 2001). It is more probable that the number of pathogens on a given produce will decline over time (rather than increase) given that the microorganism is on the outer surface of an uninjured portion of the produce and moisture is lacking.

In addition, exposure to sunlight will adversely influence the growth of microorganisms.

4.5.1 Waterborne Bacteria

Foodborne outbreaks of disease, in general, are commonly associated with bacterial microorganisms. Bacteria range in size from 0.3 to 2 μm and a variety of these single-celled microbes can be transmitted both via food and water and are capable of multiplying to large numbers outside of a host. *Campylobacter* spp. is the most commonly identified etiological agent of gastrointestinal illness in the United States (FSIS, 1997). This bacterium can be transmitted through drinking and recreational water and food, particularly poultry and milk (CAST, 1994; Wood et al., 1992; Harns et al., 1986; Hopkins et al., 1984). Not only may *Campylobacter* infections result in gastroenteritis, but infections have also been associated with Guillain–Barré syndrome, Reiter's syndrome, and reactive arthritis (Mishu and Blazer, 1993; Smith et al., 1993). Outbreaks have been linked to produce where it was speculated that cross-contamination during food preparation was the cause (CDC, 1998a; Allen, 1985).

Enterotoxigenic *E. coli* (ETEC) is transmitted by both food and water and is not an uncommon cause of diarrhea both within and outside the United States. It is speculated that this agent causes a significant amount of travelers' diarrhea in Mexico and includes symptoms of abdominal cramping and vomiting. Investigations of outbreaks involving enterotoxigenic *E. coli* revealed the consumption of raw produce as a risk factor for travelers' diarrhea due to ETEC as well as gastroenteritis among persons within the United States (Beuchat, 1996; Merson et al., 1976). Infections caused by enterohemorrhagic *E. coli* (EHEC), also known as *E. coli* O157:H7, may result in hemorrhagic colitis, and are particularly severe in the elderly and children (10 percent mortality) (Haas et al., 1999). This serotype of *E. coli* has been detected in both drinking water and recreational water and outbreaks have involved such foods as undercooked or raw ground beef, unpasteurized fruit juices, and produce including lettuce, radishes, and alfalfa sprouts (USFDA, 2001; Taormina et al., 1999; Ackers et al., 1998; CDC, 1997a, 1994).

Several of the over 2700 serotypes of *Salmonella* have been detected on fresh produce, such as sprouts, cantaloupes, and tomatoes. It is second to *Campylobacter* as the most commonly identified causative agent of gastroenteritis in the United States (FSIS, 1997). Large salmonellosis outbreaks have occurred associated with the consumption of contaminated tomatoes (Cummings et al., 2001; Lund and Snowden, 2000; Tauxe, 1997; Beuchat, 1996; Wei et al., 1995). Infections may result in diarrhea, reactive arthritis, and Reiter's syndrome (Smith et al., 1993) and the number of persons who become infected during outbreaks varies greatly from outbreak to outbreak (6 to 80 percent) (Chalker and Blaser, 1988).

Shigella spp. causes bloody diarrhea accompanied by abdominal cramps and has a low infectious dose for humans, the only host for these bacteria. Foodborne outbreaks are usually associated with food handlers (Haas et al., 1999) and have been documented for recreational waterborne outbreaks (CDC, 1990) and raw produce such as lettuce and green onions (CDC, 1999; Beuchat, 1996; Davis et al., 1988; Martin et al., 1986). Cholera – a severe diarrhea – is caused by *Vibrio cholerae* and can quickly result in death if treatment is not administered in a timely manner. More commonly associated with waterborne disease, *Vibrio cholerae* has also been associated with food, mostly seafood consumption, but also produce. In particular, mixed vegetables and cabbage have become contaminated via the use of wastewater for irrigation (Nguyen-the and Carlin, 2000; Swerdlow et al., 1992). Cases of infections with this pathogen are seen during conditions of poor sanitation and hygiene, so opportunities for contamination occur throughout produce production and preparation processes. *Yersinia enterocolitica* has also been detected on minimally processed produce where contaminated water provided the source. The natural reservoirs for this bacterium are animals (swine), but infections have been associated with bean sprouts contaminated via well water (Cover and Aber, 1989).

4.5.2 Waterborne Viruses

By definition, viruses are obligate intracellular parasites, meaning they replicate only inside a host cell. They are about 200 nm or smaller, contain only RNA or DNA (not both), and are mostly host-specific. Although most microbial foodborne illness is caused by bacterial organisms, two waterborne viruses, in particular, are associated with foodborne disease, including those associated with produce consumption. Hepatitis A virus is responsible for numerous waterborne and foodborne outbreaks worldwide, which result in the highest hospitalization ratios (as high as 6.7 percent) when compared with other waterborne viruses. Infections may result in fever, nausea, abdominal discomfort, anorexia, malaise, liver damage, and the production of jaundice. Waterborne outbreaks due to hepatitis A virus have been traced to sewage contamination (Rao and Melnick, 1986), but transmission through the consumption of contaminated shellfish (Lemon, 1985) and other foods has also been documented. Outbreaks have been associated with frozen raspberries, frozen strawberries, lettuce, raw watercress, and tomatoes (Lund and Snowden, 2000; Hutin et al., 1999; CDC, 1971, 1997b; Niu et al., 1992; Rosenblum et al., 1990; Lowry et al., 1989; Ramsay and Upton, 1989; Reid and Robinson, 1987). Recently, a large outbreak (more than 500 infected) of hepatitis A virus occurred in the United States as a result of the consumption of contaminated green onions at a restaurant chain (CDC, 2003).

Although agents of waterborne disease are generally not reportable in the United States, noroviruses are considered the major causes of U.S. waterborne illnesses due to associated health outcomes and attack rates observed when waterborne outbreaks are documented and are the primary agents responsible for adult gastroenteritis (Dolin et al., 1987; Kaplan et al., 1982). These viruses are also responsible for the many observed illness outbreaks among passengers of cruise ships. Regarding foodborne disease, outbreaks have been associated with melons, fruit salad, coleslaw, tossed salad, and fresh fruits (Lund and Snowden, 2000; Herwaldt et al., 1994; White, 1986). These viruses are routinely isolated from untreated wastewater used for irrigation, so the potential exists for produce contamination. Although water may be related to some produce contamination incidences of hepatitis A or norovirus, most often produce contamination with either virus is believed to be due to an infected food handler.

4.5.3 Waterborne Protozoa

Protozoan parasites are unicellular organisms that reproduce in a host and are capable of forming an environmentally resistant cyst or oocyst transmitted through the fecal–oral route. *Cyclospora* has been responsible for both waterborne and foodborne outbreaks of disease, including outbreaks associated with produce that became contaminated by parasite-laden water (Fleming et al., 1998; Herwaldt and Ackers, 1997). Most often, outbreaks have been associated with contaminated berries, such as raspberries and blackberries, but foods such as lettuce and basil have also been implicated (Herwaldt and Beach, 1999; Fleming et al., 1998; CDC, 1997c, d; Herwaldt and Ackers, 1997). Symptoms include watery diarrhea, abdominal cramping, anorexia, nausea, and fatigue.

Cryptosporidium is the agent responsible for several documented waterborne outbreaks, including the largest waterborne outbreak observed in the United States (Lisle and Rose, 1995). This outbreak occurred in 1993 in Milwaukee, Wisconsin, and resulted in more than 400,000 illnesses and 4000 hospitalizations (MacKenzie et al., 1994). This protozoa causes severe diarrhea, which is particularly detrimental to immunocompromised persons. Foodborne outbreaks have been associated with unpasteurized fruit juice (CDC, 1997e; Millard et al., 1994) as well as (possibly) green onions (CDC, 1998b). *Giardia* is the most identifiable waterborne agent in the United States, numerous outbreaks having been documented. Infections may result in diarrhea, abdominal cramps and bloating, and fatigue. Giardiasis has resulted from food contamination, including that of produce such as tomatoes, lettuce, and onions (Nguyen-the and Carlin, 2000; CDC, 1989). Theoretically, contaminated agricultural water could provide a route of transmission for protozoa to produce, although this has not been conclusively demonstrated.

4.6 WATER'S IMPACT ON PRODUCE QUALITY AND PUBLIC HEALTH

Studies have documented the occurrence of pathogens on different types of produce and investigations have linked such findings to foodborne illnesses (CDC, 2000b; Bean and Griffin, 1990). The number of outbreaks associated with fresh produce in the United States has increased over the last decade. It can be speculated that this is a result of a combination of factors including better surveillance and reporting of foodborne illnesses in general, an increase in fresh produce consumption (as people attempt to practice healthy lifestyles), an increase in the number of buffet-style (salad bar) restaurants accompanied by an increase of people eating outside of the home, and the increasing global market, with distributions including larger geographical areas.

Current production and processing practices for fruits and vegetables do not include a procedure to remove – or attempt to remove – any present microorganisms. Although limited due to monetary and time constraints, studies have been conducted that evaluate produce for the presence of various microorganisms and these have been described elsewhere (USFDA, 2001). In addition, epidemiological investigations have successfully identified certain pathogens as the etiological agents in produce-related outbreaks; some of these outbreaks; are mentioned in the previous section. In many produce-related outbreak situations, it appears that cross-contamination during food preparation or direct contamination from an infected food handler are often implicated as the causative factor.

As with any foodborne outbreak, linking illnesses with a specific food item and pathogen is difficult at best. In addition, determining the source (e.g., contaminated water during production or processing) presents another challenge. It is recognized that points of contact with water (as described in an earlier section) provide opportunities for (pathogenic) microbes to contaminate produce, affecting quality for the consumer. Many factors influence whether a pathogen will survive and/or multiply on fresh produce, including the type of microorganism (bacterium, virus, or protozoa), produce item and condition (injured or uninjured), and environmental factors (temperature, presence of nutrients, and so on). Environmental factors also include conditions such as the presence of moisture, which may promote the growth of some bacteria, further emphasizing the potential role of water impacting produce quality. In addition, the infiltration of wash water into fresh fruits and vegetables through pores or injured areas has also been observed, impacting microbial quality of the produce (Bartz, 1999).

Water's role in the microbial quality of produce and subsequent public health impact is not completely understood. However, some produce-related outbreak investigations have been able to obtain enough information to suspect water as the source of the contamination. Table 4.3 provides a list of some of these outbreaks where contaminated water may have been a contributing factor.

4.7 APPLICATION IN MICROBIAL RISK ASSESSMENT

4.7.1 The Risk Assessment Paradigm

Risk assessment is a tool for evaluating hazardous agents in the environment. First constructed by the National Research Council (NRC, 1983) to address chemical agents in the environment, the paradigm has since been applied to address microorganisms in both water and food (Mena et al., 2004; Mena, 2002; Lammerding et al., 2001; Haas et al., 1993, 1999; Lammerding and Paoli, 1997; Regli et al., 1991; Rose et al., 1991), including microorganisms associated with produce. This four-step framework consists of hazard identification, exposure assessment, dose–response assessment (hazard characterization), and risk characterization, and can provide a qualitative and/or quantitative estimate of the human health risks associated with exposure to a particular hazardous agent.

TABLE 4.3 Some Speculated Water-related Disease Outbreaks Involving Produce[a]

Pathogen	Produce	Location/Year	Water Contact	Reference
Cyclospora cayetanensis	Raspberries	20 U.S. States and 2 Canadian provinces/1996	Insecticide/fungicide application	Fleming et al., 1998; Herwaldt and Ackers, 1997
E. coli O157:H7	Lettuce	Montana/1995	Irrigation run-off	Ackers et al., 1998
Giardia	Lettuce and onions	New Mexico/1989	Wash water	CDC, 1989
Hepatitis A virus	Raw watercress	Tennessee/1971	Harvested in stream (near abandoned septic tanks)	CDC, 1971
Salmonella (Javiana)	Tomatoes	Several states in U.S.	Contaminated water bath used in packinghouse	Tauxe, 1997; Beuchat, 1996
Salmonella (Montevideo)	Tomatoes	Several states in U.S.	Contaminated water bath used in packinghouse	Tauxe, 1997; Wei et al., 1995
Shigella sonnei	Parsley	Several states in U.S./Canada	Use of unchlorinated water in packinghouse and as ice	CDC, 1999
Vibrio cholerae	Cabbage	Peru/1991	Untreated sewage used for irrigation	Swerdlow et al., 1992
Yersinia enterolitica	Bean Sprouts	Pennsylvania/1982	Contaminated well water used to immerse sprouts	Cover and Aber, 1989

[a]USFDA, 2001

Risk assessment is an iterative process where available information regarding the agent of interest is gathered and systematically presented. Where data are lacking, assumptions are made. The hazard identification step evaluates whether a microbial (or chemical) agent poses a human health threat based on available laboratory and field data, including information obtained from epidemiological studies. The risk assessment may end at this point if the agent is found to pose no adverse health effects to humans. The objectives of the exposure assessment step are to measure the frequency and intensity of the exposure, identify the route(s) of exposure, and characterize the population exposed. Dose–response assessment determines the relationship between the dose of the hazard and the incidence or extent (severity) of the adverse human health effect. Risk characterization presents an overall estimate (ideally quantitative) of the risk of an adverse health effect occurring based on the conclusions of the three previous steps. In addition, any uncertainties and variability incorporated in the risk assessment process are addressed here.

Risk assessment is the first step of the risk analysis process, followed by risk management and risk communication (NRC, 1994). Overall, the goal of a risk assessment is to determine or predict the public health significance associated with the occurrence of (exposure to) a particular hazard or hazards. Risk assessment results as well as the assumptions and uncertainties incorporated in the process are considered to determine appropriate (if any) management decisions. Risk managers consider risk–benefit and cost–benefit scenarios during the decision-making process and often have to compare risk assessment results of different hazards to determine where to allocate funds and resources. Risk communication is the interaction between risk managers and the public regarding the associated human health risks with a particular hazard(s).

4.7.2 Characterizing Water as a Hazard

Microbial risk assessments, or components of the process, have been conducted for various microorganisms as both waterborne disease agents and foodborne pathogens, including microorganisms discussed in this chapter (Haas et al., 1996, 2000; Holcomb et al., 1999; Teunis et al., 1999; Crockett et al., 1996; Rose et al., 1995). In addition, the application of risk assessment to the microbial quality of agricultural water has been explored (Blumenthal et al., 2000; Shuval et al., 1997). A microbial risk assessment for microorganisms that can potentially contaminate fruits and vegetables through the water route could be undertaken to characterize the likelihood and public health consequences of such an occurrence. An approach could focus on microbial-contaminated water as the hazard. Table 4.4 lists some factors to consider in such a risk assessment.

Data on the occurrence of microorganisms in agricultural water sources are needed for conducting of microbial risk assessments of water used in produce production and processing. Ideally, methods to obtain these data would be quantitative and address infectivity. Any water treatment administered, as well as the amount of water applied and retained on the produce after processing, needs to be addressed (Shuval et al., 1997). The various pathways agricultural water sources contact

TABLE 4.4 Considerations When Characterizing Water as a Hazard in Microbial Risk Assessments of Produce

Sources of water used in produce production/processing
Types of and effectiveness of water treatment applied to source water
Types of pathogens occurring in specific source water
Amount of water used during different stages of produce production/processing
Amount of water retained on produce after production/processing
Survivability of waterborne microorganisms on produce
Effectiveness of produce washing/processing by consumer

produce during production and processing should also be included in the assessment. The identification of appropriate indicator organisms has been minimally explored and could prove useful in a risk assessment approach (Hirotani et al., 2001). Furthermore, studies need to be conducted that determine the survivability of specific microorganisms on various produce, including susceptibility to sanitizers. In addition, an understanding of the dose–response relationships for specific waterborne pathogens is critical in the risk assessment process, and consumption patterns of various produce need to be characterized.

4.8 IMPLICATIONS FOR HACCP

The Hazard Analysis and Critical Control Point (HACCP) is a systematic process that was first developed in 1959 by the Pillsbury Company along with the National Aeronautics and Space Administration (NASA) and the U.S. Army Natick Laboratories to ensure safety (quality) measurements were met regarding food and water in a space environment. The process has since been implemented to minimize the possibility of contamination of food during production and processing to strive for the highest quality food product at the time of consumption (CODEX Alimentarius Commission, 1997). This approach provides another mechanism to prevent foodborne disease along with end-product testing.

The HAACP system is guided by seven principles that have been described elsewhere (CODEX Alimentarius Commission, 1997). The basic premise is to identify hazards (microbial, chemical, or physical) that may be introduced during food production/processing as well as identify the particular points where these hazards could be introduced. These "critical control points" are then targeted as areas where strategies could be implemented to avoid contamination by a particular hazard or hazards. This system also incorporates a "checks and balances" to monitor these critical control points and verify that the HACCP system is effective. This system has been applied (and sometimes required) in a variety of manufacturing settings, including the meat, poultry, dairy, and seafood industries, as well as by juice manufacturers.

When considering risk analysis, HACCP is really a risk management component based on the overall premise of the system. The most important – and perhaps most difficult – task of a HAACP system, however, is in the identification of hazards and subsequent identification of the associated critical control points. This is where risk assessment is a valuable tool to identify such hazards in a way that assists in the determination of critical control points. Risk assessment can take a "farm-to-fork" approach in its identification and characterization of hazards associated with food throughout the food production chain.

4.8.1 Incorporating Water Issues in HACCP Programs

Identifying water as a critical ingredient with the potential to impact food quality has gained much attention in recent years by food industries. Acceptable water quality is sometimes taken for granted, depending upon the water source and purpose during food production and processing. For growers of fruits and vegetables, the importance of utilizing high-quality water is recognized, but no guidelines are available to provide direction for water monitoring and interpretation of water-monitoring data. A HACCP system can provide a mechanism to consider water quality in food manufacturing settings (including in the agricultural arena) and monitor its integrity throughout the process.

A risk assessment approach could be undertaken to identify specific hazards associated with water use during food manufacturing and to pinpoint areas where control measures and monitoring of those control measures should be implemented. During the production of fruits and vegetables, for example, an initial overview of the specific processes involved could identify areas where pathogens could be introduced via water. Control points during pre- and postharvest could be targeted in a HACCP program to prevent or eliminate the occurrence of waterborne pathogens that could then be transmitted to consumers through food.

As discussed earlier in this chapter, pathogens can contact produce via the water route at various stages during pre- and postharvest. Table 4.5 provides a list of some critical control points related to water usage that could be part of a HAACP program applied in the fruit and vegetable production and processing setting. Source water

TABLE 4.5 Possible Critical Control Points Related to Water Usage (Quality) in an Agricultural Setting During the Production and Processing of Fruits and Vegetables

Source Water
Water Treatment
Irrigation Water
Contact Sprays
Wash Water
Rinse Water
Ice
Container Sanitation
Equipment/Truck Sanitation
Field Worker/Packer Hygiene

may be contaminated with pathogens by a variety of routes, such as manure contact. The amount of pathogen content in manure varies, and recommendations to prevent agricultural water from becoming contaminated are based on distance between the manure site and the source water. Agricultural water is applied throughout the pre- and postharvest stages, including for irrigation, as part of fertilizer and pesticide sprays, as wash water (for produce as well as equipment, container, and truck sanitation), as rinse water, to make (packing) ice, and for handwashing by field and packinghouse workers. The potential for agricultural water to become contaminated with pathogens is dynamic; water may be contaminated at the source or may become contaminated during one of the pre-/post-harvesting events listed in Table 4.5. The greatest challenge with implementing such a HACCP program related to water quality is the lack of specific guidelines available to growers so that monitoring data is meaningful and responsive action steps are triggered.

4.9 GLOBAL CHALLENGES

The amount of fresh fruits and vegetables imported into the United States has increased dramatically over the past decade. Sales have increased from $2.0 billion in 1987 to more than $4.0 billion in 1997 (Kaufman et al., 2000). There is great variability in the production practices of fruits and vegetables throughout the world, and USFDA has attempted to unify the structure of these practices through their implementation of the Good Agricultural Practices (GAPs) for growers in the United States as well as those outside exporting to the United States (USFDA, 2001). Implementing effective HACCP programs is challenging at best; each program needs to be developed and enforced by trained personnel with knowledge of all aspects of their particular environment and industry. Fresh produce are currently being distributed to a large number of locations and at increasing distances from point of harvest, indicating a greater need for quality control mechanisms, yet creating more challenges in protecting produce quality. It is speculated that this global distribution of produce is a contributing factor for the observed increase in foodborne illnesses in the United States associated with fruits and vegetables.

The application of water is yet another challenge when considering produce production on a global scale. More research is needed regarding the impact of water on produce quality at the different stages of pre- and postharvest. In particular, specific waterborne pathogens as potential contaminants of produce during pre- and postharvest need to be further characterized to fully understand the associated public health implications and to determine effective preventive strategies.

REFERENCES

M. Ackers, B.E. Mahon, E. Leahly, B. Goode, T. Damrow, P.S. Hayes, W.F. Bibb, D.H. Rice, T.J. Barrett, L. Hutwagner, et al., An outbreak of *Escherichia coli* O157:H7 infections associated with leaf lettuce consumption. *J. Infect. Dis.*, 177, 1588–1593 (1998).

A.B. Allen, Outbreak of campylobacteriosis in a large educational institution – British Columbia. *Can. Dis. Weekly Rep.*, 2, 28–30 (1985).

J.A. Bartz, Washing fruits and vegetables: lessons from treatment of tomatoes and potatoes with water. *Dairy, Food, and Environ. San.*, 19 (12), 853–864 (1999).

R.K.X. Bastos and D.D. Mara, The bacterial quality of salad crops drip and furrow irrigated with waste stabilization pond effluent: an evaluation of the WHO guidelines. *Water Sci. Technol.*, 31 (12), 425–430 (1995).

N.H. Bean and P.M. Griffin, Foodborne disease outbreaks in the United States, 1973–1987: pathogens, vehicles, and trends. *J. Food Prot.*, 53 (9), 804–817 (1990).

N.H. Bean, J.S. Goulding, M.T. Daniels, and F.J. Angulo, Surveillance for foodborne disease outbreaks: United States, 1988–1992. *J. Food Prot.*, 60, 1265–1286 (1994).

L.R. Beuchat, Pathogenic microorganisms associated with fresh produce. *J. Food Prot.*, 59, 204–206 (1996).

L.R. Beuchat and J.-H. Ryu, Produce handling and processing practices. *Emerg. Infect. Dis.*, 3 (4), 459–465 (1997).

U.J. Blumenthal, D.D. Mara, A. Peasey, G. Ruiz-Palacios, and R. Stott, Guidelines for the microbiological quality of treated wastewater used in agriculture: recommendations for revising WHO guidelines. *Bull. World Health Org.*, 78 (9), 1104–1116 (2000).

R.E. Brackett, Incidence, contributing factors, and control of bacterial pathogens in produce. *Postharvest Biol. Technol.*, 15, 305–311 (1999).

CAST (Council for Agricultural Science and Technology), *Foodborne Pathogens: Risks and Consequences.* CAST, Ames, Iowa, 1994.

CDC (Centers for Disease Control and Prevention), Infectious hepatitis – Tennessee. *Morbidity Mortality Weekly Rep.*, 20, 357 (1971).

CDC (Centers for Disease Control and Prevention), Epidemiologic notes and reports common-source outbreak of giardiasis – New Mexico. *Morbidity Mortality Weekly Rep.*, 38, 405–407 (1989).

CDC (Centers for Disease Control and Prevention), Waterborne disease outbreaks. U.S. Department of Health and Human Services, Atlanta, GA. *Morbidity Mortality Weekly Rep.*, 39 (SS-1), 1–57 (1990).

CDC (Centers for Disease Control and Prevention), Foodborne outbreaks of enterotoxigenic *Escherichia coli* – Rhode Island and New Hampshire, 1993. *Morbidity Mortality Weekly Rep.*, 43 (5), 81,87–88 (1994).

CDC (Centers for Disease Control and Prevention), Surveillance for foodborne-disease outbreaks – United States, 1988–1992. *Morbidity Mortality Weekly Rep.*, 45 (SS-5), 1–55 (1996).

CDC (Centers for Disease Control and Prevention), Outbreaks of *Escherichia coli* O157:H7 infection associated with eating alfalfa sprouts – Michigan and Virginia, June–July 1997. *Morbidity Mortality Weekly Rep.*, 46 (32), 741–744 (1997a).

CDC (Centers for Disease Control and Prevention), Hepatitis A associated with consumption of frozen strawberries – Michigan, March 1997. *Morbidity Mortality Weekly Rep.*, 46, 288–289 (1997b).

CDC (Centers for Disease Control and Prevention), Outbreak of cyclosporiasis – Northern Virginia – Washington DC – Baltimore, Maryland metropolitan area, 1997. *Morbidity Mortality Weekly Rep.*, 46 (30), 689–691 (1997c).

CDC (Centers for Disease Control and Prevention), Update: outbreaks of cyclosporiasis – United States, 1997. *Morbidity Mortality Weekly Rep.*, 46 (21), 461–462 (1997d).

CDC (Centers for Disease Control and Prevention), Outbreaks of *Escherichia coli* O157:H7 infection and cryptosporidiosis associated with drinking unpasteurized apple cider – Connecticut and New York, October 1996. *Morbidity Mortality Weekly Rep.*, 46, 4–8 (1997e).

CDC (Centers for Disease Control and Prevention), Outbreak of *Campylobacter enteritis* associated with cross-contamination of food – Oklahoma, 1996. *Morbidity Mortality Weekly Rep.*, 47, 129–131 (1998a).

CDC (Centers for Disease Control and Prevention), Foodborne outbreak of cryptosporidiosis – Spokane, Washington, 1997. *Morbidity Mortality Weekly Rep.*, 47, 565–567 (1998b).

CDC (Centers for Disease Control and Prevention), Outbreaks of *Shigella sonnei* infection associated with eating fresh parsley – United States and Canada, July–August, 1998. *Morbidity Mortality Weekly Rep.*, 48 (14), 285–289 (1999).

CDC (Centers for Disease Control and Prevention), Surveillance for foodborne-disease outbreaks – United States, 1993–1997. *Morbidity Mortality Weekly Rep.*, 49 (SS01), 1–51 (2000a).

CDC (Centers for Disease Control and Prevention), CDC surveillance summaries; March 17, 2000. *Morbidity Mortality Weekly Rep.*, 49 (SS-1), 1–51 (2000b).

CDC (Centers for Disease Control and Prevention), Hepatitis A outbreak associated with green onions at a restaurant – Monaca, Pennsylvania, 2003. *Morbidity Mortality Weekly Rep.*, 2 (47), 1155–1157 (2003).

R.B. Chalker and M.J. Blaser, A review of human salmonellosis. III. Magnitude of *Salmonella* infection in the United States. *Rev. Infect. Dis.*, 10, 111–124 (1988).

CODEX Alimentarius Commission, Hazard Analysis and Critical Control Point (HACCP) System and Guidelines For Its Application. Annex to CAC/RCP 1-1969, Rev.3. Food and Agricultural Organisation of the United Nations, FAO/Codex Alimentarius Commission, Rome, Italy, 1997.

T.L. Cover and R.C. Aber, *Yersinia enterocolitica. New Eng. J. Med.*, 321, 16–24 (1989).

C.S. Crockett, C.N. Haas, A. Fazil, J.B. Rose, and C.P. Gerba, Prevalence of shigellosis in the U.S.: consistency with dose-response information. *Int. J. Food Microbiol.*, 30, 87–99 (1996).

K. Cummings, E. Barrett, J.C. Mohle-Boetani, J.T. Brooks, J. Farrar, T. Hunt, A. Fiore, K. Komatsu, S.B. Werner, and L. Slutsker, A multistate outbreak of *Salmonella enterica* serotype baildon associated with domestic raw tomatoes. *Emerg. Infect. Dis.*, 7 (6), 1046–1048 (2001).

H. Davis, J.P. Taylor, J.N. Perdue, G.N. Stelma Jr., J.M. Humphreys Jr, R. Rowntree III, and K.D. Greene, A shigellosis outbreak traced to commercially distributed shredded lettuce. *Am. J. Epidemiol.*, 128 (6), 1312–1321 (1988).

R. Dolin, J.J. Treanor, and P. Madore, Novel agents of viral gastroenteritis. *J. Infect. Dis.*, 155, 365–371 (1987).

FDA (Food and Drug Administration, Center for Food Safety and Applied Nutrition, Office of Plant and Dairy Foods and Beverages), FDA survey of imported fresh produce. FY 1999 Field Assignment, January 30, 2001. Available at http://www.cfsan.fda.gov/~dms/prodsur6.html.

C.A. Fleming, D. Caron, J.E. Gunn, and M.A. Barry, A foodborne outbreak of *Cyclospora cayetanensis* at a wedding. *Arch. Intern. Med.*, 158, 1121–1125 (1998).

FSIS (Food Safety and Inspection Service), FSIS/CDC/FDA site study: the establishment and implementation of an active surveillance system for bacterial foodborne diseases in the United States, Report to Congress. U.S. Department of Agriculture, Washington, DC, 1997.

B. Garcia-Villanova Ruiz, R. Galvez Vargas, and R. Garcia-Villanova, Contamination on fresh vegetables during cultivation and marketing. *Int. J. Food Microbiol.*, 4, 285–291 (1987).

C.N. Haas, C.S. Crockett, J.B. Rose, C.P. Gerba, and A.M. Fazil, Assessing the risk posed by oocysts in drinking water. *J. Am. Water Works Assoc.*, September, 131–136 (1996).

C.N. Haas, J.B. Rose, and C.P. Gerba, *Quantitative Microbial Risk Assessment.* John Wiley & Sons, Inc., New York, 1999.

C.N. Haas, J.B. Rose, C. Gerba, and S. Regli, Risk assessment of virus in drinking water. *Risk Anal.*, 13 (5), 545–552 (1993).

C.N. Haas, A. Thayyar-Madabusi, J.B. Rose, and C.P. Gerba, Development of a dose–response relationship for *Escherichia coli* O157:H7. *Int. J. Food Microbiol.*, 1748, 153–159 (2000).

N.V. Harns, N.S. Weiss, and C.M. Nolan, The role of poultry and meats in the etiology of *Campylobacter jejuni/coli* enteritis. *Am. J. Publ. Health*, 76, 407–411 (1986).

B.L. Herwaldt and M.L. Ackers, An outbreak in 1996 of cyclosporiasis associated with imported raspberries. *New Engl. J. Med.*, 336 (22), 1548–1556 (1997).

B.L. Herwaldt and M.J. Beach, The return of *Cyclospora* in 1997: another outbreak of cyclosporiasis in North America associated with imported berries. *Ann. Intern. Med.*, 130, 210–219 (1999).

B.L. Herwaldt, J.F. Lew, C.L. Moe, D.C. Lewis, C.D. Humphrey, S.S. Monroe, E.W. Pon, and R.I. Glass, Characterization of a variant strain of Norwalk virus from a foodborne outbreak of gastroenteritis on a cruise ship in Hawaii. *J. Clin. Microbiol.*, 32, 861–866 (1994).

H. Hirotani, J. Naranjo, P.G. Moroyoqui, and C.P. Gerba, Demonstration of indicator microorganisms on surface of vegetables on the market in the United States and Mexico. *J. Food Sci.*, 67 (5), 1847–1850 (2001).

D.L. Holcomb, M.A. Smith, G.O. Ware, Y.-C. Hung, R.E. Brackett, and M.P. Doyle. Comparison of six dose–response models for use with food-borne pathogens. *Risk Anal.*, 19 (6), 1091–1100 (1999).

R.S. Hopkins, R.N. Olmsted, and G.R. Istre, Endemic *Campylobacter jejuni* infection in Colorado: identified risk factors. *Am. J. Publ. Health*, 74, 249–250 (1984).

Y.J.F. Hutin, V. Pool, E.H. Cramer, O.V. Nainan, J. Weth, I.T. Williams, S.T. Goldstein, K.F. Gensheimer, B.P. Bell, C.N. Shapiro, et al., A multistate, foodborne outbreak of hepatitis A. *New Engl. J. Med.*, 340, 595–602 (1999).

J.E. Kaplan, G.W. Gary, R.C. Baron, N. Singh, L.B. Schonberger, R. Fieldman, and H.B. Greenberg, Epidemiology of Norwalk gastroenteritis and the role of Norwalk virus in outbreak of acute nonbacterial gastroenteritis. *Ann. Intern. Med.*, 96, 757–761 (1982).

P.R. Kaufman, C.R. Handy, E.W. McLaughlin, K. Park, and G.M. Green, Understanding the dynamics of produce markets: consumption and consolidation grow. USDA, Economic Research Service, Agriculture Information Bulleting No. 758, 2000.

A.M. Lammerding, A.M. Fazil, and G.M. Paoli, "Microbial Food Safety Risk Assessment," in F.P. Downes and K. Ito, Eds, *Compendium of Methods for the Microbiological Examination of Foods*, 4th edn. American Public Health Association, Washington, DC, 2001.

A.M. Lammerding and G.M. Paoli, Quantitative risk assessment: an emerging tool for emerging foodborne pathogens. *Emerg. Infect. Dis.*, 3, 483–487 (1997).

S. M. Lemon, Type A viral hepatitis: new developments in an old disease, *New Engl. J. Med.* 313, 1059–1067 (1985).

J.T. Lisle and J.B. Ros., *Cryptosporidium* contamination of water in the USA and UK: a mini-review. *J. Water Supply: Res. Technol.–Aqua*, 44, 103–117 (1995).

P.W. Lowry, R. Levine, D.F. Stroup, R.A. Gunn, M.H. Wilder, and C. Konigsberg, Hepatitis A outbreak on a floating restaurant in Florida, 1986. *Am. J. Epidemiol.*, 129, 155–164 (1989).

B.M. Lund and A.L. Snowdon, "Fresh and Processed Fruits," in B.M. Lund, T.C. Baird-Parker and G.W. Gould, Eds., *The Microbiological Safety and Quality of Food*, Vol. 1. Aspen Publishers, Gaithersburg, MD; 2000, pp. 738–758.

W.R. MacKenzie, M. Neil, N.J. Hoxie, M.E. Proctor, M.S. Gradus, K.A. Blair, D.E. Peterson, et al., A massive outbreak in Milwaukee of *Cryptosporidium* infection transmitted through the public water supply. *New Engl. J. Med.*, 331, 161–167 (1994).

J.M. Madden, Microbial pathogens in fresh produce – the regulatory perspective. *J. Food Prot.*, 55 (10), 821–823 (1992).

D.L. Martin, T.L. Gustafson, J.W. Pelosi, L. Suarez, and G.V. Pierce, Contaminated produce – a common source for two outbreaks of *Shigella* gastroenteritis. *Am. J. Epidemiol.*, 124 (2), 299–305 (1986).

P.S. Mead, L. Slutsker, V. Dietz, L.F. McCaig, J.S. Bresee, C. Shapiro, P.M. Griffin, and R.V. Tauxe, Food-related illness and death in the United States. *Emerg. Infect. Dis.*, 5, 607–625 (1999).

K.D. Mena, "Environmental Exposure to Viruses: A Human Health Risk Assessment Methodology," in G. Bitton, Ed., *The Encyclopedia of Environmental Microbiology.* John Wiley & Sons, Inc., New York, 2002.

K.D. Mena, J.B. Rose, and C.P. Gerba, "Addressing Food Safety Issues Quantitatively: A Risk Assessment Approach," in R.C. Beier, S.D. Pillai, T.D. Phillips, and R.L. Ziprin. Eds., *Pre-Harvest and Post-Harvest Food Safety: Contemporary Issues and Future Directions.* Blackwell Publishing, Ames, Iowa, 2004.

M.H. Merson, G.K. Morris, D.A. Sack, J.E. Wells, J.C. Feeley, R.B. Sack, W.B. Creech, A.Z. Kapikian, and E.J. Gangarosa, Travelers' diarrhea in Mexico. *New Engl. J. Med.*, 294, 1299–1305 (1976).

P.S. Millard, K.F. Gensheimer, D.G. Addiss, D.M. Sosin, G.A. Beckett, A. Houck-Jankoski, and A. Hudson, An outbreak of cryptosporidiosis from fresh-pressed apple cider. *J. Am. Med. Assoc.*, 272 (20); 1592–1596 (1994).

B. Mishu and M.J. Blazer, Role of infection due to *Campylobacter jejuni* in the initiation of Guillain–Barré syndrome. *Clin. Infect. Dis.*, 17, 104–108 (1993).

R. Monge and M. Chinchilla, Presence of *Cryptosporidium* oocysts in fresh vegetables. *J. Food Prot.*, 59(2), 202–203 (1996).

C. Nguyen-the and F. Carlin, "Fresh and Processed Vegetables," in B.M. Lund, T.C. Baird-Parker, and G.W. Gould, Eds., *The Microbiological Safety and Quality of Food.* Aspen Publishers, Gaithersburg, MD, 2000, pp. 620–684.

M.T. Niu, L.B. Polish, B.H. Robertson, B.K. Khanna, B.A. Woodruff, C.N. Shapiro, M.A. Miller, J.D. Smith, J.K. Gedrose, M.J. Alter, et al., Multistate outbreak of hepatitis A associated with frozen strawberries. *J. Infect. Dis.*, 166, 518–524 (1992).

NRC (National Research Council), *Risk Assessment in the Federal Government: Managing the Process.* National Academy Press, Washington, DC, 1983.

NRC (National Research Council), *Science and Judgement in Risk Assessment.* National Academy Press, Washington, DC, 1994.

J.A. Odumeru, S.J. Mitchell, D.M. Alves, J.A. Lynch, A.J. Yee, S.L. Wange, S. Styliadis, and J.M. Farber, Assessment of the microbiological quality of ready-to-use vegetables for health-care food services. *J. Food Prot.*, 60 (8), 954–960 (1997).

C.E. Park and G.W. Sanders, Occurrence of thermotolerant campylobacters in fresh vegetables sold at farmers' outdoor markets and supermarkets. *Can. J. Microbiol.*, 38 (4), 313–316 (1992).

C.N. Ramsay and P.A. Upton, Hepatitis A and frozen raspberries. *Lancet*, 1 (8628), 43–44 (1989).

V.C. Rao and J.L. Melnick, "Monitoring for Viruses in Wastewater and Water," in *Environmental Virology.* J.A. Cole, C.J. Knowles, and D. Schlessinger, Eds., American Society for Microbiology, Washington, DC, 1986, pp. 18–40.

S. Regli, J.B. Rose, C.N. Haas, and C.P. Gerba, Modeling the risk from *Giardia* and viruses in drinking water. *J. Am. Water Works Assoc.*, 83, 76–84 (1991).

T.M.S. Reid and H.G. Robinson, Frozen raspberries and hepatitis A. *Epidemiol. Inf.*, 98, 109–112 (1987).

I. Robinson and R.P. Adams, Ultra-violet treatment of contaminated irrigation water and its effect on the bacteriological quality of celery at harvest. *J. Appl. Bacteriol.*, 45, 83–90 (1978).

J.B. Rose, C.N. Haas, and C.P. Gerba, Linking microbiological criteria for foods with quantitative risk assessment. *J. Food Safety*, 15, 121–132 (1995).

J.B. Rose, C.N. Haas, and S. Regli, Risk assessment and control of waterborne giardiasis. *Am. J. Publ. Health*, 81 (6), 709–713 (1991).

L.S. Rosenblum, I.R. Mirkin, D.T. Allen, S. Safford, and S.C. Hadler, A multifocal outbreak of hepatitis A traced to commercially distributed lettuce. *Am. J. Publ. Health*, 80 (9), 1075–1079 (1990).

B. Sheikh, R.C. Cooper, and K.E. Israel, Hygienic evaluation of reclaimed water used to irrigate food crops – a case study. *Water Sci. Technol.*, 40 (4–5), 261–267 (1999).

H. Shuval, Y. Lampert, and B. Fattal. Development of a risk assessment approach for evaluating wastewater reuse standards for agriculture. *Water Sci. Technol.*, 35 (11–12), 15–20 (1997).

J.L. Smith, S.A. Palumbo, and I. Wallis, Relationship between foodborne bacterial pathogens and reactive arthritides. *J. Food Safety*, 13, 209–236 (1993).

D.L. Swerdlow, E.D. Mintz, M. Rodriguez, E. Tejada, C. Ocampo, L. Espejo, K.D. Greene, W. Saldana, L. Seminario, R.V. Tauxe, et al., Waterborne transmission of epidemic cholera in Trujillo, Peru; lessons for a continent at risk. *Lancet*, 340 (4), 28 (1992).

P.J. Taormina, L.M. Beuchat, and L.M. Slutsker, Infections associated with eating seed sprouts: an international concern. *Emerg. Infect. Dis.*, 62, 626–634 (1999).

R.V. Tauxe, Emerging foodborne diseases: an evolving public health challenge. *Dairy, Food, Env. Sanit.*, 17 (12), 788–795 (1997).

P.F.M. Teunis, N.J.D. Nagelkerke, and C.N. Haas, Dose response models for infectious gastroenteritis. *Risk Anal.*, 19 (6), 1251–1260 (1999).

USDA (United States Department of Agriculture), *Fruit and Vegetable Agricultural Practices – 1999.* National Agricultural Statistics Service, 2001. Available at http://usda.gov/nass/pubs/rpts106.htm.

USEPA (United States Environmental Protection Agency) Office of Water, Office of Wastewater Enforcement and Compliance, *September Manual: Guidelines for Water Reuse.* U.S. Agency for International Development Report nr EPA/625/R-92/004, Washington, 1992.

USFDA (United States Food and Drug Administration), Analysis and Evaluation of Preventive Control Measures for the Control and Reduction/Elimination of Microbial Hazards on Fresh and Fresh-Cut Produce. Report of the Institute of Food Technologists. IFT/FDA Contract No. 3, 2001. Available at http://www.cfsan.fda.gov/~comm/ift3-toc.html.

C.I. Wei, T.S. Huang, J.M. Kim, W.F. Lin, M.L. Tamplin, and J.A. Bartz, Growth and survival of *Salmonella montevideo* on tomatoes and disinfection with chlorinated water. *J. Food Prot.*, 58 (8), 829–836 (1995).

K.E. White, A foodborne outbreak of Norwalk virus gastroenteritis: evidence for post-recovery transmission. *Acute Dis. Epidemiol.* (Field Services Sect), Minnesota Department of Health, 124, 120–126 (1986).

WHO (World Health Organization), *Health Guidelines for the Use of Wastewater in Agriculture and Aquaculture*, Technical Report Series 778. WHO, Geneva, Switzerland, 1989.

R.C. Wood, K.L. MacDonald, and M.T. Osterholm, *Campylobacter enteritis* outbreaks associated with drinking raw milk during youth activities. *J. Am. Med. Assoc.*, 268, 3228–3230 (1992).

D.W. York, L.W. Parsons, and L. Waler-Coleman, Agricultural reuse: using reclaimed water to irrigate edible crops in Florida. In Proceedings of the 2000 Florida Water Resources Conference, April 17–19, 2000, Tampa. FWEA, FS/AWWA, FW&PCOA.

CHAPTER 5

Food Worker Personal Hygiene Requirements During Harvesting, Processing, and Packaging of Plant Products

BARRY MICHAELS

B. Michaels Group Inc. Consulting
Palatka, Florida

EWEN TODD

National Food Safety and Toxicology Center, Michigan State University,
East Lansing, MI, USA

Microbial Hazard Identification in Fresh Fruit and Vegetables, Edited by Jennylynd James

5.1 INTRODUCTION

In recent years, there have been a number of widely publicized foodborne illness outbreaks associated with consumption of fruits and vegetables. Considering the diverse variety of produce products, the manner in which they are grown, harvested, packaged, and distributed, it is not surprising that numerous risks for contamination exist (Lund and Snowden, 2000; Rangarajan et al., 2000; Michaels et al., 2004a, b). As a result of farm and packing house investigations of outbreaks by U.S. federal agencies, infected workers and poor worker hygiene are often implicated (Levine et al., 1996; FDA, 1998). It is important, therefore, for the produce industry to develop an understanding of the microbiological hazards and issues related to produce worker personal hygiene. A detailed understanding of produce-related outbreaks demonstrates the kinds of hazards and risky practices that contributed to the illnesses. The products connected to outbreaks attributed to infected workers include mamey, coconut, strawberries, green onions, leaf lettuce, and basil; these were contaminated with a wide spectrum of microbial agents, bacteria, viruses, and protozoan parasites (Katz et al., 2002; Taylor et al., 1993; Ramsay and Upton, 1989; Niu et al., 1992; Cook et al., 1995; Dentinger et al., 2001; Hilborn et al., 1999; Davis et al., 1988; Naimi et al., 2003). However, a key factor associated with these outbreaks is poor personal hygiene. The term poor personal hygiene is rather inexact, and it is the aim of this chapter to describe the various failings that comprise this classification.

5.2 MICROBIAL HAZARDS RELEVANT TO WORKER HYGIENE

5.2.1 Foodborne Outbreaks Linked to Infected Produce Workers

Contact of fecal material, animal or human, with produce, even if not visible, is a source of contamination that has been associated with past outbreaks of foodborne illness arising from fruit and vegetable consumption. This is particularly true if the fecal source is diarrheal in nature. Persons with infectious diseases increase the risk of microbial hazards because they can unintentionally contaminate fresh produce, water supplies, and other workers. However, excretors of pathogens may not demonstrate any symptoms (recovering, asymptomatic, or carrier state), yet may still be the cause of outbreaks. Numerous pathogens infecting humans exhibit periods of asymptomatic shedding with periods as short as a few hours to as long as ten years or more with some agents (FDA, 2000, 2001; Michaels et al., 2004b). Specific agents that infected produce workers excrete include *Escherichia coli* O157:H7, *Salmonella*, *Shigella*, *Staphylococcus aureus*, *Vibrio* spp., hepatitis A virus, and the parasites *Cryptosporidium* and *Cyclospora* (Table 5.1). Types of produce included fruits, leaves, and roots, all involving extensive handling. Producers must make an assessment of the potential hazards associated with the type of crop grown, the handling practices involved, and reported instances of foodborne outbreaks associated with each. Although much can be accomplished by following general guidelines of good sanitary practices, some of the hazards

TABLE 5.1 Produce Outbreaks Associated with Infected Workers

Date	Produce	Infectious Agent	No. of Cases	Produce Origin	Reference
1987	Raspberries (frozen)	Hepatitis A virus	92	United Kingdom	Reid and Robinson, 1987
1989	Canned mushrooms	*Staphylococcus aureus*	99	China	Levine et al., 1996
1990	Strawberries	Hepatitis A virus	53	United States	Niu et al., 1992
1991	Frozen coconut milk	*Vibrio cholera* O1	3	Thailand	Taylor et al., 1993
1994	Green onions (scallions)	*Shigella flexneri*	72	California	Cook et al., 1995
1996	Leaf lettuce	*E. coli* O157:H7	49	United States	Hilborn et al., 1999
1997	Strawberries (frozen)	Hepatitis A virus	250	California	CDC, 1997
1997	Green onions (scallions)	*Cryptosporidium parvum*	55	United States	CDC, 1998
1997	Basil	*Cyclospora cayetanensis*	341	United States	Pritchett et al., 1997
1998	Green onions (scallions)	Hepatitis A virus	43	United States/California	Dentinger et al., 2001
1998	Mamey (sapote fruit pulp)	*Salmonella typhi*	13	Guatemala	Katz et al., 2002
1999	Parsley (chopped)	*Shigella sonnei*	486	United States	CDC, 1999
2003	Parsley (chopped)	Enterohemorrhagic *E. coli*	77	United States	Naimi et al., 2003

can only be determined by experience from outbreaks and by conducting some kind of risk assessment. Operators should follow acceptable general and product-specific standards for personal hygiene, food handling, and sanitation. It should be recognized that both management and workers are responsible for the appropriate handling of fresh fruit and vegetables to prevent the possibility of product contamination. Specific personal hygiene and sanitation requirements depend on the product harvested and processed.

5.2.2 Factors Involved in Infected Worker Contamination of Produce

Produce involved in outbreaks has often been extensively handled by infected workers and then subjected to temperature abuse, which allowed pathogenic bacterial growth to take place. Other problems identified, particularly in developing countries, have been workers coming from areas of high endemic rates of disease. These areas have poor hygiene education, a lack of appropriate washing and toilet facilities, and lack of sanitary and hygienic supplies (Michaels et al., 2004b). For any kind of food processing or preparation operation, worker hygiene starts with fastidious emphasis on facility sanitation, personal cleanliness, sanitary behavior, knowledge of worker health status, and elimination of risky handling practices. This is particularly true for fruits and vegetables that do not have a final decontamination step. Placement and maintenance of toilet facilities and the development of effective training programs for each type of employee are important ingredients in prevention. Toilet facilities must be accessible and monitored to prevent soil and water contamination as well as cross-contamination to animals and workers (Zagory, 1999). Workers must be taught proper hygiene technique and Standard Sanitary Operating Procedures (SSOP) as hand hygiene knowledge cannot be assumed (University of Maryland, 2002a). These hygienic procedures used by producers and processors may be termed Good Agricultural Practices (GAPs) and Good Handling Practices (GHPs) (University of Maryland, 2002a).

The term *personal hygiene* encompasses a variety of practices used to reduce the risk of disease transmission by food workers. It includes activities both at work and at home prior to the food-handling employees coming to work. Good hygiene habits practiced by agricultural workers are critically important in decreasing the opportunity for pathogens to contaminate fruit and vegetables. Although hand hygiene is the most fundamental aspect of personal hygiene, many other elements are of importance. Hands are, however, what move vast quantities of produce from growing sites to packing and processing operations. If employees do not wash their hands before starting work and after using the toilet, potential human pathogens can be transferred from workers' hands to fruit and vegetable products. In addition, some hygiene requirements are designed to prevent physical contamination of produce with foreign material such as stones, glass fragments, and objects worn by workers, such as necklaces, bracelets, or earrings. Worker cooperation is crucial in eliminating contamination of fresh fruit and vegetables destined directly for consumption by consumers or to be further processed for retail or food service.

5.2.3 Water Availability and Sanitary Status

The failure to use properly chlorinated water and barriers to transmission, such as gloves and various other appropriate utensils, has also been identified as a risk factor in these and other outbreaks (Table 5.2; Michaels et al., 2004a). A sanitary water source for consumption and hand washing reduces the likelihood of enteric infections contaminating the hands of workers.

Common sources of potable water usually include:

1. Ground water originating from below the ground surface and typically pumped up and out for use (i.e., well water) or flowing naturally to the surface (i.e., spring water or artesian well);
2. Treated surface water originating from lakes, rivers, creeks, canals, and reservoirs (i.e., ponds and water impoundment areas);
3. Treated municipal system, which comes from a city or municipality's water treatment plant.

As seen in Figure 5.1, untreated surface water is a more likely source of human pathogens than groundwater. This is because of the real possibility of direct contamination with animal feces, sewage, or land run-off from higher locations. It is important for producers and packers to have a clear understanding of the likely contamination routes of pathogens into raw water catchments being used. Fields with cattle or other animals that are close to a water source or gulls roosting near reservoirs are indications of potential contamination. Drinking water should always be of higher microbial quality than that used for agricultural processes because of the risk of infecting workers or indirectly contaminating produce at the point of handling. For this reason, it is highly recommended that water used for human consumption or hand hygiene be from municipal sources when at all possible. This can be accomplished by piping water from a municipal supply or by tank truck employing adequately cleaned and sanitized water-holding vessels. When neither of these solutions is a practical option, water treatment systems need to be installed to ensure safety.

Primary water treatment processes used to render water supplies free of pathogens include filtration, disinfection, and treatment to remove organic and inorganic contamination (USEPA, 2001). Initial processes are often used to clean up the water by removing solids and turbidity prior to filtration. This is done by addition of chemicals with rapid and vigorous mixing, followed by coagulation, flocculation, and/or sedimentation. Filtration through rapid sand, slow sand, diatomaceous earth, or cartridges, are also used to remove the remaining solids and larger pathogens like *Giardia* and *Cryptosporidium*, but not viruses and most bacteria. The three most commonly used disinfection technologies are chlorine, ozone, and chloramines (USEPA, 2001). Other disinfectants that may be used include chlorine dioxide and ozone. Unfortunately, chlorine-based formulations are not very effective against *Cryptosporidium*, which has been implicated as the cause of waterborne outbreaks and some foodborne episodes associated with fresh produce.

TABLE 5.2 Factors Contributing to Outbreaks Attributed to Produce Caused by Infected Workers

Date	Produce	Lack of Adequate Water Quality	Workers with Limited Hygiene Education	Poor or No Toilet Facilities	Poor, Limited or No Hand Hygiene Facilities	Bare Hand Contact	Lack of Contact Surface Cleaning	Lack of Child Care for Workers	Reference
1989	Canned mushrooms				×	×			Levine et al., 1996
1990	Strawberries					×			Niu et al., 1992
1991	Frozen coconut milk					×			Taylor et al., 1993
1994	Green onions					×		×	Cook et al., 1995
1996	Leaf lettuce	×			×	×			Hilborn et al., 1999
1996	Raspberries	×		×	×	×	×		CDC, 1997
1997	Strawberries (frozen)			×		×			CDC, 1997
1997	Green onions					×		×	CDC, 1998
1997	Basil					×			Pritchett et al., 1997
1998	Green onions			×	×	×		×	Dentinger et al., 2001
1998	Mamey					×			Katz et al., 2002
1999	Parsley (chopped)	×	×	×	×	×			Naimi et al., 2003

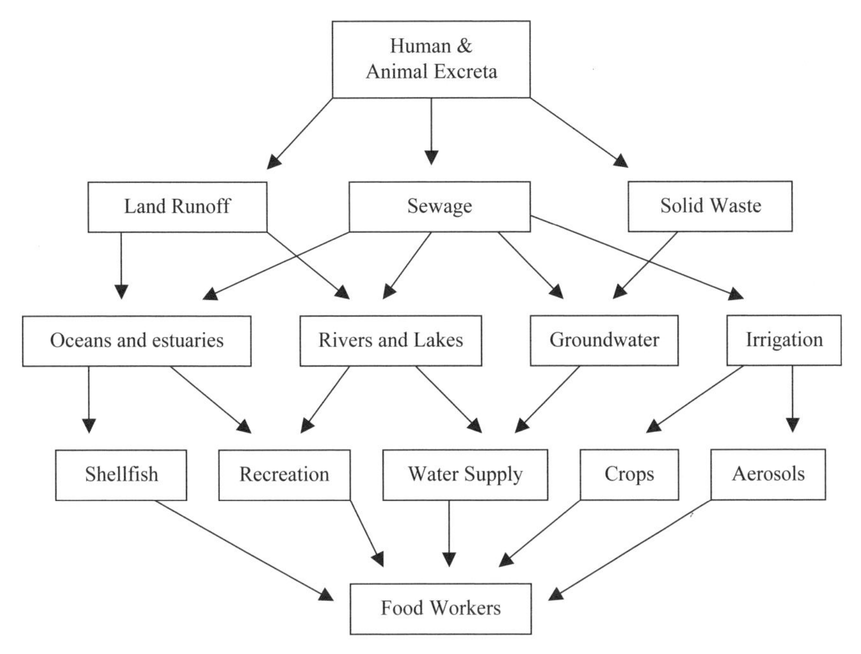

Figure 5.1 Routes of enteric pathogen transmission (adapted from Rao and Melnick, 1986).

5.3 ROLE OF LAWS, REGULATIONS, AND GUIDELINES RELATED TO WORKER HYGIENE

5.3.1 Role of Regulatory Authorities

Various local, state/provincial, and national regulations are in place around the world governing standards for worker health and hygiene during the growth, harvesting, processing, and marketing of foods, including produce destined for human consumption (ASBH, 1968; CDC, 1997; EUREPGAP, 2003; CHC, 2004). Operators need to be fully aware of all applicable regulations, and be able to implement them effectively. Management handling of on-site sanitary facilities directly affects worker health and hygiene, and ultimately the potential safety of the finished product. In the United States, placement, construction, and disposal of human waste are governed also by the Environmental Protection Agency (EPA) regulations (USEPA, 2003). Typically, strict field sanitation laws require growers to provide toilets, hand-washing water, and potable drinking water for agricultural field workers. All of these regulations are put into place in recognition of the fact that the health and hygiene of employees who handle fresh produce have a direct impact on the safety of food product for the consumer. Regulations may even go as far as requiring producers to provide regular food and worker safety training and documenting the training received by each employee. Therefore, in the strictest jurisdictions, workers who come into contact (directly or indirectly)

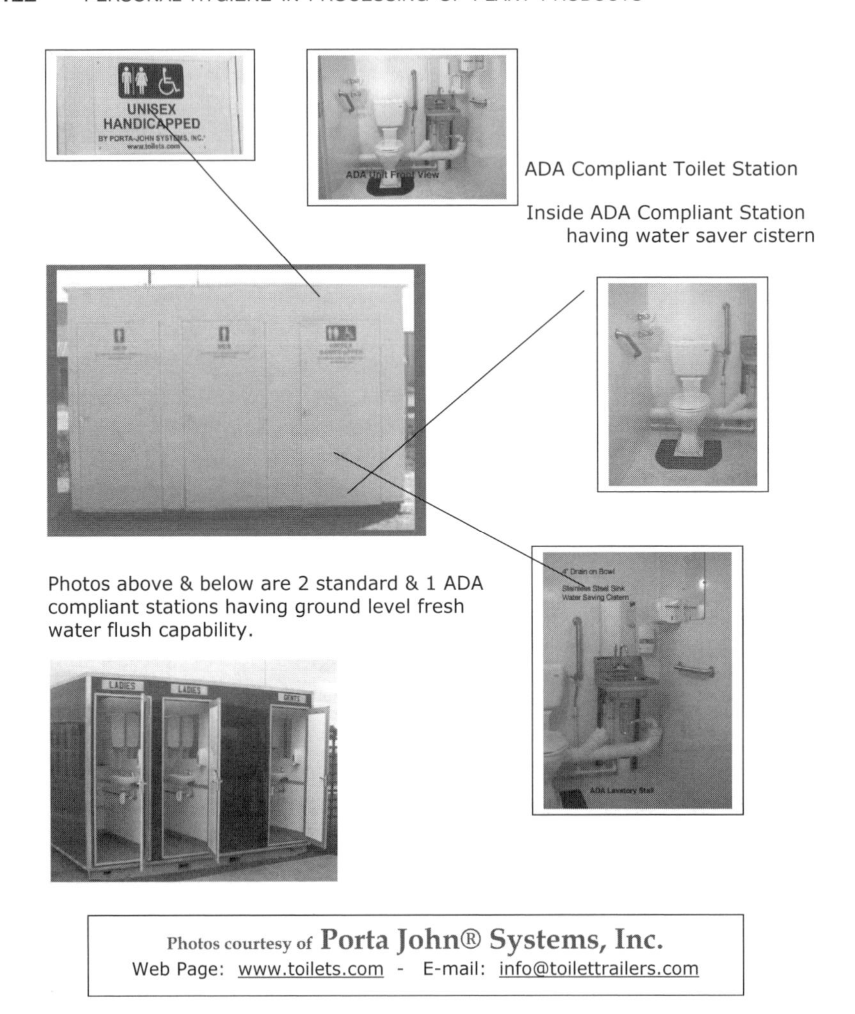

Figure 5.2 Portable toilet configurations suitable for field use.

with fresh fruit or vegetables must comply with specific hygiene and health provisions established to prevent food product contamination. Figure 5.2 shows portable toilet systems compatible with the American with Disabilities Act (ADA) requirements and suitable for field usage. Each contains an individual sink and water saver cistern, and is equipped with soap, paper towels, and toilet paper.

Because of the potential for contamination of produce by human wastes in the field or in packing facilities, the U.S. Occupational Safety and Health Act (OSHA, 2005a) regulations were enacted covering proper toilet facility management (OSHA, 2005b). Therefore, at least in the United States, packing and field facilities need to be operated according to the laws and regulations described in the Code of Federal Regulations (CFR).

5.3.2 Role of Industry Organizations in Establishing Standards

Numerous grower groups, cooperatives, and trade organizations, having suffered economic loss from food-safety scares or negative publicity, have organized to develop internal standards of practice. Typically, a group is connected to a particular crop, geographical region, or growing style, and members agree upon voluntary standards of practice. These industry-specific groupings focus on problems or processes unique to them and by pooling resources can take safety and quality programs to the next level by an economy of scale. A program such as the California Strawberry Commission quality assurance program (QAP) ensures those producers take all the necessary steps to minimize product contamination (Boyd, 2000). Many such trade organizations have produced multilingual manuals, instruction programs, worksheets, and training programs for their managers and workers.

5.3.3 Role of Third-Party Inspection Systems

Fruit and vegetable products are used as essential ingredients in many processed and ready-to-eat foods provided by food-service entities. In food processing and food service, day-to-day process (safety) control and assurance of safety for incoming raw material are important. There are often individuals within organizations whose sole job is to conduct plant/product inspections and auditing. The bigger the company the more difficult it is for company personnel to get around to all of the suppliers and this is where third-party inspectors are employed. More and more outside auditors are used to inspect and make recommendations for product safety and/or product quality process improvements. This is especially true of retail food operations and food-service entities. They rely on receiving safe, high-quality product from suppliers and the only means they have to enforce their product requirements is through audits. There are a number of U.S. companies that specialize in this kind of auditing, such as Silliker Laboratories, Cook and Thurber, Sani-best Inspection Services, the American Sanitation Institute (ASI), and the American Institute of Baking (AIB). Companies that focus on auditing have product-specific expertise and may work for many different clients. They may be empowered by the companies that employ them to constantly improve quality and/or safety by continuously setting higher standards. Product-specific quality problems are often addressed to increase shelf life, ease of handling, or organoleptic aspects of product. Various certifying organizations have been established to set standards for third-party auditors to ensure that they maintain equally high levels of competency. Auditors for these companies must become certified in HACCP systems and the collection and examination of data relevant to product and process safety and quality. Finally, companies can band together, sharing audit information that helps improve product safety and quality. They can choose to recognize and improve audits to reduce duplication and increase audit effectiveness. This increases efficiency at all levels. For the producer this means that instead of having to entertain dozens of auditors in the course of a year, three or four inspections and audit visits suffice.

5.4 ESTABLISHING AN EFFECTIVE PERSONAL HYGIENE PROGRAM

5.4.1 Role of Management

A dedicated management establishes rules and procedures for the maintenance of food safety goals, ensuring that sanitary practices are soundly established and effectively enforced. Senior management should choose personnel who understand the entire food-safety program and can communicate job-specific training effectively to workers. This includes knowledgeable supervisors to lead workers in hygiene education and awareness training. These supervisors must be capable of teaching workers proper sanitary procedures and oversee the workers' hygienic and sanitary preparation and handling of fruit and vegetables. Having responsible management that leads by example encourages workers to practice good personal hygiene habits and work using approved procedures. Producers must ensure supervisors are knowledgeable and responsible to identify and promote good personal hygiene and sanitary working procedures. Training is critical for both managers and other employees, because good hygienic practices are not typically included in any basic education. Managers are then able to instruct, monitor, and coach staff on proper procedures for the harvesting, handling, and packing of fruit or vegetable products being produced at each facility, on a regular basis, commensurate with operational requirements. They need to identify potential food contamination situations and take any necessary corrective actions to prevent or reduce any behavior that may result in the contamination of fruit and vegetables. This is especially important at the commencement of harvest, prior to seasonal startup of packing lines, or when a new worker is introduced into the process. The reason for this attention is to prevent produce contamination and reduce the likelihood of illnesses in consumers and potential recalls.

5.4.2 Safe and Sanitary Water Supply

Basic System Requirements. An approved water supply is the basis for worker health maintenance during working hours and is critically important for carrying out appropriate personal hygiene procedures (Gast and Holt, 2000; CDC, 1997; CHC, 2004).

Contaminated drinking water may contain pathogenic organisms such as enterohemorrhagic *Escherichia coli*, *Salmonella*, and *Shigella*. Other microbial contaminants may include viruses such as hepatitis A and norovirus as well as protozoa parasites such as *Giardia lamblia*, *Cryptosporidium parvum*, and *Cyclospora cayetanesis*. The presence of these organisms or microbial indicators such as coliforms, enterococci, or coliphage in water is interpreted as fecal contamination (Sobsey et al., 1998). Even though commonly found coliform bacteria are not in themselves normally considered harmful, their presence is a biological marker for water that may be contaminated with pathogens (Doyle et al., 1997). This is borne out by records of waterborne disease outbreaks in which very low levels of coliforms have been found. If the total coliform test on a sample of drinking water is positive (1 or more coliforms per 100 ml of water), either a fecal coliform test or an *E. coli* test should be performed to determine if the coliform bacteria found are of animal or human fecal origin (USEPA, 1986). Positive results on either of these two tests are a strong indication that the water may be contaminated with fecal material. If a positive result is found, an investigation of the water treatment and distribution system is advised (USEPA, 1986). If necessary, the level of disinfectant should be increased under the direction of a water-treatment specialist, to ensure water is safe for human ingestion, or water should be boiled, disinfected, or filtered as a precaution (Bryan et al., 1996; Pontius, 1996). As in the United States, most produce-growing countries have regulations concerning the microbiological standards for drinking water, which should also include maximum permissible levels of chemical contaminants, including heavy metals.

Potable water needs to be available not only for drinking and washing hands, but also for washing and processing produce. Whatever water-treatment system is employed, it is necessary for the producer or processor to verify the quality of the water to ensure and document that it is adequate for use for drinking and hygiene. Newer technology employs the use of chlorine injection units, microbiological filtration, and UV light treating units.

Requirements for Drinking Water Handling. Even if potable water is available on site, it can become contaminated through improper storage or cross-contamination. Therefore,

1. Water supply systems should be in good condition and monitored throughout each shift and not just at the beginning of the day's work;
2. Water should be stored in clean, previously sanitized containers and tanks;
3. Water containers should be washed and sanitized on a daily basis;
4. Water storage containers should be closed at all times;
5. Containers should be kept away from the sun and excessive heat; and
6. Disposable, single-use cups should be provided.

If drinking water is stored or transported in tanks or other storage devices prior to consumption or use in hand hygiene, it is important to periodically clean and sanitize these storage devices on a frequent basis. Records indicating cleaning and sanitizing schedules, persons responsible, and chemicals utilized should be kept as part of the GMP/SSOP procedure.

Constant monitoring is needed to assure that supply systems for drinking water in the fields and packing areas are in good condition and operating properly. This water should be stored in clean tanks that are washed and sanitized on a daily basis. When cleaning ports on top of water tanks are easily accessible by workers, these should be kept locked to prevent workers from using the tanks to wash hands or clothing by dipping (Ayers, 2001). Where practical, water storage containers should be kept out of direct sunlight and excessive heat, because contamination levels could increase as conditions favor growth of microorganisms. Sanitary water fountains or disposable cups should be available for drinking. If disposable drinking cups are used, workers should be instructed to use a different cup each time. Sanitary water must be supplied to hand hygiene stations to ensure that when hands are washed, the water supply does not become a source for hand contamination.

Testing of Water Supply and Responsibility for Maintenance. The following testing and maintenance programs should be followed to promote water safety:

1. Management must designate a supervisor responsible for potable water system safety, maintenance, and testing, whether piped or tanked to the required locations.
2. Microbiological and physical evaluation should be performed on facility drinking water on a periodic basis when the water is being stored or treated on a site, other than approved municipal water delivery systems.
3. Organoleptic evaluations of water for odor, taste, and color should be performed on a daily basis as part of the facility sanitation monitoring procedure.
4. Failure of water quality tests or indications that the quality is not adequate should result in water replacement to reduce the chances of worker infection. Senior management should be notified of the problem to be able to determine if finished product might be in jeopardy.

Because of the risk of contamination, frequent physical and microbiological evaluations must be performed on drinking and hand-hygiene station water when the water is stored or treated on site. If any of the organoleptic water tests indicate poor quality, the water should be replaced to reduce the chances of worker infection, and senior management should be notified so that they are able to ascertain how this occurred. Detailed records of these testing or monitoring evaluations should be part of the GMP program and kept as evidence of the effectiveness of the water treatment and systems. When municipal water is used, records from the city or municipal water system should be kept as documentation of the water quality of water being used.

5.4.3 Effective Worker Training

Produce growers, packers, and processors need to develop sanitation training programs that fit their particular operation and types of employees involved. This is particularly important for areas where edible products or ingredients are handled, prepared, or stored and where equipment or utensils are washed and/or sanitized. This includes personal hygiene of field workers, packaging plant employees, and inspection staff. New, seasonal, part-time, full-time, and management staff should have practical training in basic sanitary practices and hygiene. Management is responsible for establishing an ongoing educational program that informs all employees of their responsibilities in all aspects of sanitation in the execution of their duties. Management further has the responsibility of verifying that employees are consistently meeting the requirements of the job. It is required that the exact type of training will vary based on employees' work responsibilities. Training can vary from formal classroom instruction to one-on-one presentation of job requirements and demonstrations. For example, proper and effective hand washing needs demonstration by the instructor and actual try-outs by the employees with the comment that this is a critical aspect of food production and employees will be observed carrying out the correct procedure. All agricultural workers should grasp the consequences of poor personal hygiene and how it affects themselves, their families, and the safety of the produce consumer.

New and Seasonal Employee Training Program. New employees need to know the basics of sanitation and the reasons behind this. They should understand food-safety hazards as they relate to each job or task the worker performs. In addition, specific training for handling and use of agricultural chemicals and other potentially hazardous cleaning chemicals needs to be completed. This is done to prevent chemicals from finding their way into the product and prevent exposure of workers to potentially toxic compounds. In the United States, employee safety training should be based on the OSHA (1928) requirements that are applicable to worker health and training requirements. Many other nations have similar occupational safety agencies with regulations specifying training requirements that must be followed to remain in compliance.

For new workers, training should be provided at the beginning of their employment term. For seasonal or part-time workers, if a complete training program is not deemed necessary, workers should, at the very least, receive good personal hygiene and food-handling training as it relates to their specific task in the operation. Training must be provided systematically and must be fully understood by the entire work crew. Visual aids, written instructions, check lists, or other appropriate means must be developed to aid the instruction supervisor in teaching workers the basic training program. Demonstrations of handling procedures are usually more effective than verbal instructions. Feedback to the trainer from workers is important to assess the overall effectiveness of the training. As with any food-safety program, commitment of administration to the program is essential in achieving success. Detailed information about toilet and hand-washing location and use needs to be conveyed. It cannot be assumed that workers know how to use toilets or carry out basic hand hygiene, especially if there are language and culture differences. Employees with diarrhea or other symptoms of disease must declare this and be sent home or assigned different tasks that do not allow contact with food or equipment. Every step must be taught to avoid having employees only being partially informed. After training personnel give detailed instruction, new employees should shadow experienced workers for a period necessary to carry out their job safely and effectively. "Wash Hands" signs must be well displayed in all washrooms, lunchrooms, break, and smoking areas. Applicable documentation and records regarding each employee and the training they have received should be kept on file.

To be effective, all training programs must be accurately delivered (both in oral and written forms) at the language and comprehension level of the workers on site. It is management's responsibility to consider the language needs of workers who may not be able to read their own language or understand the language being used by management on the farm, in the packing house or in the processing plant. For workers with a mother tongue different from that of the operation, translated material and use of visual aids are strongly recommended in order to facilitate worker comprehension. In some cases, training materials need to be provided in several different languages because of the mixed nature of work forces. In order to guarantee that employees comprehend and implement the training, trainers must be aware of language/dialects of the employees and should consider other cultural peculiarities, aversion and ingrained habits of trainees, when developing the training program. The actual level of knowledge an employee needs to demonstrate sanitary practices varies according to the type of operation (production, harvesting, or processing), produce type, and the responsibilities and activities of the employee. Trainers must effectively address special cultural encumbrances to personal hygiene such as the widespread belief that the process of hand washing or washing hands with cold water causes arthritis (Noritake, 2004; Puryear, 2000). In many developing countries, toilet paper is not placed in the toilet bowl. From field reports it seems that toilet paper is often dried for reuse or reusable cloths are washed, dried, and reused (Ayers, 2001; Michaels et al., 2001a). Issues like this need to be clearly addressed in training multicultural work crews and it is sometimes necessary to employ special trainers who are aware of these cultural nuances and have developed approaches of overcoming obstacles.

Refresher Training. Periodic refresher or follow-up training sessions are needed to ensure that workers continue to understand what has to be done, and especially if there is a change in established procedures. These refresher courses will be partly based on the observations that are made during routine verifications of employee performance to ensure the effectiveness of the program. When deviations from the taught personal hygiene and food-handling practices are noticed, workers should be coached to follow the correct procedures. When deviations in standard operating procedures are found to be widespread, then the training program must be assessed, problems in comprehension or procedure reevaluated, and, if necessary, modified to correct such deviations in the future. Persistent lack of compliance with the SSOPs by employees should result in disciplinary action, because the reputation of the company is at risk.

Worker Training Record Keeping. All training (formal or informal) must be kept on record to identify each worker's participation in training and the level of training they have achieved at that facility. Proper documentation should be kept, recording all medical reports, and worker gastrointestinal disease reports. Using this approach the health of all personnel can be assessed and necessary corrective actions can be put into place to minimize the risk of product contamination. Should an adverse foodborne event occur linked to product from these premises, then the records may also be useful to facilitate traceback of a disease outbreak, as well as showing due diligence in following GMP procedures.

5.4.4 Exclusion of Ill Workers

A set procedure must be established when gastrointestinal infection is suspected in workers. Legally, if health authorities have reasonable cause to suspect the possibility of disease transmission from any food-processing plant employee, the health authority is required to secure a morbidity history of the suspected employee. Health authorities could also instigate an investigation as necessary, and take appropriate action to prevent or reduce the development of a foodborne outbreak.

Public health authorities may require any or all of the following measures:

1. The immediate exclusion of the employee from all food processing plants;
2. The immediate closure of the food processing plant concerned until, in the opinion of the health authority, no further danger of disease outbreak exists;
3. Restriction of the employee's services to some area of the establishment where there would be no danger of transmitting disease;
4. Examination of medical and laboratory records of employees.

Assuring worker health increases worker productivity and aids in preventing potential microbial contamination of fruits and vegetables. An infected employee (whether symptomatic or not) can easily contaminate fruit or vegetable products with human microbial pathogens. If good hygiene such as hand washing after sneezing, touching hair or other parts of the body, or using the restroom is not practiced, food contamination is possible. These pathogens can then be transmitted to consumers who handle or eat the contaminated produce. For this reason, U.S. state and federal regulations found in the FDA Model Food Code call for the exclusion of infected workers (FDA, 2001).

Supervisors monitoring employee health should be familiar with the symptoms of infectious diseases and have all sick employees report possible illness to them prior to beginning work. Workers who are sick, especially those with diarrhea, should be excluded from handling produce. Persons not showing outward signs or symptoms of disease can in many instances still transmit microbial pathogens. Many pathogens can colonize the human gut without clear evidence of disease. Many pathogens also undergo a period of presymptomatic shedding. In both cases, pathogenic microorganisms can be spread to others via contaminated food. These infected employees are much more difficult to detect, and even regular testing of stool specimens for pathogens does not guarantee clean products or equipment surfaces.

Physical symptoms that identify an employee with the potential for causing microbial contamination of produce include (FDA, 2005; Gast and Holt, 2000; Rangarajan et al., 2000; University of Maryland, 2002)

- Severe abdominal pain or Diarrhea;
- Vomiting;
- Fever, Nausea or Dizziness;
- Exposed cuts, sores or open wounds;
- Hepatitis A or yellow jaundiced skin and eyes.

Employees with gastrointestinal illness, cuts on hands, or open wounds can contaminate fruits and vegetables during handling and food-preparation practices. Symptoms that identify an employee as having the potential for transmitting microbial contamination to produce include diarrhea, vomiting, dizziness, abdominal cramps, exposed or open wounds, hepatitis, or yellow jaundiced skin. Food workers having upper respiratory infection should also be assigned duties not involved in food handling because of infectious secretions possibly harboring spreading multiple pathogens. (FDA, 2005; Gast and Holt, 2000; Rangarajan et al., 2000; University of Maryland, 2002).

Workers should be trained to report gastrointestinal illness symptoms to supervisors. Sick infected employees should not participate in food-handling activities that involve direct contact with produce until they have obtained clearance from a licensed healthcare provider stating they are no longer infectious. (FDA, 2005; Gast and Holt, 2000; Rangarajan et al., 2000; University of Maryland, 2002.)

As part of the training program, supervisors should instruct workers in the recognition of disease symptoms and report any possible appearance of symptoms to designated supervisory personnel. Once notified, it is the duty of supervisors to assign workers with symptoms of disease to activities that do not involve contact with the produce products. It is extremely important that supervisors be provided with the necessary training on foodborne pathogens and disease symptoms enabling them to make judgments regarding the best course of actions for managing ill employees. Workers once removed from produce-handling tasks as a result of illness should not be allowed to return to those jobs until a licensed healthcare professional provides written medical documentation stating that they are free of the infectious agents capable of being transmitted to food. Alternatively, a healthcare provider could provide written notice that symptoms experienced by the worker are the result of a chronic noninfectious condition or are unrelated to a foodborne intestinal disease. Workers excluded from handling produce may be assigned other jobs or given sick leave. (FDA, 2005; Gast and Holt, 2000; Rangarajan et al., 2000; University of Maryland, 2002.) Foodborne outbreaks can be extremely, costly and the economic consequences of a contaminated product, including recalls and lawsuits, far outweigh the establishment of a benefits package for workers. (Sockett and Todd, 2000)

5.4.5 Health Maintenance Programs

Ideally, agricultural workers should have access to a healthcare system. Employers should provide fruit and vegetable handlers with a training program on good food-handling and hygiene practices. The possibility of produce contamination is directly related to the quality of the worker-training program. This training should be reinforced constantly identifying potential food contamination situations and taking any necessary corrective actions. Producers must provide food-safety training for management and all other employees.

5.4.6 Sanitary Worker Habits: Worker Hygiene Practices

Ultimately, the day-to-day responsibility for reducing the risk of product contamination during primary production, packaging, or processing resides with agricultural workers. For that reason they need to be educated as to risks related to poor personal hygiene. Workers must be taught to maintain a uniformly high degree of personal cleanliness that is consistent with the process being performed, and in such a way it becomes automatic. Whether it involves field work or the processing plant, standards must be maintained such that each employee knows what is expected of them. During training sessions, producers and processors can provide training materials and educational resources that train employees on what is required, but, at the end of the day, implementation starts with the worker and how they prepare for work. The overall effectiveness of the program is based on how well workers understand and implement personal hygiene and safe handling practices.

Management must provide workers with detailed information about expected daily personal hygiene practices, and the importance of these practices in maintaining safe food products and their jobs. During daily worker monitoring, those workers not living up to minimum requirements must be disciplined by being dismissed from service until requirements are met.

The practices important in maintaining good personal hygiene habits include

- Regular bathing;
- Wearing task-appropriate clean clothes;
- Keeping fingernails clean and trimmed short;
- Only using designated toilet facilities at all times;
- Washing hands in a proper and effective manner after any possible contamination;
- Using hairnets or headgear if required in packaging or processing areas;
- Never spitting on the premises, whether it be in the field or packing house;
- Refraining from eating food, gum, candy, or tobacco products in harvesting, packaging, or processing areas.

In the fields, portable toilet units should be maintained in a condition that encourages agricultural worker usage (Fig. 5.2). Workers must never spit on anything, anywhere, anytime. Workers must never be allowed to pilfer or eat any of the produce during production. Eating food, drinking beverages, chewing gum or candy, or using tobacco products (including chewing tobacco and snuff) must be prohibited in production areas and only allowed in areas designated for this.

Eating and drinking, except for water from fountains in production, should be confined to the designated lunchroom areas.

Hands, including fingernails, should be clean and washed at all times after using the lavatory facilities. Hand-washing facilities should be located at the entrances to production areas and used by everyone entering the food-production/handling area. Making these visible enables their use to be monitored, and additional safeguards with respect to hand washing are provided. Scratching of the head, face, and so on, or the placing of fingers in or around the mouth or nose are not acceptable practices. Care of the hands includes the treatment and appropriate covering of all cuts and sores by the plant nurse or personnel trained in first aid and designated to be responsible for its administration (CHC, 2004).

5.4.7 Worker Hygiene

Hand Hygiene: When, Where, and How. Hand hygiene is the single most effective means of preventing infectious disease transmission (Semmelweiss, 1861; CDC, 1997; FDA, 2000, 2001, 2003). The importance of proper hand-washing techniques should not be dismissed when dealing with produce harvest and processing. Although most of the scientific work on the subject of hand hygiene has been developed based on work in the medical field, much of what has been learned is directly applicable to food handling. Generally speaking, the scientific facts do not change just because field workers or processing plant personnel are involved in the hand-hygiene process. First, one should not assume workers know how to wash their hands properly. The hygiene culture from which they come should be understood, workers should be taught proper hand-washing techniques and expected levels of personal hygiene as a prerequisite for work.

All workers should be required to wash their hands at the following times:

- Before beginning work each day;
- After all visits to the washroom;
- After handling contaminated or potentially contaminated materials;
- After a break or eating a snack or meal;
- Where feasible, when involved in field packing, packinghouse operations, or in direct contact or close proximity to fruit and vegetables;
- After sneezing, coughing, or blowing the nose;
- After touching or scratching the skin or wounds;
- After touching the hair, face, mouth, or nose;
- After smoking;
- After any other absence from the workstation;
- After touching dirty surfaces, equipment, and utensils;
- After handling dirty raw materials, trash on the floor, waste material or garbage;
- After performing maintenance on any equipment;

- After touching or handling fertilizers, pesticides, chemicals, or cleaning materials; and
- Any other situation that may lead to contamination of the hands.

Pathogens can be transmitted easily to consumers who eat contaminated produce (CDC, 1997; IAFP, 1999; Rangarajan et al., 2000; CFIA, 2003; Michaels, 2004a). Kerr et al. (1993) frequently found *Listeria* spp. on hands of food workers who were deemed to not adequately wash hands compared to clerical worker controls or food workers without *Listeria* spp. on hands. Considering the risk of transmission to food products, Good Hygienic Practices (GHPs) should be applied by all agricultural workers having direct contact with produce. This includes washing hands after sneezing or using the restroom as well as other situations described above.

A recommended hand-washing procedure includes the following steps:

1. Moisten hands with warm water.
2. Use 1–3 ml soap or a gentle antimicrobial soap with proven efficacy having either the active ingredient triclosan or other disinfectant.
3. Ensure contact period with antimicrobial soap is 15–20 s.
4. Friction should be used during the washing process (palm, top of hands, rubbing in between fingers vigorously). Thumbs and tips of fingers are frequently missed during hand washing, as well as the fact that left hands are often washed better than right hands. The hand-washing technique used should address these known hand-washing deficiencies.
5. A good lather should be developed through vigorous rubbing during the soaping process. Hands should be washed up to the elbow or to include exposed skin beyond the sleeve of protective work clothing, such as a smock or uniform.
6. A fingernail brush or fingernail cleaning tool should be used to clean under fingernails, if necessary. Disposable, fresh, or preferably sterile, fingernail cleaning implements should be used.
7. The friction employed during the washing process aided by surfactants in the soap will loosen bacteria from hands, as well as flakes of skin that may harbor bacteria.

The hand-washing process consists of wetting hands, the addition of soap, and performing the hand-washing process, followed by rinsing with water and drying. Details of the hand-washing process, including hand-washing frequency, duration, and aggressiveness, should be presented in the facility SSOPs. It should be noted that investigators have found that as frequency, duration, and aggressiveness of hand washing increase, so does skin damage. This results in infection and colonization (Price, 1938; Marples, 1965; Brunn and Solberg, 1973; Ojajarvi et al., 1977; Gawkrodger et al., 1986; Roth and James, 1988) or increased microbial (pathogen) counts on hands (Ojajarvi et al., 1977; Noble and Pitcher, 1978; Larson, 1984). This occurs through the removal of stratum corneum or alteration of stratum corneum permeability, combined with overexposure to soap surfactants causing skin defatting (Kirk, 1966; Mitchell and Rawluk, 1984; Goffin et al., 1996), dry skin, chapping and pain, with cracking and fissures. Dry skin can cause increased shedding of skin squames and attached microflora (Meers and Yeo, 1978; Rotter, 1984; Larson et al., 1986), resulting in higher microbial counts in the immediate environment, and degrading the environmental aerobiology in the facility. Continuous exposure to many produce types is known to cause allergic contact dermatitis (Fisher, 1986; Van der Valk and Mailbach, 1996; Michaels and Ayers, 2000). For all of these reasons, use of gentle soap and avoidance of aggressive cleaning compounds/hygiene implements or skin damaging work is a key to maintaining worker skin health (Michaels and Ayers, 2000).

All of the negative skin consequences described above support proper and effective use of gloves in place of total reliance on hand washing. Many glove types have irritant or sensitizing properties (Fisher, 1986; Fay, 1991; Michaels, 2001). Latex protein allergies are a class of these sensitizing chemicals found in gloves. In addition, many produce allergens cross-react with latex allergens (Michaels, 2004a,b). It is therefore important to be watchful of skin reactions related to latex glove use. Because of risks to workers and consumers, latex gloves are generally not recommended for use by food handlers (Michaels, 2001, 2005).

Hand Soaps or Detergents. For effective hand washing, the microbial principles are fairly clear (Semmelweiss, 1861; Taylor, 1978a, b). These consist of loosening dead skin cells and attached resident/transient microorganisms with soaps and detergents (Price, 1938; Larson and Lusk, 1985) through the mechanical action of hands (Bibel, 1977; Pether and Gilbert, 1971; Blackmore, 1989; Larson and Lusk, 1985; Michaels and Ayers, 2000), rinsing well by dilution with copious amounts of running water, followed by drying with paper towels (Knights et al., 1993; Redway et al., 1994; Gould, 1994; Michaels et al., 2001b, 2003a; Michaels, 2002; Michaels and Ayers, 2002). Soaps used prior to the Second World War (WWII) were of the type used for thousands of years made from boiling fat and wood ashes together (Fishbein, 1945). After WWII, synthetic surfactants came into widescale use. The soap and/or synthetic detergents used to wash hands can have a profound effect on skin physiology. A significant relationship exists between exposure to soaps and detergents and eczematous hand conditions (Larson, 1985). Detergents in hand soaps can remove skin lipids and damage the protective cornified epidermis (Kirk, 1966). The presence of the outermost epidermal layers, the stratum corneum, and skin lipids all help protect the skin from overdrying and overwetting. The lipid barrier can be completely destroyed by overwashing, but can be repaired by application of lipid compounds in the form of hand lotions (Kirk, 1966). This can be accomplished, for example, with emollients present in alcohol-based instant hand sanitizers. This avoids the risk of lotion product contamination.

Antimicrobial Soaps and Sanitizers. Antimicrobials in soap may be of some help in reducing microbial counts on hands. However, it has been suggested that exposure time is so short that it is doubtful there is a significant effect (Broughall et al., 1984; Larson, 1984; Graham, 1990; Bowell, 1992). Several different antimicrobial ingredients are available for use in soaps employed for hand washing in the food-processing/service environment (Crisley and Foter, 1965; Paulson, 1992; Larson, 1989; Block, 1991). Triclosan is hydrophobic, substantive, and has been reported to have anti-inflammatory properties. For this reason is recommended as an antimicrobial soap active ingredient. Bar soaps are usually avoided because of the possibility of contamination through multiple worker use (Bannan and Judge, 1965; Heiss, 1969; Kabara and Brady, 1984), either by prolonged water contact or through in use contamination. Although antimicrobial soap is preferred over bland soap or soap lacking an antimicrobial ingredient, even if the soap carries an E2 rating (USDA, 1998), exposure time during hand washing is generally considered too short (Larson et al., 1986; Graham, 1990; Michaels and Ayers, 2000) for any appreciable decrease in microbial counts on hands. This can be further complicated by the presence of organic matter, dirt, grease, or soil (Green, 1974; Michaels and Ayers, 2000; Michaels et al., 2003b; Lin et al., 2003), which reduces the effectiveness of some active ingredients (Burke et al., 1971; Borgatta et al., 1989). Use of alcoholic instant hand sanitizer, so successfully employed in the medical field, is only to be employed after hand washing takes place because it is necessary to remove organic soils to retain effectiveness (FDA, 2003; Michaels et al., 2003b; Lin et al., 2003).

Soap production facilities used by food processors should be operated under a hold/release system to prevent contaminated product being shipped to food preparation facilities. Frequency of hand washing appears to be of greater importance than one-time efficacy (Paoli et al., 2003; Michaels et al., 2004b) and in that context, use of antimicrobial soap in itself is less important than maintenance of clean hands, skin health, and microbial integrity of the water supply and soap product used. Some antimicrobial ingredients are irritants in high concentrations or potential sensitizers, and can cause allergic contact dermatitis of the same type frequently experienced with irritating/sensitizing glove types (Van der Valk and Mailbach, 1996; Fisher, 1986; Borgatta et al., 1989; Fay, 1991; Goh, 1989; Behrens et al., 1994). For this reason, thorough rinsing of soaps from skin is essential in retaining skin health (Kirk, 1966; Taylor, 1978b; Larson, 1985; Michaels and Ayers, 2000).

Rinsing and Water Temperature. Rinse well in water at an adequate flow rate (3–4 l/min) adjusted to a comfortable temperature range. During the rinse, hands and fingertips should be positioned down into sink with rinse progressing from elbow or forearm to fingertip. In this way water is drained from fingertips into sink. It is important to remove all soap residues from hands during rinsing, as irritation can develop following prolonged exposure to soap components.

Rinsing of hands must be accomplished by rubbing the hands together, much as is done in the washing process. Of primary importance is the hygienic quality of the water supply, because contaminated water can compromise the entire hygiene process with the introduction of microbial contaminants to the hands. Properly performed, rinsing will result in removing transient bacteria and soil loosened by soap, along with dead skin cells loosely attached to the surface (Semmelweiss, 1861; Taylor, 1978a, b; Kirk, 1966).

Most food-safety scientists have specified that water temperatures between 110°F and 120°F should be used for rinsing. Price (1938) showed that water temperature did not make a difference in resident flora removal when aggressively using a brush. Recent work with bland (nonantimicrobial) soap indicated that there is no significant difference in resident flora removal rates between washing and rinsing with 70°F and 120°F water. In subsequent experiments with antimicrobial soaps, there was little difference between using wash water at 85°F or water at 110°F (Michaels et al., 2001c; Michaels et al., 2002). Washing and rinsing hands at excessively low temperatures, equivalent to that found in a refrigerated packing or storage room, is uncomfortable and may result in poor hand washing compliance, leaving more residual flora on hands than normally would be removed with water at slightly higher temperatures (Michaels and Ayers, 2000).

A poor rinse clearly runs the risk of leaving both resident and transient contaminants on hands (Taylor, 1978b; Larson and Lusk, 1985; Michaels and Ayers, 2000). Thorough rinsing is also extremely important in that it removes potential irritants and sensitizers originating in food, soaps, metals, and facility sanitizers and disinfectants (Fisher, 1986; Michaels and Ayers, 2000). If rings are worn, then thorough rinsing will help remove both allergens derived from food or spices and sanitizer/disinfectants that can react with ring metal. It also can help to remove microorganisms trapped under the rings (Fisher, 1986). Rinsing with excessively hot water (above 120°F) runs the risk of scalding or damage to the skin, irritation, pain, defatting (Kirk, 1966; Mitchell and Rawluk, 1984; Goffin et al., 1996), skin cracking and fissures (Adams, 1990; Goffin et al., 1996), and possible colonization (Ojajarvi et al., 1977; Roth and James, 1988; Borgatta et al., 1989; Van der Valk and Mailbach, 1996), all resulting in reduced hand washing with subsequent increases in microbial counts on hands (Ojajarvi, 1981; Larson, 1984; McGinley et al., 1988).

An ideal handwash station uses faucets, soap, and paper towel dispenser that operate automatically (hands free) or through use of the knee, foot, or elbow (Eberhart et al., 1983; Hattula and Stevens, 1997; Michaels and Ayers, 2000; Harrison et al., 2003a, b; Michaels et al., 2001d). These are unavailable in restrooms and many food-preparation facilities, increasing the risks of cross-contamination through use of sinks having contaminated faucet handles (Crane and Shevlin, 1964; Sheldon, 1994; Michaels et al., 2001d; Mermel et al., 1997). These contact surfaces receive contamination from all users when the faucet is turned. If a wet hand is used to turn off the faucet, then the next user picks up contamination deposited by one user and hands are contaminated.

The water flow rate is important for several main reasons. If too low a pressure (2.5 l/min) is set, the stream tends to be intermittent. A continuous stream is required, with water pressure helping to remove microbial agents, soil, and residual soap. At too high a flow rate (>4 l/min), splash back from the sink can occur. This can result in microbial contamination splashing on to clothes, contact surfaces, and hands (Hattula and Stevens, 1997).

Hand Drying Process. Hands should be dried in a sanitary manner using methods resulting in frictional removal of remaining transient organisms. Hand wetness has been identified as an important determinant in cross-contamination. Numerous experimental studies have shown that wet hands or surfaces can transfer more microbial contaminants than dry hands or dry surfaces (Marples and Towers, 1979; Mackintosh and Hoffman, 1984; Patrick et al., 1997; Rheinbaben et al., 2000). Although in many produce-handling situations it is difficult to avoid wet hands, as hands sometimes even stay wet on a continuous basis, use of dry hands or glove surfaces can help reduce chances for cross-contamination. For hand drying, single-use paper towels, above all other methods, have been proven to effectively remove both transient and normal flora along with loose skin cells (Eberhart et al., 1983; Blackmore, 1989; Gould, 1994; Rangarajan et al., 2000; Michaels et al., 2001b, 2003a; Michaels and Ayers, 2002). Friction provided during the drying process can be an effective final microbial removal step. In fact, friction seems to be a critical aspect of hand hygiene where viruses are concerned. Several studies have shown friction somewhat more important than surfactant or antimicrobial soap action (Bidawid et al., 2000; Guzewich and Ross, 1999; Michaels and Ayers, 2000; Lin et al., 2003).

Hands should be dried in a way that avoids recontamination. Touching of paper towel dispenser button, cranks, or levers, or use of hot air dryers can recontaminate hands, resulting in cross-contamination. One should avoid contact with contaminated surfaces, such as sink or faucet handles, which can result in cross-contamination. If a faucet requiring full grasp of the faucet handle is used, then the water should be left to run until after rinsing hands. After drying hands, one should grasp the faucet handle with the used paper towel and turn off the water.

Hand drying (FDA, 2001) is a critical moment in the hygiene process of a food-processing/service facility and is important in reducing the likelihood of microbial contamination of food. It is the final part of what has been considered the basic personal hygiene process. HACCP measures of hand-drying effectiveness include speed of drying, degree of dryness, effective removal of contaminating microorganisms, and prevention of cross-contamination (Guzewich and Ross, 1999). Several systems currently allowed for use in food-processing and service environments (FDA, 2001) contain flawed aspects lacking in one or more of these measures of effectiveness. With use of hot/warm air dryers, hands can become contaminated (Eberhart et al., 1983; Blackmore, 1987, 1989; Knights et al., 1993; Redway et al., 1994) with coagulase-positive *Staphylococcus aureus*, Enterobacteriaceae from toilets, or pathogens form aerosols in the food-processing environment (Darlow and Bale, 1959; Bound and Atkinson, 1966; Newsom, 1972; Gerba et al., 1975). The speed of hand drying is relevant to the hazard analysis of the hygiene process, because systems that cannot effectively serve sequential users (as in a food-processing environment) or take an inordinate period of time to accomplish the task cause users to wipe hands on clothes. With both continuous cloth towels and hot/warm air dryers, users have been observed drying hands on clothing (Blackmore and Prisk, 1984; Matthews and Newsom, 1987; Knights et al., 1993; Michaels and Ayers, 2000).

Paper towels are generally considered to be the most hygienic means of drying the hands. Friction is present when paper towels are used, and to a lesser extent with cloth towels, but lacking with hot/warm air dryers. Friction during the hand-drying process can remove bacteria and microorganisms from precisely those areas of the hands missed during the hand-washing process (Taylor, 1978a, b; Gould, 1994; Michaels et al., 2003a). There have been limited reports (possibly used as an excuse for not washing hands) that coarse paper towels cause dry skin and reduce hand-washing effectiveness. Experience does not indicate this to be a problem unless towels are excessively course and harsh detergents are used. Credible investigators (Larson et al., 1986) indicate that the friction created during the hand-washing process (rubbing hands together and use of paper towel) is not as damaging to the stratum corneum as detergents. This is supported by the work of Kirk (1966). If harshness is experienced, both soap and paper towels should be reexamined for appropriateness.

Despite the sanitary advantages offered by paper towels, problems are encountered as a result of the potential contamination of hands from common use of paper towel dispenser mechanisms, consisting of buttons, cranks, and levers (Michaels et al., 2001a, b; Hattula and Stevens, 1997). Paper towel exits for folded towels also can become contaminated with potential pathogens (Michaels et al., 2001b; Hattula and Stevens, 1997; Griffith et al., 2003; Harrison et al., 2003a, b), especially if towels and dispensers are poorly matched and subject to occasional or frequent jamming. This results in cross-contamination of hands, and possible food or food contact surface contamination. Movement activated automatic paper towel dispensers avoid hand contact with levers or knobs.

Nail Hygiene. Hand washing is considered a basic procedure that children learn at an early age. However, each person has a different background and a different concept of proper hand washing. Therefore, personnel should be well trained in these practices no matter how basic they sound. In this context, fingernail hygiene is an important component of hand hygiene training, having both at-home and at-work components.

Workers in their home environment should, before coming to work, consistently use the fingernail brush. In this respect it represents a personal item; however, when nail regions become heavily contaminated with accumulated dirt and food soils while on the job, clean and appropriate fingernail brushes should be available for use at the hand hygiene station. The nail region can represent a reservoir of normal flora, transient contamination, and potential pathogens. Over 95 percent of the bacteria on hands can be found in these regions and toilet paper poke-through can deposit pathogens during bouts of diarrhea. These parts of the hands, in addition to being missed during normal hand washing, are not cleaned effectively during normal hand wash or by normal hand wash combined with use of alcoholic hand sanitizers. It has been shown that only through use of a fingernail brush can this microbe-rich region of the hand be cleaned (Lin et al., 2003).

Selection of fingernail brush should be made with care and overuse should be cautioned. Improperly designed fingernail brushes can damage skin, fingertip, and nail region, resulting in infection. Disposable fingernail brushes are available from several suppliers and use of these implements can help prevent puncture transmission of blood-borne infectious agents (HIV, hepatitis B & C, and so on), fungal agents, and opportunistic bacteria such as *Candida* spp. and *Pseudomonas* spp. (Michaels and Ayers, 2000). When not using fingernail brushes of the disposable kind, these implements need to be sanitized or sterilized between uses, otherwise they become contaminated with the microbial agents described above (Michaels and Ayers, 2000). Use of fingernail brushes on premises requires management to have a sufficient quantity of nailbrushes available for their entire workforce. Chlorine sanitizer treatment should be capable of restoring sanitary status once contaminated, but the sanitizer supplier should be consulted for recommendations as to adequate procedures and test strips used to measure sanitizer strength.

5.4.8 Working Apparel

Clothing. In determining what working apparel should be required, practical considerations should be given as to whether the work is harvesting, packing, or processing. Judgment will have to be used by plant employees and inspection staff when interpreting these guidelines. The main intent of any working apparel policy is to protect the edible produce from all possible sources of contamination, including material soiled during the different operations. In order to prevent potential contamination, suitably clean, washed, and well-maintained work clothing must be provided to workers while on the job. Work clothing provided by the producer should be worn alone or over any piece of worker clothing that may come into contact with food products. The wearing and storage of work clothing must be in accordance with sanitary norms. Use of work clothing in packing areas may be less restrictive than for that worn in processing areas. For produce processing, work clothing should not be worn into washrooms, lunchrooms, or any areas other than the processing floor. Work clothes should not be stored in lockers used for street clothing or outside the plant. In those instances that workers must enter other areas of the plant, which are considered incompatible with the more critical processing activities (e.g., barn or slaughter area, maintenance shed, and so on) appropriate measures must be taken to prevent cross-contamination. This can be accomplished by wearing color-coded or clearly identified smocks or cover garments. Employees in ready-to-eat product areas should, when leaving the process area, remove their outer garments and hang them in specifically designated areas to prevent any possible contamination of ready-to-eat food. All of these details must be codified in SSOPs as part of the GMP plan. If workers involved in picking or harvest wear their own clothes, then specific standards should be communicated clearly to workers with regard to clothing type (overalls, shirt, hat, and so on) and cleanliness.

Clothing that is not considered acceptable for a specific type of work should be specifically explained to employees prior to beginning of work.

Footwear. Various types of footwear may be required depending on the conditions encountered in the work environment. First and foremost, footwear should be of a design and construction equal to the challenge encountered in the work environment. Because much agricultural work involves wet work, including washing and rinsing, waterproof work boots are often required. According to data from the National Safety Council (NSC), there are over 100,000 injuries each year in the United States involving feet and toes, representing 19 percent of all work-related disabling injuries (NSC, 2000). An analysis of Florida State Workers' Compensation records revealed that falls accounted for around 12 percent of the recorded injuries in horticultural production occupations (Becker et al., 1991).

In most agricultural occupations the shoes or boots are very important in supporting the ankle, providing slip resistance and protecting against crushing injuries (Ellis and Lewis, 1988). ANSI sets standards for shoes and boots. Ideal work shoes should meet those standards. A typical ANSI rating could be 1-75 C-25. This translates into the toe withstanding 75 foot pounds of impact and 2500 pounds of compression.

Chevron or cleated soles are by far the best for slippery terrain because of the squeezing action they provide on surfaces. Softer soles are better when encountering slippery indoor conditions, whereas the harder, less resilient, cleat-type sole is usually preferred for tougher outdoor use. Except around wet environments, leather covering for the foot and ankle is preferred and could be specified as a prerequisite for obtaining work.

In wet environments, rubber is considered satisfactory for most conditions. However, where chemicals, oils, greases, or pesticide are commonly encountered, boots made of neoprene, polyvinyl chloride (PVC), or a blend of PVC and polyurethane should be employed because of their chemical resistance.

Jewelry and Personal Adornments. Personal adornment and accessories to clothing, such as jewelry, badges, and buttons, create a potential risk of product contamination and, at times, create a risk to the safety of the employee when moving equipment is involved. All such items should not be worn in food-preparation and packaging areas. In these same areas, items such as pens, pencils, thermometers, and so on, should not be kept in coat/shirt pockets where they may accidentally fall into products. Jewelry items that cannot be removed such as continuous loop earrings, weddings bands, medic alert bracelets or necklaces, and so on, also create a potential risk to the safety of the product and must be adequately covered and accounted for at the beginning and end of the shift.

Facial or body adornments that cannot be removed, or which workers will not remove based on religious or cultural reasons, may have the potential to contaminate food products. For that reason they must be properly attached, removed, or adequately covered when they are worn in food-preparation and packaging areas. There is no doubt that these adornments can pose food-safety hazards; therefore, proper use of head cover, gloves, or surgical masks are acceptable mitigation strategies. Use of "band aids", as are often used to cover rings, is not considered adequate and in fact these easily become dislodged, leading to the potential for further product contamination.

Hairnets and Headgear. Hairnets and headgear provide protection from contamination of food product and may also play an important role in protecting the agricultural worker. Protective headgear, which includes safety hats and bump caps, and hard hats if used properly, could prevent most head injuries commonly encountered in agricultural work. Signs should be posted so that workers know when to wear them. Certain areas may need to be posted to make sure everyone entering those areas is using them. Jobs requiring protection may include operating and repairing machinery, trimming trees, entering or leaving buildings with low doors, or working under low ceilings.

High-impact plastic hard hats are commonly employed in a wide range of agricultural jobs. They are lightweight, offering a good level of sun, impact, and water protection. Hats with a webbing suspension system provide workers with good scalp ventilation in hot climates, and many types are sold with liner inserts for winter weather conditions. Hard hats should be inspected on a seasonal basis. Hats that appear faded, dull, chalky, or brittle are obviously showing signs of degradation and should be replaced before safety is compromised.

Hard hats used on the farm should be cleaned regularly with mild soap and water. Cleaning and adjusting the suspension straps is particularly important in preventing scalp problems and making sure the hat fits squarely on the head. Hats should not be used for any purpose other than head protection. They are not to be used for holding or storing liquids or other objects; they are not step stools, hammers, or water buckets. Hard hat manufacturers warn that workers should not wear anything under them unless it is one of their approved liners designed and manufactured specifically for that purpose. They also warn that scratches or cuts that penetrate deeply into the surface can cause hard hats to weaken and fail upon impact. Manufacturers recommend replacing hard hats every few years if they are exposed to extreme temperatures, sunlight, or chemicals, and caution against painting or repairing hats, as flaws can be masked or incompatible paints can soften the underlying material. Damaged hats should be discarded and they should not be stored in direct sunlight when not in use.

In processing areas all persons should wear head coverings. All persons entering areas where ready-to-eat food products are open to exposure must wear appropriate head coverings. This is necessary to ensure that food product is not contaminated through the introduction of falling hair. In order to accomplish this objective, hairnets or hats must cover all exposed hair. If hairnets are used, management must ensure that mesh is small enough to prevent loose hair from ending up in food. Most sanitarians require that where ready-to-eat food products are involved, hairnets be used to cover any exposed facial hair with the exception of trim moustaches. As with work clothing, head coverings must always be clean and maintained in good repair.

Dust Mask and Goggles. Wind-blown dust is one of the inevitable aspects of field planting and harvesting. Workers in dusty environments should be protected from dust particles entering their eyes and lungs. Unvented goggles will protect eyes from dust.To prevent lungs from the dust, masks can be worn, but they must fit properly to prevent particles from penetrating the mask or getting in around it. Several types and styles of dust particle masks are available, although, as a rule, all of them are made with fabric that traps the particles before they can be inhaled into the lungs.

The most comfortable masks are made from stretchy material and fit snugly over the mouth and nose. Testing the effectiveness of the mask is easy. If discolored mucus is observed after use, the mask is not working. Masks that fail to filter out airborne dust result in greater levels of spitting and coughing and, in addition to health damage to the worker, can lead to compromising food safety (Rangarajan et al., 2000). Respirators afford the most protection but are cumbersome to use over periods of 6 to 8 hours. Even so, working in close proximity to some pesticides may require a respirator; this is determined by consulting the label on the pesticide products that are being used.The label will also specify the cartridge needed for that pesticide if a respirator is required.

Gloves. Bare hand contact with ready-to-eat food has often been linked to outbreaks of foodborne illness (Michaels et al., 2004b), resulting in regulatory authorities implementing no bare hand contact requirements (FDA, 2000, 2001, 2003). When properly used, gloves are an effective way of preventing contamination and protecting the employee. However, gloves can become a means of spreading pathogens when they are not appropriately cleaned and disinfected or changed after a potential contamination incident (e.g., using the restroom or answering the phone). Properly used, gloves can mitigate risk, but often when gloves are used in place of bare hands, risk is simply transferred, with no net gain in risk reduction (Michaels, 2001; Michaels et al., 2004b). Improperly used gloves or use of gloves that are not up to the task intended can cause risk of outbreak to be greatly amplified (Michaels et al., 2004b). It should be clearly understood by workers and supervisors that the use of gloves is not a substitute for hand washing or other hygienic practices (Guzewich and Ross, 1999; FDA, 2003; Michaels, 2005).

The 2001 FDA Food Code recognizes that various grades of glove are available for use by food-service facilities and considers them to be either multiuse or single-use utensils (FDA, 2001). Material durability, strength, and cleanability are the key factors in distinguishing multiuse from single-use utensils. Multiuse utensils are required to be durable, nonabsorbent, and resistant to attack by corrosive facility sanitizers, with sufficient strength to withstand repeated use, washing, and sanitizing treatments without damage, distortion, and decomposition. Both multiuse and single-use glove types are required to be safe for contact with food. Although regulations do not allow migration of deleterious substances, colors, odors, or tastes to food, many single-use glove types have been documented to leach allergens and plasticizers onto food (Michaels et al., 2004b). For this reason, where produce handling is taking place, regulators or food-safety inspectors may want a say in what gloves are used for the handling of specific food products. Regulatory authorities may require letters of safety assurance (Michaels, 2004a, 2005).

Although single-use gloves must be used for one task and no other purpose, being discarded when damaged or soiled, multiuse gloves can be cleaned and sanitized as often as necessary to reduce the risk of cross-contamination. Thicker multiuse gloves have significant application in produce handling because these gloves are capable of protecting workers from insect/arachnid bites and physical hazards associated with specific produce types. As long as a multiuse glove remains intact, in good repair, and can be cleaned and sanitized appropriately with sufficient frequency to prevent cross-contamination, it is permitted for use in food-processing and service operations. Gloves that are durable and resistant to puncture, over extended periods of wear can be considered impermeable covers necessary to cover skin wounds, cuts, and boils. Extended-wear gloves that have the requisite strength and durability to withstand repeated washing and sanitizer cycles have advantages over flimsy single-use types and bare hands with respect with the degree with which they can be easily cleaned or sanitized. However, they may feel uncomfortable to some workers because of the retention of heat and sweat, and may be taken off at inappropriate times. It should be noted that cleaning and sanitizing gloves has been found to be up to 1000 times more effective than cleaning or sanitizing hands (Michaels et al., 2004b). Gloves can be cleaned more often and more effectively than hands, without suffering the negative effects of sanitizers on bare skin.

Gloves, just as any other food-handling utensil, can be reused, preventing bare hand contact with ready-to-eat food and reducing risk. It is the duty of the facility food-safety manager to make sure staff knows how and when to clean and sanitize regularly and that this be part of a facility's Standard Sanitary Operating Procedure (SSOP). In some produce-handling applications, slash-resistant gloves are employed during cutting. Public health authorities prefer these to be covered by smooth, durable, nonabsorbent gloves or a single-use glove. Single-use gloves should be used for only one task and no other purpose. They should be discarded when damaged or soiled, or when interruptions occur in an operation (FDA, 2001).

When using single-use gloves, frequent replacement of gloves can help cleanliness and reduce the potential for growth of microorganisms in wet/dirty rubber gloves. Gloves should be changed at any time that bare hands should be washed. This includes after using the restroom, smoking or eating, taking a break, covering coughs or sneezes, touching skin or wounds, touching floors or other dirty surfaces, or handling agricultural chemicals or cleaning materials.

Gloves can become a means of spreading pathogens when they are not appropriately disinfected or changed after a potential contamination or if punctured. If hands sweat heavily inside gloves, contamination in the nail region can be liquefied and glove puncture can release large amounts of microbial contamination (Cole and Bernard, 1964; Michaels, 2001; Michaels et al., 2004b). It should be clearly understood by workers and supervisors that the use of gloves is not a substitute for hand washing or other good hygiene practices.

The use of cloth gloves is usually discouraged except for applications where further produce cleaning and/or sanitizing steps are employed. Cloth gloves rapidly become soiled and subsequently could easily cause contamination of food products. For this reason, cloth gloves are often paired with covering by accepted impervious gloves of the latex, vinyl, nitrile, or neoprene types (Michaels, 2005). There are also composite glove products made by coating cloth gloves with impervious coatings of the same materials. The use of cloth gloves to handle exposed, ready-to-eat food products in food service is not permitted under FDA guidelines (FDA, 2001). This fact should help guidance to processors and packers with respect to use of various glove products in their operations.

5.5 CONCLUSIONS

Numerous outbreaks of foodborne illness caused by bacteria, viruses, and parasites linked to fruit and vegetable products have often implicated infected or contaminated food workers. A cluster of contributing factors, often described as poor personal hygiene, have been identified. Personal hygiene is considered a prerequisite program within an HACCP system designed to reduce or eliminate risk of food contamination and potential outbreak situations. Workers must be taught proper hygiene, including those activities that take place even before they arrive at work.

Selection and placement of sanitary devices related to personal hygiene must be carefully chosen to fit the individual facility, food product, and location. Worker health and welfare, food and environmental safety are impacted by decisions made on the farm, in the packing house or on processing lines. For this reason, many of these activities fall under regulatory authority, with specific aspects covered by regulations.

Because there are so many different produce types requiring unique handling requirements it is incumbent on management to do a hazard analysis of the various food products and processes used within their facility. Management has the overriding responsibility to determine what products and procedures need to be used and to provide training to make sure that hazards are controlled within their system. The control points used to reduce or minimize hazards should be identified, monitored on a regular basis, and records kept to verify control.

A critical component for maintenance of worker health and production of safe food products is having a sanitary water supply. Water used for drinking, washing hands, and cleaning produce must be of high sanitary status, tested on a regular basis to ensure microbiological and chemical levels are met. Workers must be monitored to ensure they are not working while ill or infected. Sanitary work habits need to be developed in agricultural workers through detailed training programs designed to overcome all possible obstacles to product safety. To execute effective toilet and hand hygiene, adequate facilities and supplies need to be available to workers when needed. Soap, water, toilet paper, and paper towels form the basis for hygiene, but are worthless if workers do not know how and when to use them. Fingernail brushes can be important for maintenance of clean hands, but are not without their own hazards, as these can become easily contaminated through multiple worker use, and they can increase the risk of transfer of pathogens to hands. Working apparel and personal protection equipment such as cover garments, goggles/dust masks, shoes, and gloves all have specific requirements for their safe use. Gloves can both significantly reduce or magnify risk of product contamination. Therefore using gloves should be carefully monitored to ensure they positively impact worker and product safety.

All the commonly used SSOPs described and reviewed in this chapter are considered GMPs and GAPs. They are the universally accepted approach designed to reduce the likelihood of illness to consumers and maintenance of brand/product value. Of the vast amount of produce grown on farms, harvested, and packaged, only a small percentage causes outbreaks. This is a tribute to all those developing a worker hygiene program, correcting errors when they occur, and being responsible for delivering safe products to the consumer. Produce safety is not an accident, but is a process always under construction, requiring vigilance by management to ensure that new products and solutions do not create their own unexpected set of new problems. HACCP systems on the farm and in the processing and packaging plants allow management to anticipate potential negative consequences in a highly complex, technologically driven supply chain and to be prepared for new product innovations demanded by the market. Between these extremes are the workers responsible for getting product to market and an effective set of plant product- and process-specific controls.

REFERENCES

R.M. Adams, *Occupational Skin Disease*. Harcourt Brace Jovanovich, Inc., New York, 1990.

Arkansas State Board of Health (ASBH), Rules and Regulations Pertaining to Frozen Food Processing Plants. Executive Session, Hot Springs, Arkansas, 1968.

T. Ayers, ABC Research Corp. Gainesville (FL), personal communication, June 10, 2001.

E.A. Bannan and L.F. Judge, Bacteriological studies relating to handwashing. 1. The inability of soap bars to transmit bacteria. *Am. J. Publ. Health Nations Health*, 55, 915–922 (1965).

W.J. Becker and T.A. Wood, An Analysis of Agricultural Accidents in Florida for 1991. National Agricultural Safety Database, 1991. Available at http://www.cdc.gov/niosh/nasd/nasdhome.html, accessed January 2005.

V. Behrens, P. Seligman, L. Cameron, C.G.T. Mathias, and L.J. Fine. The prevalence of back pain, hand discomfort, and dermatitis in the U.S. working population. *Am. J. Public Health* 84 (11), 1780–1785 (1994).

D.J. Bibel, Ecological effects of a deodorant and a plain soap upon human skin bacteria. *J. Hyg. (Lond)*, 78 (1), 1–10 (1977).

M. Blackmore, The journal of infection control nursing. Hand-drying methods. *Nurs. Times*, 83, 71–74 (1987).

S. Bidawid, J.M. Farber, and S.A. Sattar, Contamination of foods by food handlers: experiments on hepatitis A virus transfer to food and its interruption. *Appl. Environ. Microbiol.*, 66 (7), 2759–2763 (2000).

M.A. Blackmore and E.M. Prisk, Is hot air hygienic? A comparison of the efficiency of hot air, cotton and paper towels. *The Home Economist*, 4, 14–15 (1984).

M.A. Blackmore, A comparison of hand drying methods. *Catering & Health*, 1, 189–198 (1989).

S.S. Block, *Disinfection, Sterilization, and Preservation*. Lea & Febiger, Pennsylvania, 1991.

L. Borgatta, M. Fisher, and N. Robbins, Hand protection and protection from hands: Handwashing, germicides and gloves. *Women & Health*, 15, 77–81 (1989).

W.H. Bound and R.I. Atkinson, Bacterial aerosol from water closets. A comparison of two types of pan and two types of cover. *Lancet*, i, 1369–1370 (1966).

B. Bowell, Operation clean-up. Hands up for cleanliness. *Nurs. Stand*, 6 (15–16), 24–25 (1992).

V. Boyd, Strawberry growers adopt voluntary program to pre-empt future scares. *The Grower*, 2 (3), 7–8 (2000).

J.M. Broughall, P. Bird, B. Jackson, and C. Marshman, An automatic monitoring system for measuring handwashing frequency in hospital wards. *J. Hosp. Infect.*, 5 (4), 447–453 (1984).

J.N. Brunn and C.O. Solberg, Hand carriage of gram-negative bacilli and *Staphylococcus aureus*. *Br. Med. J.*, 2, 580–582 (1973).

F.L. Bryan, O.D. Cook, K. Fox, J.J. Guzewich, D. Juranek, D. Maxson, C. Moe, R.C. Swanson, and E.C.D. Todd, *Procedures to Investigate Waterborne Illness*, 2nd edn. International Association of Milk, Food and Environmental Sanitarians, Inc., Des Moines, IA, 1996.

J.P. Burke, M. Finland, H.M. Gezon, D. Ingall, and J.O. Klein, Proteus mirabilis infections in a hospital nursery traced to a human carrier. *N. Engl. J. Med.*, 284, 115–121 (1971).

Canadian Food and Inspection Agency (CFIA), Severe Acute Respiratory Syndrome (SARS). Notice to Food Processors, Distributors and Importers, 2003. Available at http://www.inspection.gc.ca/english/ops/secure/20030403c.shtml, accessed August 2004.

Canadian Horticultural Council (CHC), *Personal Hygiene and Sanitary Working Procedures*, 3rd edn, On-Farm Food Safety Guidelines for Fresh Fruit and Vegetables in Canada. CHC, Ottawa, 2004.

Centers for Disease Control and Prevention (CDC), Hepatitis A associated with consumption of frozen strawberries – Michigan. *Morb. Mortal. Wkly. Rep.*, 46, 288–295 (1997).

Centers for Disease Control and Prevention (CDC), A foodborne outbreak of cryptosporidiosis – Spokane, Washington. *Morb. Mortal. Wkly. Rep.*, 47 (27), 565–567 (1998).

Centers for Disease Control and Prevention (CDC), Outbreaks of *Shigella sonnei* infection associated with eating fresh parsley – United States and Canada, July–August 1998. *Morb. Mortal. Wkly. Rep.*, 48 (14), 285–289 (1999).

W.R. Cole and H.R. Bernard, Inadequacies of present methods of surgical skin preparation. *Arch. Surg.*, 89, 215–222 (1964).

K.A. Cook, T. Boyce, C. Langkop, K. Kuo, M. Swartz, D. Ewert, E. Sowers, I. Wells and R. Tauxe. Scallions and shigellosis: a multistate outbreak traced to imported green onions. *Epidemic Intelligence Service 44th Annu. Conf.* Mar. 27–31, CDC, Atlanta, GA, p. 36 (1995).

C.B. Crane and F.B. Shevlin, Washing and hand-drying in schools. *The Medical Officer*, 95 (1964).

F.D. Crisley and M.J. Foter, *The Use of Antimicrobial Soaps and Detergents for Hand Washing in Food Service Establishments*. Division of Environmental Engineering and Food Protection, Public Health Service, U.S. Department of Health, Education, and Welfare, pp. 278–284, 1965.

H.M. Darlow and W.R. Bale, Infective hazards of water closets. *Lancet*, i, 1196–1200 (1959).

H. Davis, J.P. Taylor, J.N. Perdue, G.N. Stelma, J.M. Humphreys, R. Rowntree, III, and K.D. Greene, A shigellosis outbreak traced to commercially distributed shredded lettuce. *Am. J. Epidemiol.* 128, 1312–1321 (1988).

C.M. Dentinger, W.A. Bower, O.V. Nainan, S.M. Cotter, G. Myers, L.M. Dubusky, et al., An outbreak of Hepatitis A associated with green onions. *J. Infect. Dis.*, 183, 1273–1276 (2001).

M.P. Doyle, L.R. Beuchat, and T.J. Montville, *Food Microbiology: Fundamentals and Frontiers*. American Society for Microbiology Press, Washington, DC, 1997.

R. Eberhart, M.L. Goetz, and J. LaVillaureix, The bacteriological risk relative to the utilization of hot air dryer to dry hands. *VIIe Journess Regionales D'Hygiene Hospitaliere*, 36–38 (1983).

J.N. Ellis and H.B. Lewis, *Introduction to Fall Protection*. American Society of Safety Engineers, Des Plaines, IL, 1988.

EUREPGAP, Fruit and Vegetables, 2003. Proceedings of Fourth EUREPGAP Conference, Madrid, Spain, September 18–25, 2003.

M.F. Fay, Hand dermatitis. The role of gloves. *AORN J.*, 54, 451–467 (1991).

M. Fishbein (ed.), *Medical Uses of Soap: A Symposium*. J.B. Lippincott Company, Philadelphia, 1945.

A.A. Fisher, *Contact Dermatitis*. Lea & Febiger, Philadelphia (PA), 1986.

Food and Drug Administration (FDA), Guide to Minimize Microbial Food Safety Hazards for Fresh Fruits and Vegetables. Available at http://vm.cfsan.fda.gov/~acrobat/prodguid.pdf, accessed September 2005.

Food and Drug Administration (FDA), *CFP 2000 Backgrounder: No Bare Hand Contact*. U.S. Department of Health and Human Services, Public Health Service Food and Drug Administration, Washington, DC, 2000.

Food and Drug Administration (FDA), *Food Code*. U.S. Department of Health and Human Services, Public Health Service Food and Drug Administration, Washington, DC, 2001.

Food and Drug Administration (FDA), *Hand Hygiene in Retail and Food Service Establishments. FDA Food Service Facts*. U.S. Department of Health and Human Services, Public Health Service Food and Drug Administration, Washington, DC, 2003. Available at http://cfsan.fda.gov/~comm/handhyg.html, accessed September 2005.

K.B. Gast and K. Holt, Minimizing Microbial Hazards for Fresh Fruits and Vegetables, Health and Hygiene. Kansas State University Pub Number: MF2481. Available at http://www.oznet.ksu.edu/library/hort2/mf2481.pdf, accessed June 2005.

D.J. Gawkrodger, M.H. Lloyd, and J.A. Hunter, Occupational skin disease in hospital cleaning and kitchen workers. *Contact Dermatitis*, 15, 132–135 (1986).

C.P. Gerba, C. Wallis, and J.L. Melnick, Microbiological hazards of household toilets: droplet production and the fate of residual organisms. *Appl. Microbiol.*, 30, 229–237 (1975).

V. Goffin, C. Pierard-Franchimont, and G.E. Pierard, Sensitive skin and stratum corneum reactivity to household cleaning products. *Contact Dermatitis*, 34, 81–85 (1996).

C.L. Goh, Contact sensitivity to topical antimicrobials. (II). Sensitizing potentials of some topical antimicrobials. *Contact Dermatitis*, 23 (3), 166–171 (1989).

D. Gould, The significance of hand-drying in the prevention of infection. *Nurs. Times*, 90, 33–35 (1994).

M. Graham, Frequency and duration of handwashing in an intensive care unit. *Am. J. Infect. Control*, 18 (2), 77–81 (1994).

S. Green, Hand hygiene in practice. *Food Manufacture*, October, 19–20, 63 (1974).

C.J. Griffith, R. Malik, R.S. Cooper, N. Looker, and B. Michaels, Environmental surface cleanliness and the potential for contamination during handwashing. *Am. J. Infect. Control*, 31 (2), 93–96 (2003).

J. Guzewich and M.P. Ross, *White Paper: Evaluation of Risks Related to Microbiological Contamination of Ready-to-Eat Food by Food Preparation Workers and the Effectiveness of Interventions to Minimize those Risks*. Food and Drug Administration, Center for Food Safety and Applied Nutrition, College Park, Maryland, September 1999.

W.A. Harrison, C.J. Griffith, T. Ayers, and B. Michaels, Bacterial transfer and cross-contamination potential associated with paper-towel dispensing. *Am. J. Infect. Control*, 31 (7), 387–391 (2003a).

W.A. Harrison, C.J. Griffith, B. Michaels, and T. Ayers, Technique to determine contamination exposure routes and the economic efficiency of folded paper-towel dispensing. *Am. J. Infect. Control*, 31 (2), 104–108 (2003b).

J.L. Hattula and P.E. Stevens, A descriptive study of the handwashing environment in a long-term care facility. *Clin. Nurs. Res.*, 6; 363–374 (1997).

F. Heiss, Cade-test with a new soap. *Asthet. Med. (Berl)*, 18 (12), 223–232 (1969).

E.D. Hilborn, J.A. Farrar, M.K. Glyn, J.L. Hadler, M.A. Lambert-Fair, J.H. Mermin, P.A. Mshar, R. Mshar, L. Slutsker, M. Swartz, A. Voetsch, and C. Wojtkunski, A multistate outbreak of *Escherichia coli* O157:H7 infections associated with consumption of mesclun lettuce. *Arch. Intern. Med.*, 159 (15), 1758–1764 (1999).

IAFP, *Procedures to Investigate Foodborne Illness*, 5th edn. International Association for Food Protection, Des Moines, IA, 1999.

J.J. Kabara and M.B. Brady, Contamination of bar soaps under "in-use" conditions. *J. Environ. Pathol. Toxicol. Oncol.*, 5 (4–5), 1–14 (1984).

D.J. Katz, M.A. Cruz, P.D. Florella, R.M. Hammond, J.A. Suarez, and M.J. Trepka, An outbreak of typhoid fever in Florida associated with imported frozen fruit. *J. Infect. Dis.*, 15 (186), 234–239 (2002).

K.G. Kerr, D. Birkenhead, P.M. Hawkey, J. Major, and K. Seale, Prevalence of Listeria spp. on the hands of food workers. *J. Food Prot.*, 56 (6), 525–527 (1993).

J.E. Kirk, Hand washing. Quantitative studies on skin lipid removal by soaps and detergents based on 1500 experiments. *Acta Derm. Venereol. Suppl. (Stockh.)*, 1, 183 (1966).

B. Knights, C. Evans, S. Barrass, and B. McHardy, *Hand Drying: An Assessment of Efficiency and Hygiene of Different Methods*, A survey carried out by the Applied Ecology Research

Group, report issued by University of Westminster for the Association of Makers of Soft Tissue Papers, London, UK, 1993.

E.L. Larson, Current handwashing issues. *Infect. Control*, 5 (1), 15–17 (1984).

E. Larson, Handwashing and skin: Physiologic and bacteriologic aspects. *Infect. Control* 6, 14–23 (1985).

E. Larson, Guideline for use of topical antimicrobial agents. *Am. J. Infect. Control*, 16, 253–266 (1988).

E. Larson, Handwashing: it's essential – even when you use gloves. *Am. J. Nurs.*, 89 (7), 934–939 (1989).

E.L. Larson and E. Lusk, Evaluating handwashing technique. *J. Adv. Nurs.*, 10 (6), 547–552 (1985).

E.L. Larson, K.J. McGinley, G.L. Grove, J.J. Leyden and G.H. Talbot, Physiologic, microbiologic, and seasonal effects of handwashing on the skin of health care personnel. *Am. J. Infect. Control*, 14, 51–59 (1986).

W.C. Levine, R.W. Bennett, Y. Choi, R.A. Gunn, P.M. Griffin, K.A. Hendricks, K.J. Henning, D.P. Hopkins, and J.R. Rager, Staphylococcal food poisoning caused by imported canned mushrooms. *J. Infect. Dis.*, 173 (5), 1263–1267 (1996).

C.M. Lin, F. M. Wu, M.P. Doyle, B.S. Michaels and K. Williams, A comparison of hand washing techniques to remove *Escherichia coli* and calicivirus under natural or artificial fingernails. *J. Food Prot.*, 66, 2296–2301 (2003).

B.M. Lund and A.L. Snowden, "Fresh and Processed Fruits," in B.M. Lund, T.C. Baird-Parker, and G.W. Gould. Eds., *The Microbiological Safety and Quality of Food*, Vol., I. Chapter 27. Aspen, Gaithersburg (MD), 2000.

C.A. Mackintosh and P.N. Hoffman, An extended model for transfer of micro-organisms via the hands; differences between organisms and the effect of alcohol disinfection. *J. Hyg. Cambr.*, 92, 345–355 (1984).

M.J. Marples, *The Ecology of the Human Skin.* Thomas Publishing Co., Springfield, IL, 1965.

R.R. Marples and A.G. Towers, A laboratory model for the investigation of contact transfer of microorganisms. *J. Hyg. (Camb.)*, 82, 237–248 (1979).

J.A. Matthews and S.W.B Newsom, Hot air electric hand driers compared with paper towels for potential spread of airborne bacteria. *J. Hosp. Infect.*, 9, 85–88 (1986).

K.J. McGinley, E.L. Larson, and J.J. Leyden, Composition and density of microflora in the subungual space of the hand. *J. Clin. Microbiol.* 26, 950–953 (1988).

P.D. Meers and G.A. Yeo, Shedding of bacteria and skin squames after handwashing. *J. Hyg. (Camb.)*, 81, 99–105 (1978).

L.A. Mermel, S.L. Josephson, J. Dempsey, S. Parenteau, et al., Outbreak of *Shigella sonnei* in a clinical microbiology laboratory. *J. Clin. Microbiol.* 35, 3163–3165 (1997).

B. Michaels, Are gloves the answer? *Dairy, Food & Environ. Sanit.*, 21 (6), 489–492 (2001).

B. Michaels, Handwashing: An effective tool in the food safety arsenal. Food Quality, Sept./Oct., 45–53 (2002).

B. Michaels, Understanding the glove risk paradigm: Part I. *Food Safety Mag.*, 10 (3), 24–27 (2004a).

B. Michaels, Understanding the glove risk paradigm: Part II. *Food Safety Mag.*, 10 (4), 15–20 (2004b).

B. Michaels, Gloves: There's more to it at hand. *Food Quality Mag.*, February/March, 71–75 (2005).

B. Michaels and T. Ayers, Hazard Analysis of the Personal Hygiene Process. Proceedings of the 2nd International Conference on Food Safety, October 11–13, 2000, Savannah, GA, National Sanitation Foundation, pp. 191–200.

B. Michaels and T. Ayers, Handwashing (and Drying) – The Most Effective Means of Reducing the Risk of Infection. Proceedings of the Third International Conference on Food Safety, May 24–25, 2002, Porto, Portugal, pp. 151–168.

B. Michaels, M. Celis, T. Ayers, and V. Gangar, A microbial survey of toilet paper and associated performance variables related to its role in reducing communicable disease transmission. *J. Food Prot.*, 64 (Suppl. A), 92 (2001a).

B. Michaels, V. Gangar, T. Ayers, E. Meyers, and M.S. Curiale, The Significance of Hand Drying after Handwashing, 3rd International Conference on Culinary Arts and Sciences, April 17–20, 2001, Cairo, Egypt. Al-Fayoum (Egypt): Bournemouth University/ University of Cairo, 2001b, pp. 294–303.

B. Michaels, V. Gangar, A. Schultz, M. Arenas, M. Curiale, T. Ayers, and D. Paulson, Handwashing water temperature effects on the reduction of resident and transient (*Serratia marcescens*) flora when using bland soap. *Dairy Food & Environ. Sanit.*, 21 (12), 997–1007 (2001c).

B. Michaels, B. Smith, and M. Pierson, Pathogenic and indicator bacteria associated with handwashing and drying contact surfaces. *J. Food Prot.*, 64 (Suppl. A), 95 (2001d).

B. Michaels, V. Gangar, A. Schultz, M. Arenas, M. Curiale, T. Ayers, and D. Paulson, Water temperature as a factor in handwashing efficacy. *Food Service Technol.*, 2, 139–149 (2002).

B.S. Michaels, V. Gangar, T. Ayers, H. Guider, M.S. Curiale, and C. Griffith, Optimization of Paper Towel Hand Drying During Normal Handwashing Processes. Proceedings of APIC 30th Annual Conference & International Meeting, June 8–12, 2003, Association for Professionals in Infection Control and Epidemiology; San Antonio, TX. Washington, 2003a, p. 14.

B. Michaels, V. Gangar, C.M. Lin, and M. Doyle, Use limitations of alcoholic instant hand sanitizer as part of a food service hand hygiene program. *Food Service Technol.*, 3 (2), 71–80 (2003b).

B. Michaels, T. Ayers, and E. Todd, Produce Worker Personal Hygiene Issues. First World Congress on Organic Food: Meeting the Challenges of Safety and Quality for Fruits, Vegetables, and Grains, March 29–31, 2004, Michigan State University, East Lansing, MI, 2004a.

B. Michaels, C. Keller, M. Blevins, G. Paoli, T. Ruthman, E. Todd, and C.J. Griffith, Prevention of food worker transmission of foodborne pathogens: risk assessment and evaluation of effective hygiene intervention strategies. *Food Service Technol.*, 4 (2), 31–49 (2004b).

K.G. Mitchell and D.J.R. Rawluk, Skin reactions related to surgical scrub-up: Results of a Scottish Survey. *Br. J. Surg.*, 71, 223–224 (1984).

National Safety Council (NSC), NSC *Injury Facts*, 2000 edn. National Safety Council, Itaska, IL, 2000.

T.S. Naimi, J.M. Bartkus, J.M. Besser, D.J. Boxrud, C.W. Hedberg, H. Kassenborg, G. Krause, et al., Concurrent outbreaks of *Shigella sonnei* and enterotoxigenic *Escherichia coli* infections associated with parsley: implications for surveillance and control of foodborne illness. *J. Food Prot.*, 66 (4), 535–541 (2003).

S.W.B. Newsom, Microbiology of hospital toilets. *Lancet*, ii; 700–703 (1972).

M.T. Niu, J.K. Gedrose, B.K. Khanna, M.A. Miller, L.B. Polish, B.H. Robertson, C.N. Shapiro, J.D. Smith, and B.A. Woodruff, Multistate outbreak of Hepatitis A associated with frozen strawberries. *J. Food Dis.*, 166, 518–524 (1992).

W.C. Noble and D.G. Pitcher, Microbial ecology of the human skin. *Adv. Microbiol. Ecol.*, 2, 245–289 (1978).

S. Noritake, Wilshire Technologies Corp. Carlsbad, CA, personal communication, July 30, 2004.

Occupational Safety and Health Association (OSHA), General Environmental Controls; Sanitation, 29 CFR 1910.141, 2005a. Available at http://www.osha-slc.gov/OshStd_toc/OSHA_Std_toc.html. accessed June 2005.

Occupational Safety and Health Association (OSHA), General Environmental Controls; Sanitation, 29 CFR 1910.110, 2005b. Available at http://www.osha-slc.gov/OshStd_toc/OSHA_Std_toc.html, accessed June 2005.

J. Ojajarvi, The importance of soap selection for routine hand hygiene in hospital. *J. Hyg. (Lond.)*, 86, 275–283 (1981).

J. Ojajarvi, P. Makela, and I. Rantasalo, Failure of hand disinfection with frequent hand washing: a need for prolonged field studies. *J. Hyg. (Camb.)*, 79, 107–119 (1977).

G. Paoli, T. Ruthman, and B. Michaels, Modeling Infectious Disease Transmission Using GoldSim. GoldSim Newsletter, Fall 2003. Available at http://www.goldsim.com/eNews/0310/Fall2003_article.htm. accessed June 2005.

D.R. Patrick, T.E. Miller, and G. Findon, Residual moisture determines the level of touch-contact associated transfer following handwashing. *Epidemiol. Infect.*, 119 (3), 319–325 (1997).

D.S. Paulson, Evaluation of three handwash modalities commonly employed in the food processing industry. *Dairy, Food & Environ. Sanit.*, 12 (10), 615–618 (1992).

J.V.S. Pether and R.J. Gilbert, The survival of Salmonella on finger-tips and transfer of the organisms to food. *J. Hyg. (Camb.)*, 69, 673–681 (1971).

F.W. Pontius, Guidelines for boil-water advisories. *J. Am. Water Works Assoc.*, 88 (12), 18–102 (1996).

P.B. Price, The bacteriology of normal skin; a new quantitative test applied to a study of the bacterial flora and the disinfectant action of mechanical cleansing. *J. Infect. Dis.*, 63, 301–318 (1938).

R. Pritchett, C. Gossman, V. Radke, J. Moore, E. Busenlehner, K. Fischer, K. Doerr, C. Winkler, M. Franklin-Thomsen, J. Fiander, J. Crowley, et al., Outbreak of cyclosporiasis – Northern Virginia – Washington, D.C. – Baltimore, Maryland, Metropolitan Area. *MMWR*, 46 (30), 689–691 (1997).

K. Puryear, Some migrant workers believe that washing hands in cold water could cause arthritis. *Food Chem. News*, August 28, 2000, p. 32 (2000).

C.N. Ramsay and P.A. Upton, Hepatitis A and frozen raspberries. *Lancet,* 1 (8628), 43–44 (1989).

A. Rangarajan, E.A. Bihn, R.B. Gravani, M.P. Pritts, and D.L. Scott, Food Safety Begins on the Farm; Good Agricultural Practices for Fresh Fruits and Vegetables, Cornell Dept of Horticulture GAPs Program Brochure (2000). Available at http://www.hort.cornell.edu/commercialvegetables/issues/foodsafe.html, accessed June 2005.

V.C. Rao and J.L. Melnick, "Human Viruses in Sludges, Soils and Groundwater," in V.C. Rao and J.L. Melnick, Eds., *Environmental Virology*. American Society of Microbiology, Washington, DC, 1986, pp. 55–71.

K. Redway, B. Knights, Z. Bozoky, A. Theobald, and S. Hardcastle, Hand Drying: A Study of Bacterial Types Associated with Different Hand Drying Methods and with Hot Air Dryers. Report prepared by the Applied Ecology Research Group, University of Westminster for the Association of Makers of Soft Tissue Papers, London, UK, 1994.

T.M. Reid and H.G. Robinson, Frozen raspberries and hepatitis A. *Epidemiol. Infect.*, 98, 109–112 (1987).

F. Rheinbaben, S. Schunemann, T. Gross, and M.H. Wolff, Transmission of viruses via contact in a household setting: experiments using bacteriophage straight phiX174 as a model virus. *J. Hosp. Infect.*, 46, 61–66 (2000).

R.R. Roth and W.D. James, Microbial ecology of the skin. *Ann. Rev. Microbiol.*, 42, 441–464 (1988).

M.L. Rotter, Hygienic hand disinfection. *Am. J. Infect. Control*, 5, 18–22 (1984).

I.P. Semmelweiss, *Die Aetiologie, der Bergriff and die Prophylaxis des Kindbettfiebers*. CA Hartleben, Pest, Wein und Lepizig, 1861.

J.E. Sheldon, Combating infection. 25 tips on hand washing. See if you know them all. Nursing, 24, 20 (1994).

M.D. Sobsey, M.J. Casteel, H. Chung, G. Lovelace, O.D. Simmons, F. Hsu, and J.S. Meschke, Innovative Technologies for Waste Water Disinfection and Pathogen Detection. In Proceedings of: Disinfection 1998 – The Latest Trends in Wastewater Disinfection: Chlorination vs. UV Disinfection, Baltimore, MD, 1998.

P. Sockett and E.C.D. Todd, Economic impact of foodborne disease. In *The Microbiological Safety and Quality of Food*. Vol 2. B.M. Lund, T.C. Baird-Parker, and G.W. Gould (eds.), Aspen Publishers Inc., Gaithersburg, MD, pp. 1563–1588, 2000.

J.L. Taylor, An evaluation of handwashing techniques – 1. Nurs. Times, 74, 54–55 (1978a).

J.L. Taylor, An evaluation of handwashing techniques – 2. *Nurs. Times*, 75, 108–110 (1978b).

J.L. Taylor, J. Tuttle, T. Praukul, K. O'Brien, T.J. Barrett, B. Jolbarito, Y.L. Lim, D.J. Vugia, J.G. Morris, Jr, R.V. Tauxe, and D.M. Dwyer. An outbreak of cholera in Maryland associated with imported commercial frozen fresh coconut milk. *J. Infect. Dis.*, 167:1330–1335, 1993.

University of Maryland, *Improving the Safety and Quality of Fresh Fruits and Vegetables: A Training Manual for Trainers*. University of Maryland, USA, 2002.

U.S. Department of Agriculture (USDA), *List of Proprietary Substances and Nonfood Compounds Authorized for Use Under USDA Inspection and Grading Programs*. USDA Food Safety and Inspection Service, Washington, DC, 1998.

U.S. Environmental Protection Agency (EPA), Ambient Water Quality Criteria for Bacteria, EPA Office of Water Regulations and Standards, EPA 832-B-92-005, 1986.

U.S. Environmental Protection Agency (EPA), Method 1623: Cryptosporidium and Giardia in water by filtration/IMS/FA. EPA 821-R-01-025, Office of Water, Office of Science and Technology, Engineering and Analysis Division, U.S. Environmental Protection Agency, Washington, DC, 2001.

U.S. Environmental Protection Agency (EPA), Domestic Septage Regulatory Guidance, A Guide to the EPA 503 Rule. EPA, Office of Water Regulations and Standards, 832-B-92-005 September 1993.

U.S. Environmental Protection Agency (EPA), Ultraviolet disinfection guidance manual. EPA 815-D-03-007, Office of Water, Office of Science and Technology, Engineering and Analysis Division, U.S. Environmental Protection Agency, Washington, DC, 2003.

P. Van der Valk and H.I. Mailbach, Eds., *The Irritant Contact Dermatitis Syndrome*. CRC Press, New York, 1996.

D. Zagory, Sanitation Concerns in the Fresh-Cut Fruit and Vegetable Industry. University of California, Davis Food Processors Sanitation Workshop, 1990. Available at http://www.davisfreshtech.com/articles_freshcut.html, accessed September 2005.

CHAPTER 6

Overview of Hazards in Fresh-Cut Produce Production: Control and Management of Food Safety Hazards

MARIA ISABEL GIL and MARIA VICTORIA SELMA

Research Group on Quality, Safety and Bioactivity of Plant Foods, Department of Food Science and Technology, CEBAS-CSIC, Campus Universitario, Espinardo, Murcia, Spain

Microbial Hazard Identification in Fresh Fruit and Vegetables, Edited by Jennylynd James

6.1 FRESH-CUT PRODUCTS

Early terminology referred to minimal processing, which was described as "handling, preparation, packaging and distribution of agricultural commodities in a fresh-like state" (Shewfelt, 1987). The term has evolved into "lightly processed" (Abe and Watada, 1991) and now "fresh-cut" is currently accepted (Watada et al., 1996). The International Fresh-cut Produce Association (IFPA) defines fresh-cut products as any fruits or vegetables that have been trimmed and/or peeled and/or cut into 100 percent usable product that is bagged or prepackaged to offer consumers high nutrition, convenience, and flavor while still maintaining its freshness (IFPA, 2002).

Fruits and vegetables have natural barriers, like skins and rinds, which minimize the chances that any surface contamination could be transferred to the internal edible portions. The physiology of fresh-cut vegetables and fruits is essentially the physiology of wounded tissue (Cantwell, 1995, 1996). This type of processing, involving abrasion, peeling, slicing, chopping, or shredding, differs from traditional processing in that the tissue remains viable (or "fresh") during subsequent handling. Thus, the behavior of the tissue is generally typical of that observed in plant tissues that have been wounded or exposed to stress conditions (Cantwell, 1992; Brecht, 1995). The disruption of tissue and cell integrity that results from fresh-cut processing increases enzymatic and respiratory processes. Wounded tissues undergo accelerated deterioration and senescence. The visual symptoms of deterioration of fresh-cut produce include flaccidity from loss of water, changes in color (especially increased oxidative browning at the cut surfaces) (Saltveit, 1997) and microbial contamination (Brecht, 1995; King and Bolin, 1989; Varoquaux and Wiley, 1994). Nutrient losses may also be accelerated when plant tissues are wounded (Klein, 1987). Minimizing the negative consequences of wounding in fresh-cut vegetables and fruits will result in increased shelflife and greater maintenance of nutritional quality. Moreover, cutting circumvents the physical barrier, allowing juices to leak and promote microbiological growth. Large microbiological populations, including potentially human pathogens, develop on cut produce items. Microbial decay of fresh-cut fruit products may occur much more rapidly than in vegetable products because of the high levels of sugars found in most fruit tissues. However, the acidity of fruit tissue usually helps suppress microbial growth (Gorny, 1996). The main focus of this chapter will be the microbiological safety of fresh-cut products as one of the major concerns for the fresh-cut industry.

6.2 FOOD SAFETY SYSTEMS IN THE FRESH-CUT PRODUCE INDUSTRY

It has become necessary for the fresh-cut produce industry to initiate comprehensive sanitation programs to ensure the microbiological safety of its products.

The most comprehensive, science-based program to date for reducing pathogen contamination in fresh-cut products is the Hazard Analysis Critical Control Point

Program (HACCP). The HACCP approach focuses on controlling pathogens at their source rather than trying to detect them in finished products. HACCP addresses the root causes of food-safety problems in production, storage, transportation, and so on, and it is preventive (FDA, 1994). The effective sanitation programs that include Good Agricultural Practices (GAPs), Good Manufacturing Practices (GMPs), Sanitation Standard Operating Procedures (SSOPs) and HACCP ensure that fresh-cut fruits and vegetables are among the safest food products available. They are preventive programs to keep produce from being contaminated by pathogenic bacteria, viruses, and parasites as well as physical and chemical contaminants. Good Agricultural Practices provide guidelines on how to minimize potential biological hazards between the field and the fresh-cut processing plant. Good Manufacturing Practices are broadly defined criteria to follow for minimizing product contamination in the processing plant. They are the first lines of defense in controlling pathogen buildup in the product or environment of a fresh-cut processing plant (Hurst, 2002).

A comprehensive Sanitation Program contains the required elements to ensure a clean and sanitary environment. These programs should be well documented with written Standard Operating Procedures (SOP), understood by all employees, and periodically audited to ensure that critical processing steps are accomplished. The use of HACCP helps to minimize the potential hazards that may be associated with fresh-cut produce processing. The format for developing a HACCP plan is, in most cases, product and process specific. A generic model can serve as a useful tool to demonstrate how to create a HACCP plan for a specific product (Hurst, 2002). However, the model is not applicable for every facility and these features make each HACCP program a plant-specific plan. Each involves seven basic principles.

- *Principle 1*: Conduct a hazard analysis. Flow diagram steps of a process determine where significant hazards exist and what control measures should be instituted.
- *Principle 2*: Determine the critical control points (CCPs) required to control the identified hazards. CCPs are any steps where hazards can be prevented, eliminated, or reduced to acceptable levels.
- *Principle 3*: Establish critical limits that must be met to ensure that CCPs are under control.
- *Principle 4*: Establish procedures to monitor CCPs. These are used to adjust the process to maintain CCP control.
- *Principle 5*: Establish corrective actions to be taken when there is a deviation identified by monitoring of a CCP.
- *Principle 6*: Establish verification procedures for determining if the HACCP system is working correctly.
- *Principle 7*: Establish effective record-keeping procedures that document the HACCP system.

There are a number of interventions that can be implemented to avoid postharvest contamination of produce with human pathogens. Various government agencies and industry associations have published comprehensive guidelines for the produce industry. A description of GMPs appears in the U.S. Code of Federal Regulations (21CFR 110, 1994). The Food and Drug Administration has issued a Guide to Minimize Microbial Food Safety Hazards for Fresh Fruits and Vegetables (FDA, 1998). The IFPA has been very active in providing food-safety information, publishing the Food Safety Guidelines for the Fresh-cut Produce Industry (IFPA, 2002).

Hurst (2002) defined the fresh-cut risk factors in two categories. One category concerns the factors or conditions contaminating fresh produce with indigenous pathogens during cultivation or at harvest including poor agronomic practices, use of contaminated water for crop irrigation or mixing chemical sprays, application of improperly composted manure as fertilizer, and lack of training among field workers on good personal hygiene. A second category of microbiological risk concerns any of the processing operations in the fresh-cut plant. This will be the focus of this chapter.

6.3 FRESH-CUT PRODUCE PRODUCTION

Various steps are included in the preparation of fresh-cut products, each of which are specific unit operations (Fig. 6.1). Each unit operation must be performed properly to ensure that finished product quality, shelf life, and food safety are satisfactory (Gorny, 1996). The operations involved in preparing fruits and vegetables as fresh-cut products generally increase the rates of deterioration and may also play a role in the spoilage mechanisms (Cantwell, 1996). The unit operations in the fresh-cut product preparation have been described as in the following subsections (Gorny, 1996):

6.3.1 Raw Material Receiving and Storage

Inspection is the first operation on receiving the raw materials as part of quality control to achieve a standard product quality. Rejecting inferior quality product on the receiving dock and not allowing it into cold storage or the processing room can avoid many problems (Gorny, 1996). The main quality criteria are appearance, including overall freshness, absence of insects, physiological, and microbial diseases, absence of necrotic tissue, and compliance with regulations on pesticide residues and nitrate content (Varoquaux and Mazollier, 2002). The accepted product should be quickly moved into the appropriate temperature storage room or directly to the processing room (Fig. 6.1).

6.3.2 Preliminary Washing and Sorting

Washing, and even scrubbing, some commodities and then rinsing whole fruit in water may be helpful to remove dirt and reduce microbial populations before

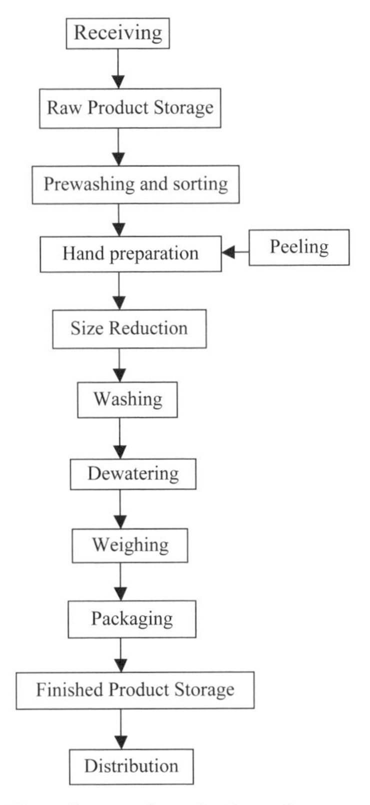

Figure 6.1 Flow diagram for a fresh produce processing plant.

cutting. Sometimes, using only a bath in which bubbling is maintained by a jet of air, permits turbulence to eliminate practically all traces of earth and foreign matter without bruising the product. Many hardy produce items are brush-washed with oscillating brushes to scrub surfaces for the physical removal of soil and microorganisms. This is often done in conjunction with a detergent followed by a rinse of potable water. Brushing also removes a portion of the natural waxy cuticle on the produce surface that acts as a barrier to microorganisms. Some produce items that may be damaged by brushes are washed in a bath or under a spray. This may or may not include gentle agitation and/or detergents to aid in removal of soils. Washing efficiencies vary with commodity, type of washing system, type of soil, contact time, detergent, and water temperature.

Any fruits or vegetables that show signs of decay will be removed immediately from the processing line to help reducing microbial contamination. Grading and sorting comprise the last separation stage before processing. Separation of raw materials into size and weight provides uniformity and standardization of the finished product. The most important grade factors are size, shape, color, firmness, bruises, and cut surfaces (Yildiz, 1994). Some grading is carried out manually by trained personnel, who are able to assess a number of grading factors simultaneously. Automated grading has the advantages of speed, reliability, and low labor cost.

6.3.3 Peeling

Before cutting, many fruits or vegetables are subject to the removal of the outer layer, which is referred to as peeling, paring, skinning, husking, shelling, and so on. Industrial peeling of large volumes of products can be accomplished mechanically on root vegetables (potatoes, beets, carrots, turnips, and onions). Most fruit require peeling to remove tough fibrous skin before cutting. Hand peeling will, in most cases, provide the highest quality product and the highest yield, although it is slow, costly, and wasteful. Hand peeling is extremely labor-intensive, but it is the only option for some commodities. Abrasive peelers are used commonly in the fresh-cut industry for items such as potatoes, carrots, and onions. These peelers utilize abrasive surface rollers to remove the outer skin from the product. The skin is washed away with a fine spray of water. Abrasive peeling is very damaging to commodities and can result in surface scarring, leading to chalking or white blush in peeled carrots, for example.

6.3.4 Size Reduction/Cutting

All unwanted parts of the plant, including most of the outer green leaves, cores, stems, and seeds, must be trimmed and discarded from the bulk commodities before further size reduction is carried out. This is also the area for sorting out decay or defects. Trimming is done manually, although it may be partly mechanized, at least for broad leaf lettuces. Any products exhibiting decay should be discarded, as they will inoculate the cutting surface with microbes, and potentially allow for the inoculation of each successively trimmed fruit or vegetable. If defects and decay are not sorted out at this point, one defect will be cut into many small pieces and be dispersed in a large amount of product, making it almost impossible to sort out (Gorny, 1996).

Size reduction describes all means by which fruits and vegetables are cut or broken into smaller, uniform pieces of definite shape and size. Wounding of plant tissue hastens the deterioration of fresh produce by disrupting membranes and allowing enzymes to intermix with substrates. In addition, cuts and punctures allow for microbial contamination of products as well as moisture loss (Shewfelt, 1987).

The product may be sliced or shredded using very sharp blades. Hand knives and stationary coring units are used effectively for this operation. When fresh horticultural products are sliced mechanically, diced, and strip cut, high-speed machines constructed of 18-8 stainless steel are used. Slicing knives, circular knives, spindle, or crosscut knives are used for dicing. The most satisfactory cutting device is a knife of extreme sharpness and as thin as structurally possible (Yildiz, 1994). Replacing and/or sharpening knives on a regular basis is highly recommended. Early studies showed the effects of knives used for cutting on the stability of cut lettuce (Bolin et al., 1977). It has been demonstrated that knife sharpness has a significant impact on the shelf-life of fresh-cut lettuce products (Bolin and Huxsoll, 1991; Bolin et al., 1977).

Cutting and grinding equipment must be thoroughly washed after each operation. An area of GMPs that is often overlooked in fresh-cut processing facilities is the maintenance of equipment and in particular cutting equipment (Gorny and Kader, 1996). Equipment used in cutting, peeling, sorting, packaging, or other operations on fresh-cut produce might be a source of human pathogens if fragments of contaminated fruits and vegetables become lodged within the equipment. So microbial growth can occur or biofilms containing human pathogens can become established on food contact surfaces (Sapers, 2003). It is possible to produce aseptically diced or sliced products by using sterile knives and aseptic conditions (Yildiz, 1994).

6.3.5 Size and Defect Sorting

Size sorting is done to assure that finished product piece size is within acceptable limits to meet customer needs. The most effective way to accomplish this is by the use of shaker screen sizers, which allow undersized small pieces (fines) to pass through a vibrating screen. Pieces greater than the size of the aperture of the shaker screen will continue on to the next steps in processing (Gorny, 1996).

6.3.6 Washing/Cooling

Properly washing and cooling of fresh-cut products immediately after cutting is one of the most important steps in fresh-cut processing. In most processing lines, the product immediately drops into a washing tank after shredding. This step is named a prewasher, and washes away exudates and juices that would otherwise rapidly pollute the disinfecting tank. The prewashing process depends on the nature of the horticultural products, the need for maximum dirt removal with minimum water, avoidance of injury to the product, and rate of output (Yildiz, 1994). Thorough washing with water alone to remove the free cellular contents that are released by cutting was found to be effective in prolonging the shelf-life of cut carrots (Bolin and Huxsoll, 1991).

Water is one of the key elements in the quality of the fresh-cut products. Three parameters have to be controlled in washing fresh-cut fruits and vegetables: quantity of water used (5–10 l/kg of product), temperature of water to cool the product, and concentration of the sanitizer (Yildiz, 1994). Rinse water temperature has been recommended to be as cold as possible for the product being rinsed, and 0°C (32°F) has been the optimal water temperature for most products. However, infiltration of washwater into intact fruit has been demonstrated with several fruits and vegetables, and is thought to have contributed to an outbreak of salmonellosis associated with fresh marked tomatoes. Washwater contaminated with microorganisms, including pathogens, can infiltrate the intercellular spaces through pores when conditions are right. Internal gas pressures and surface hydrophobicity usually prevent uptake of water. However, when the water temperature is much lower than the produce temperature, the internal gas contracts, thereby creating a partial vacuum that will draw in water through pores, channels, or punctures and be sufficient to draw water into the fruit (Bartz, 1999). Commercial experience has shown that tomato dump tank water temperature should be about 5°C higher than the fruit temperature to minimize infiltration (Sapers, 2003). In other studies, it is advised that the temperature of the washwater be, ideally, at least 10°C higher than that of the fruit (Beuchat, 1998; Lund and Snowdon, 2000).

Soaking or moving brines, using water sprays, rotary drums, rotating brushes, or shaker washers can wash fresh-cut products. Rinse water may be sprayed on fresh-cut products from above via nozzles as they proceed on a conveyor belt. Fresh-cut products may also be submerged through the rinse system, which has the advantage of gently agitating the product pieces and more effectively removing dirt and other debris. Rotary drum washers are used for cleaning apples, pears, potatoes, and beets. High-pressure water is sprayed over the product and it never comes in contact with dirty water. In wire-cylinder leafy vegetable washers, medium-pressure sprays of fresh water are used for washing spinach, lettuce, parsley, and leeks (Yildiz, 1994).

Water knives are a noncontact cutting method that uses a concentrated stream of high-pressure water. It is an ideal system for a product that requires a clean environment. The water knife cutting does not cause an increase of temperature along the cut edge and washes the cell exudates away by the stream produced.

6.3.7 Dewatering

After washing, dewaterers, centrifuges, screens, and dehumidifiers are used to remove water. The method of choice in the fresh-cut industry to dewater product is centrifugation. The time and speed of centrifugation are key parameters to be adjusted for each product. Excessive centrifugation can result in cellular damage and cause products to leak cellular fluids after packaging.

Many fresh-cut products are too delicate to withstand centrifugal drying, and forced air in a semifluidized bed may be used to strip water away from products. It is particularly effective on product pieces that have smooth surfaces. Any forced air used in such an operation must be filtered so as not to contaminate products (Gorny, 1996).

6.3.8 Weighing and Packaging

The final operation in the processing of fresh-cut products takes place in the assembly and packaging room. The assembly room is the most critical zone in the processing chain and aseptic techniques are employed. Operators working in the assembly room must wear protective clothing like aprons, sleeve covers, masks, hair caps, and gloves. Inside the assembly room, a positive air pressure is maintained with filtered air. After inserting the correct amount of product into the package, the packs are sealed. Unperforated, polymeric films are used in an effort to maintain product quality, while extending shelf-life. The use of polymeric packaging material with appropriate prepackaging cooling/preparation and sanitation treatments is a major tool used to achieve adequate shelf-life for fresh-cut products (Schlimme and Rooney, 1994).

In general, methods and treatments that decrease aerobic respiration rate (without inducing anaerobic respiration), decrease microbial populations or retard microbial growth rate. These methods retard moisture loss from produce tissue, minimize mechanical damage to tissue, inhibit or retard enzyme-catalysed softening and discoloration reactions, and delay ripening/maturation/senescence, and are employed to extend the shelf-life of fresh processed fruits and vegetables (Schlimme and Rooney, 1994). When fresh-cut products are packaged, the atmosphere within the package may be evacuated or flushed with a mixture of gases to establish more rapidly a desirable modified atmosphere. The correct combination of packaging material, produce weight, and gas composition within a package are critical components, which must be determined for each product to maintain product quality and extend product shelf-life (Gorny, 1996). Packaging cannot correct for unsanitary product handling, temperature abuse, or poor-quality raw materials. Proper sealing of bags is critical in maintaining product quality. Bags with imperfect seals will have higher oxygen concentrations and accelerate browning. Seal integrity as well as side and bottom seals on preformed bags should be checked often.

6.3.9 Storage and Transportation

Proper control of storage and transportation will extend the shelf-life of fresh-cut fruits and vegetables. Fresh-cut products arrive at terminal market facilities, including food distribution centers, wholesalers, retailers, and foodservice operations. Fresh-cut fruits and vegetables must be held continuously at refrigeration temperatures and guarded from temperature abuse during distribution and retailing. One important safety consideration is the potential for foodborne human pathogens to grow relatively undetected in produce during refrigerated distribution, or storage.

6.4 POTENTIAL MICROBIAL RISK OF FRESH-CUT PRODUCT OPERATION

6.4.1 Pathogens in Fresh-Cut Produce

Food biological safety is still the top priority of the fresh-cut industry. Although fruits and vegetables are among the safest foods available, pathogens may be present, and a number of outbreaks of foodborne disease have been traced to fresh-cut products. The pathogens most frequently associated with fresh-cut products are discussed below. Recent produce food-safety issues related to fresh-cut fruits and vegetables are shown in Table 6.1 (Gorny, 2004).

Listeria monocytogenes. *Listeria monocytogenes* is a pathogenic bacteria that can cause the serious disease “listeriosis.” If the organism manages to infect the

TABLE 6.1 Recent Produce Food Safety Issues

Romaine lettuce: Spokane, WA, July 2002 (27 cases of *E. coli* O157:H7)
Tomatoes: August–September 2002 (21 states, 512 cases, *Salmonella*)
Lettuce: San Diego, CA, October 2003 (20 cases *E. coli* O157:H7)
Spinach: San Mateo, CA, October 2003 (32 cases/2 deaths *E. coli* O157:H7)
Green onions: August 2003 (500 ill/4 deaths, hepatitis A)
Cantalope: 2000, 2001, 2002 (2 deaths, *Salmonella*)
Tomatoes: July 2004 (60 cases *Salmonella javiana*)
Green onions: August 2004 (*Salmonella*)

central nervous system, the disease carries a high mortality rate. Such serious infection in humans is relatively rare, but current estimates indicate that approximately 20 percent of the population has heightened susceptibility, with the very young, the very old, and the immunocompromised being particularly at risk (Mead et al., 1999). Furthermore, pregnant women need to take greater care to avoid infection, as it can lead to an intrauterine infection causing abortion and severe infections in neonates (Doganay, 2003).

One of the factors that makes *L. monocytogenes* particularly difficult to control in fresh-cut fruits and vegetables is its ability to grow at refrigeration temperatures on packaged fresh products (Farber et al., 1998; Guerra et al., 2001). The minimum temperature for growth is reported to be −0.4°C (Walker and Stringer, 1987); some strains even seem able to grow down to about −1.5°C (IIR, 1998). Growth can occur in the presence and almost absence of oxygen.

L. monocytogenes has been isolated from minimally processed fresh fruits and vegetables. Previous work has shown that *L. monocytogenes* survives or grows on many processed vegetables, such as ready-to-use lettuce and endive salads (Steinbruegge et al., 1988; Beuchat and Brackett, 1990; Carlin and Nguyen-the, 1994; Gombas et al., 2003), shredded cabbage (Kallender et al., 1991), and minimally processed artichokes (Sanz et al., 2003) at refrigeration temperatures. *L. monocytogenes* has also been isolated from chilled, ready-to-use fruits and vegetables at rates ranging from 0 percent (Petran et al., 1988; Farber et al., 1989; Gohil et al., 1995; Fenlon et al., 1996) to 44 percent (Sizmur and Walker, 1988; MacGowan et al., 1994; Doris and Seah, 1995).

In the U.S. microbiological criteria there is a "zero tolerance" policy for *L. monocytogenes*, making no distinction between foods contaminated at high and low levels. A maximum concentration of 10^2 CFU g^{-1} at the point of consumption for foods that does not support the growth of *L. monocytogenes* is part of the Canadian Regulatory policy. This criterion is also accepted in several European countries (Gravani, 1999). In spite of policies differing from the U.S. policy, the incidence of listeriosis reported in these industrialized countries is not noticeably different from the incidence in the United States (ICMSF, 2002). Some researchers have concluded that a nonzero tolerance policy would encourage the development of measures to minimize the growth of the organism in foods. One example is the application of multiple-hurdle technologies (Chen et al., 2003). As recognized in

FDA risk assessment, the food-processing environment represents a key point of food contamination; therefore, concentrated efforts to control *L. monocytogenes* there must be made, beginning with good manufacturing practices, standard sanitation operation procedures, and implementation of properly devised HACCP programs (CFSAN, 2001; Roberts and Wiedmann, 2003).

Clostridium botulinum. The *C. botulinum* species compromises a group of Gram-positive, endospore-forming obligate anaerobic bacteria that form a powerful neurotoxin (Varnum and Evans, 1991). Botulism in adult humans is caused by the consumption of food containing the toxin produced by the growth of *C. botulinum*. The severity and fatality rate of botulism have been a significant concern to food processors and consumers since the late 1800s and, as a result, discussion will be restricted to this species. Human botulism is normally attributed to subspecies antigenic types A, B, E, and, occasionally, type F (Dodds and Austin, 1997).

Endospores of *C. botulinum*, which provide protection from adverse conditions, are ubiquitous and can potentially contaminate plant food products during growth, harvesting, and processing (Rhodehamel, 1992). Under suitable environmental conditions, with temperatures higher than 5°C, low oxygen conditions, and a pH above 4.6, this microorganism may produce a deadly toxin. The risk of *C. botulinum* on ready-to-eat MAP fresh-cut fruits and vegetables has also been investigated extensively by a number of research groups in recent years (Solomon et al., 1990; Larson et al., 1997; Austin et al., 1998; Hao et al., 1998; Gorny, 2003). In several surveys, *C. botulinum* has not been detected in fruit and vegetables, but in some investigations a high incidence was found (Lund and Peck, 1994). In a survey of modified-atmosphere-packaged vegetables, one package each of shredded cabbage, chopped green pepper, and Italian salad mix tested positive for type A. A salad mix also contained type A and B spores. The results showed an incidence of 0.36 percent (Lilly et al., 1996). In the case of mushrooms, a much lower incidence of *C. botulinum* was reported (Hauschild et al., 1978; Notermans et al., 1989).

Escherichia coli O157:H7. Despite the commensal status of the majority of strains, pathogenic strains, particularly enterohemorrhagic *E. coli* O157:H7, is recognized as an emerging foodborne pathogen. Gastroenteritis and hemorrhagic colitis are classical symptoms, while complications including thrombocytopenic purpura and hemolytic uremic syndrome have also been documented (Martin et al., 1986; Michino et al., 1998; Law, 2000). The latter potentially leads to renal failure and death in 3–5 percent of juvenile cases (Karmali et al., 1983; Griffin and Tauxe, 1991).

Fresh vegetables such as lettuce, alfalfa sprouts, and coleslaw have been identified as a vehicle for *E. coli* O157:H7 outbreaks (Beuchat, 1996; CDC, 1997; Smith De Waal et al., 2002; Wu et al., 2002). Researchers have shown these organisms can survive at low pH. In addition, packaging under modified atmosphere has no effect on survival or growth of *E. coli* O157:H7 on shredded lettuce, sliced cucumber, and minimally processed artichokes (Abdul-Raouf et al., 1993; Beuchat, 1996; Sanz et al., 2003).

***Shigella* Species.** The genus *Shigella* is composed of four species, *S. dysenteriae*, *S. boydii*, *S. sonnei*, and *S. flexneri*, all pathogenic to humans. Several outbreaks of shigellosis have been attributed to the consumption of contaminated raw vegetables such as lettuce salads, shredded cabbage, and green onions (Beuchat, 1996). Although relatively fragile, some strains of *Shigella* are able to tolerate acidic conditions less than pH 6. Furthermore, *S. sonnei* can survive at refrigeration temperatures under low O_2 packaging conditions.

***Salmonella*.** The genus *Salmonella* is composed of over 2700 serotypes. Some pathogenic species of *Salmonella* include *S. typhimurium*, *S. enteritidis*, *S. senftenberg*, *S. saint-aul*, and *S. montevideo*. Symptoms of salmonellosis include diarrhea, nausea, abdominal pain, vomiting, mild fever, and chills (ICMSF, 1996).

Salmonellae are mesophiles, with optimum temperatures for growth of 35–43°C. The growth rate is substantially reduced at <15°C, and for most *Salmonellae* is prevented at <7°C. The minimum growth temperature of *Salmonellae* is reported to be 5.2°C (ICMSF, 1996). Salmonellae are facultatively anaerobic, capable of survival in low-O_2 atmospheres. *Salmonella* have been isolated from several fresh intact and fresh-cut fruits and vegetables (Saddik et al., 1985; Doyle, 1990, 2000; Tauxe, 1991). Several outbreaks of salmonellosis were associated with the consumption of bean sprouts and fresh fruits, mainly melons (Beuchat, 1996).

***Aeromonas* Species.** The ability of *Aeromonas hydrophila* and *A. sobria* to cause human infection has not been fully confirmed. *A. hydrophila* causes a broad spectrum of infections (septicemia, meningitis, endocarditis) in humans, often in immunocompromised hosts, and *Aeromonas* spp. have been associated epidemiologically with travellers' diarrhea. *A. hydrophila* is a psychrotroph and grows slowly at 0°C, but temperatures of 4–5°C will support growth in foods. It is a facultative anaerobe, capable of growing in atmospheres containing low concentrations of oxygen. Experiments on unprocessed vegetables stored under controlled and constant atmospheres showed that *A. hydrophila* was unaffected at 4°C and 15°C by 3–10 percent CO_2 and 8–10 percent O_2 (Berrang et al., 1989a).

A. hydrophila has been isolated from many sources, including fresh fruits and vegetables (Brandi et al., 1996). The literature contains numerous reports of the presence of *A. hydrophila* on vegetables. Berrang et al. (1989a) found *A. hydrophila* to be present on fresh asparagus, broccoli, and cauliflower. Furthermore, *A. hydrophila* has been isolated in commercial vegetable salads (Marchetti et al., 1992; García-Gimeno et al., 1996) and mayonnaise salad samples (Knochel and Jeppesen, 1990).

***Yersinia enterocolitica*.** *Y. enterocolitica* causes gastrointestinal symptoms, potentially mediated through the activity of a heat-stable enterotoxin, that may develop into suppurative and autoimmune complications (Robins-Browne, 1997). Raw vegetables are frequently contaminated by *Y. enterocolitica* (up to

43–46.1 percent), but most of the isolates have been identified as nonpathogenic on the basis of their serotypes (Delmas and Vidon, 1985; De Boer, 1986).

***Campylobacter* Species.** *Campylobacter jejuni* has emerged as a major human gastrointestinal pathogen (Ketley, 1997). Some species of *Campylobacter* are a leading cause of bacterial enteritis. The disease symptoms associated with these *C. jejuni* vary from gastrointestinal symptoms to the Guillain–Barré Syndrome, the latter being characterized by symmetrical ascending paralysis (Ho et al., 1995). Although consumption of contaminated food of animal origin is largely responsible for infection, *C. enteritidis* has also been associated with consumption of contaminated fruits and vegetables (Beuchat, 1996). Studies reported by Castillo and Escartin (1994) indicate that *C. jejuni* can survive on sliced watermelon and papaya for a sufficient time to be a risk to the consumer.

Viral Pathogens. The Norwalk and hepatitis A viruses are the most commonly documented viral food contaminations and are in the top ten causes of foodborne disease outbreaks in the United States (Cliver, 1997). Although it is unlikely that viruses can grow in or on foods, their presence on fresh produce as vehicles for infection is of concern. Hepatitis A infection has been linked to the consumption of salad items, contaminated lettuce and uncooked diced tomatoes (Karitsky et al., 1984; Sattar et al., 1994), and recently to green onions (see Table 6.1).

Protozoan Pathogens. Common parasites that occur on fresh vegetables include *Giardia lamblia*, *Entamoeba histolytica*, and *Ascaris* spp. Infection with *G. lamblia* may be acquired through person-to-person transmission or ingestion of fecally contaminated water or food (Barnett et al., 1995). In one outbreak of giardiasis, epidemiologic evidence has implicated an asymptomatic food handler as the probable source of *G. lamblia* and raw sliced vegetables as the vehicle of transmission (Francis et al., 1999).

6.4.2 Contamination During Processing, Packaging, and Storage

Most pathogens are not native to fresh-cut produce; however, more and more evidence shows that fresh-cut produce can harbor potential pathogens. Raw fruits and vegetables may be contaminated with pathogens when they enter the processing chain (Nguyen-the and Carlin, 1994). During fresh-cut fruit and vegetable production, a series of processing steps, including handling, equipment contact (slicing), washing, packaging, and storage are carried out. Each of these treatments may affect microbial colonization and growth. The development of food-poisoning microorganisms depends on the properties of the microorganisms and vegetables, as well as the effects of processing, packaging, and storage.

Handling. Contamination during processing may occur via unhygienic personnel or inadequately cleaned equipment. When handling is done inappropriately, it can damage fresh produce, rendering it more suitable for growth of spoilage and pathogenic microorganisms (Brackett, 1994). Outbreaks of *Salmonella* infections were traced to precut watermelons in 1955, 1979, and 1991 (CDC, 1979; Blostein, 1993). In 1991, more than 400 cases of *Salmonella poona* infections were linked to presliced cantaloupe that originated in Texas or Mexico (CDC, 1991). Therefore, scrupulous hygiene must be applied at this level of processing.

Equipment Contact. Cutters and slicers can be potent sources of contamination, because they usually provide inaccessible sites that harbor bacteria and increase surface area for contamination and growth (Garg et al., 1990). Furthermore, liquid losses during this stage promote the dissemination and growth of microorganisms. Moisture and exudates on cut surfaces and on surfaces of utensils and equipment provide excellent media for rapid growth of microorganisms. Therefore, during the processes of slicing or shredding, there may be an increase in the number of bacteria present. Brackett (1994) showed that exposing vegetables to various types of cutting result in a six- to sevenfold increase in microbial populations.

Shredding and slicing procedures can increase the number of mesophilic bacteria from 10^3-10^4 to 10^5-10^6 CFU g^{-1} for a range of vegetables (Garg et al., 1990) and from 10^4-10^5 to 10^6 CFU g^{-1} for lettuce and chicory salads (Jöckel and Ötto, 1990). Furthermore, during this step, the risk of pathogen contamination such as *L. monocytogenes* and *Salmonella* can be increased. For instance, *L. monocytogenes* was found in 19 percent of minimally processed food vegetables examined in United Kingdom, whereas only 1.8 percent of the individual ingredients were contaminated. The authors concluded that substantial contamination by *L. monocytogenes* occurred during chopping, mixing, and packaging (Velani and Roberts, 1991). Lainé and Michard (1988) and Nguyen-the and Carlin (1994) reported that *L. monocytogenes* is regularly found in shredded cabbage and on shredders in the processing line. Fresh Products Northwest recalled its crunch pack fresh sliced apple packages because the product might have been contaminated with *L. monocytogenes* (Safety Alerts, 2001a). Boskovich Farms Fresh Cut division recalled halved red bell peppers, as the product may have been contaminated with *L. monocytogenes* (Safety Alerts, 2001b). These cases strongly suggest that consumption of fresh-cut fruits and vegetables can result in foodborne disease outbreaks. In 1981, a *L. monocytogenes* outbreak was associated with coleslaw (Hurst, 1995).

Salmonellae can become established and multiply in the environment and equipment of a variety of food-processing facilities, potentially cross-contaminating raw foods during processing (ICMSF, 1996). In 1990 and 1993, an outbreak of *Salmonella* spp. infection was epidemiologically linked to consumption of sliced tomatoes (Wood et al., 1991; CDC, 1993).

Washing and Sanitizing Treatments. Fresh-cut products are not subject to thermal or other preservative treatments designed to reduce or eliminate microbial load, unlike some other processed foods. Therefore, an important step in the processing of fresh-cut fruits and vegetables is thorough washing. This is often accompanied by dipping in an antimicrobial solution, to reduce microbiological load. Apart from the elimination of soil residues and plant debris, washing with tapwater has a limited effect on microflora. Washing of ready-to-eat salad and spinach leaves, for example, reduced total counts and *Enterobacteriaceae* by 0.2 to 0.5 log cycles (Garg et al., 1990; Jöckel and Otto, 1990) and no reduction of total counts on parsley were reported (Käferstein, 1976). A reduction of one log cycle in the counts of *Bacillus cereus* on bean sprouts was obtained after successive washing with tapwater (Harmon et al., 1987).

Strict hygiene is critical during washing and disinfection in order to avoid contamination of produce. Chlorinated water from the washing bath on a line of processing chicory salads was found to contain 10^3 bacteria ml^{-1} (Nguyen-the and Prunier, 1989). This is low compared with the 10^5 to 10^6 CFU g^{-1} found on the product. However, transfer of microorganisms can probably occur through the washing bath because of the reuse of washwater. Thus, washing baths may result in the transfer of microorganisms. Processing does not significantly alter the composition of the mesophilic microflora, which are similar in endproducts and unprocessed fruits and vegetables (Nguyen-the and Prunier, 1989; Magnusson et al., 1990; Delaquis et al., 2004). However, processing could increase contamination by some foodborne pathogens. Pathogens, if present on raw vegetables, may not be fully eliminated by commercial disinfection procedures, while natural competitive organisms are reduced or removed. As a result, disinfection may produce conditions favoring the growth of pathogens.

Packaging and Storage. The effect of atmospheric composition on foodborne pathogen growth on fresh-cut fruits and vegetables has not been studied widely. The important interaction between modified atmosphere packaging (MAP) and microbial food safety must always be considered. Research efforts to fully understand these relationships are currently under way. Modified atmosphere packaging of refrigerated fresh-cut products could represent a risk for public health for different reasons. Oxygen decreases rapidly in stored fresh-cut fruits and vegetables. This creates progressively anaerobic conditions favorable for the development of foodborne pathogens such as *C. botulinum* and *S. sonnei.* Furthermore, these atmospheric conditions, as well as refrigeration temperatures, may inhibit the development of some spoilage aerobic microorganisms. Natural competitors of pathogens that could tolerate low temperature and oxygen concentrations are inhibited (Daniels et al., 1985; Farber, 1991; Bennik et al., 1995). Moreover, inhibition of natural flora could increase the shelf-life of products, allowing the development of pathogens without noticeable changes in quality.

Fresh-cut fruits and vegetables are normally stored at refrigerated temperatures that may select for psychrotrophic microorganisms. Storage temperature is the single most important factor affecting the growth of microorganisms in fresh-cut fruits and vegetables. Two factors should be taken into account to explain the effect of temperature in addition to its direct action on growth rate. First, temperature determines the respiration rate of the product and, therefore, changes in the gaseous atmosphere in the package affecting the behavior of microorganisms. In addition, temperature could influence the rate of senescence of fresh-cut fruits and vegetables, modifying the environment for microorganisms (Nguyen-the and Carlin, 1994).

Researchers have shown when storage temperature decreases, growth of the mesophilic bacteria is reduced in shredded chicory salads (Nguyen-the and Carlin, 1994), carrot slices (Buick and Damoglou, 1987), and shredded lettuce (Bolin et al., 1977; Beuchat and Brackett, 1990). Storage of fresh-cut fruits and vegetables at adequate refrigeration temperatures will limit pathogen growth to those that are psychrotrophic. Conversely, temperature abuse during storage will markedly reduce the lag and generation times, permitting development of mesophilic pathogens and more rapid growth of psychrotrophic pathogens.

Clostridium botulinum. Under temperature abuse conditions, *C. botulinum* proteolytic strains can grow, having toxigenic potential at temperatures between 10 and 50°C (Pierson and Reddy, 1988; Hauschild and Dodds, 1993). Nonproteolytic strains are psychrotrophs and hence capable of growth and production of toxins under refrigeration conditions in vegetables (Carlin and Peck, 1996).

Recent research efforts have examined the *C. botulinum* (proteolytic A and B spores types) risk factor for various fresh-cut MA-packed produce. In general, overt gross spoilage of fresh-cut produce occurs well before toxin is produced on shredded cabbage, shredded lettuce, broccoli florets, sliced carrots, and rutabaga (Gorny, 2003). An FDA microbiology group observed an incidence rate of 0.41 percent in MA-packaged fresh-cut lettuce, cabbage, and carrots (Rhodehamel et al., 1993). The low incidence of *C. botulinum* proteolytic spores on produce and their restricted growth at 10°C or below further reduces the risk of botulism contributed by MAP of produce. However, if MA-packages are subjected to temperature abuse, growth and toxin production by *C. botulinum* could occur.

Toxin production by proteolytic strains has been observed in some cases when the headspace still contained between 1 and 2 percent oxygen (Sugiyama and Yang, 1975; Hotchkiss et al., 1992). When packaging film was pierced to increase its permeability, oxygen concentrations in packages of inoculated mushrooms never dropped below 2 percent (2.2–7.0 percent at the end of storage) and toxin production was not observed (Sugiyama and Yang, 1975). These results were all obtained after a few days at room temperature, which was a temperature abuse for products that should be kept refrigerated. Solomon et al. (1990) found that an inoculum load of 96 spores/g of cabbage produced toxin on day 4 when held at 22–25°C. At a level between 6 and 60 spores/g, the product did not become toxic for up to six days. The product was deemed organoleptically unacceptable on day 7. Therefore, these authors concluded that MAP shredded cabbage challenged with high concentrations of *C. botulinum* proteolytic spores (>100/g) did support botulinal toxin production.

A particular concern is the possibility that psychrotrophic, nonproteolytic strains of *C. botulinum* types B, E, and F are able to grow and produce toxins under MAP conditions. Little is known about the effects of modified atmosphere storage conditions on toxin production by *C. botulinum*. The possibility of inhibiting *C. botulinum* nonproteolytic strains by incorporating low levels of O_2 in the package does not appear to be feasible. Psychrotrophic strains of *C. botulinum* are able to produce toxins in an environment with up to 10 percent O_2 (Ahvenainen,

2003). Fresh produce items are recognized to harbor spores of *C. botulinum*, particularly proteolytic strains of type A and B. However, although thermally processed vegetables have been associated with numerous outbreaks of botulism (Pierson and Reddy, 1988), MAP of fresh vegetables (no thermal process) has not been confirmed as a cause of botulism. One exception is that of mushrooms (Sugiyama and Yang, 1975). Solomon et al. (1990) did implicate MAP cabbage in a botulism outbreak in Florida in 1987. They assumed that the cabbage had been packaged in a modified atmosphere, and the laboratory methodology used to establish the link between MAP and botulism was not typical of a MAP system used for produce.

Listeria monocytogenes. The growth of *L. monocytogenes* in fresh-cut fruits and vegetables packaged under modified atmospheres has been the focus of several studies. Some investigations demonstrated possible growth of *L. monocytogenes* in MAP fresh-cut vegetables. The results depended very much on the type of products and storage temperature. *L. monocytogenes* grows similarly in shredded lettuce packaged under air or 97 percent N_2 and 3 percent O_2 (Beuchat and Brackett, 1990), and in other vegetables packaged under air or 97 percent N_2 and 3 percent O_2 (Berrang et al., 1989b; Beuchat and Brackett, 1991). Conversely, Ringlé et al. (1991) found the bacteria did not grow at 4 or 8°C on shredded chicory salads packaged under air in a semipermeable film, but did grow when the product was packaged under N_2 in a high-barrier film. Similarly, Francis and O'Beirne (1997) reported that nitrogen flushing combined with storage at 8°C enhanced the growth of *L. monocytogenes* on shredded lettuce. In work by Carlin et al. (1996), chicory leaves inoculated with *L. monocytogenes* were stored at 10°C in air, or under 10, 30, or 50 percent CO_2 with 10 percent O_2. *L. monocytogenes* grew better as the concentration of CO_2 increased. Therefore, MAP does not seem to be an efficient way to reduce multiplication of *L. monocytogenes*.

Shigella sonnei. *S. sonnei* growth is not inhibited by low-oxygen packaging conditions. For instance, *S. sonnei* counts on shredded cabbage at 24°C increased by a factor of 3 to 4 log cycles after one day, irrespective of the packaging conditions (vacuum packaged, packaged under 30 percent N_2 and 70 percent CO_2, or packaged aerobically). However, during the subsequent days of storage, the bacteria disappeared more rapidly in the vacuum-packaged cabbage, and cabbage packaged under modified atmosphere (Nguyen-the and Carlin, 1994). At 0–6°C, *S. sonnei* did not grow in any samples, but survived better in the vacuum-packaged cabbage and in cabbage packaged under modified atmosphere.

Aeromonas hydrofila. *A. hydrophila* is able to multiply in fresh-cut vegetables stored in refrigerated conditions (4–5°C) under MAP. García-Gimeno et al. (1996) have reported the behavior of *A. hydrophila* on vegetable salads under a controlled atmosphere at 4 and 15°C. *A. hydrophila* survived but did not grow on samples stored at 4°C. In contrast, it grew on salads stored at 15°C, from 10^3 to 10^8 CFU g^{-1} in the first 24 h, with a subsequent decline in population.

On nutrient media, growth of *A. hydrophila* was reduced under 100 percent CO_2 when compared with air (Golden et al., 1989). On other food commodities, such as meat products, *A. hydrophila* has been detected at low temperatures in a variety of vacuum-packaged fresh products (Hudson et al., 1994; Doherty et al., 1996; Özbas et al., 1996). Devlieghere et al. (2000) developed a model, predicting the influence of temperature and CO_2 on the growth of *A. hydrophila*. In a study by Berrang et al. (1989a), regarding the effects of gas atmospheres on fresh asparagus, broccoli, and cauliflower storage at 4 and 15°C, fast proliferation of *A. hydrophila* was observed at both temperatures, but growth was not significantly affected by gas atmosphere. However, other researchers, using a solid surface model system, observed a decrease on the growth rate of *A. hydrophila* as a result of increasing CO_2 concentration (Bennik et al., 1995).

Yersinia enterocolitica. *Y. enterocolitica* is generally recognized as one of the most psychrotrophic foodborne pathogens. A minimum growth temperature of 4°C is generally accepted, but growth at 1°C has also been observed (Schiemann, 1989). Although raw vegetables are frequently contaminated by *Y. enterocolitica*, information about how atmospheric conditions on fresh-cut vegetables affect the growth of this microorganism is extremely limited. On meat products, CO_2 minimally retards the growth of *Y. enterocolitica* at refrigerated temperatures (Ahvenainen, 2003). Oxygen also seems to play an inhibiting role on the growth of *Y. enterocolitica* (Garcia de Fernando et al., 1995).

Escherichia coli O157:H7. *E. coli* O157:H7, as a mesophilic bacteria, is not able to grow in fresh-cut products under good refrigeration. However, temperature abuse may facilitate growth of *E. coli* O157:H7 on processed vegetables (for example, shredded vegetables) because of the substantial amounts of nutrients provided through processing (Abdul-Raouf et al., 1993). CO_2 had no inhibitory effect on growth of *E. coli* O157:H7 on shredded lettuce stored at 13 or 22°C. Potential growth was increased in an atmosphere of $O_2/CO_2/N_2$ 5/30/65, compared with growth in air (Abdul-Raouf et al., 1993; Diaz and Hotchkiss, 1996). *E. coli* O157:H7 populations survive on produce stored at 4°C and proliferate rapidly when stored at 15°C. Reducing the storage temperature from 8 to 4°C significantly reduced growth of *E. coli* O157:H7 on MAP vegetables. However, viable populations remained at the end of the storage period at 4°C (Richert et al., 2000).

6.5 STRATEGIES TO REDUCE OR ELIMINATE HAZARDS IN FRESH-CUT PRODUCTS

6.5.1 Temperature

Storage of produce at adequate refrigeration temperatures will limit pathogen growth. Mesophilic pathogens, such as *Salmonella* and *E. coli* O157:H7, are unable to grow where temperature control is adequate (i.e., 4°C). However, both of them can survive in chilled conditions for extended periods (ICMSF, 1996; Piagentini et al., 1997; Francis and O'Beirne, 2001). Furthermore, if temperature abuse occurs, they may then grow on a range of vegetables (Richert et al., 2000). This illustrates the essential requirement for strict temperature control, particularly in the case of extended shelf-life products devoid of additional growth-limiting hurdles.

Storage at refrigerated temperatures probably selects for psychrotrophic microorganisms, *L. monocytogenes*, *Y. enterocolitica*, nonproteolytic *C. botulinum*, and *A. hydrophila* being the most notable. In spite of the ability of *L. monocytogenes* to grow at refrigeration temperatures, reducing storage temperature will extend the lag phase and reduce the rate of growth (Beuchat and Brackett, 1990; Carlin et al., 1995). For instance, some researchers have shown that *L. monocytogenes* populations remained constant or decreased on packaged vegetables stored at 4°C, but at 8°C, growth of *L. monocytogenes* was supported on all vegetables, with the exception of coleslaw mix (Francis and O'Beirne, 2001). Therefore, even mild temperature abuse during storage permits more rapid growth of psychrotrophic pathogens (Berrang et al., 1989a; Carlin and Peck, 1996; García-Gimeno et al., 1996; Farber et al., 1998; Conway et al., 2000; Rodriguez et al., 2000). Strict control of refrigeration temperature throughout the chill-chain is crucial for maintaining microbiological safety.

In the United Kingdom, according to Food Hygiene Regulations, retail packs of prepared salad vegetables should be maintained at temperatures below 8°C. Guidelines for handling chilled foods, published by the UK Institute of Food Science and Technology (IFST, 1990) recommend a storage temperature range of 0–5°C for prepared salad vegetables. Some vegetables may suffer damage if kept at the lower end of this temperature range.

On the other hand, survival of hepatitis A virus on lettuce has been found significantly lower at room temperature than at 4°C (Bidawid et al., 2001). This is in agreement with previous studies that found larger survival of viruses on vegetables stored at refrigeration temperatures (Bagdasaryan, 1964; Badawy et al., 1985). The behavior of protozoan parasites on refrigerated produce is not well known. However, the increase in incidence of produce-linked outbreaks caused by these organisms indicates that more research in this area is necessary.

Because of difficulties in maintaining low temperatures in the chill-chain, additional barriers to control the growth of pathogens (particularly psychrotrophs) are required. A combination of beneficial intrinsic factors, processing factors, and extrinsic inhibitory factors results in considerable improvement in the microbial safety of the product. Suitable combinations of growth-limiting factors at subinhibitory levels can inhibit certain microorganisms through the "Hurdles Concept" (Scott, 1989). Theoretically, each hurdle contributes to the eventual inhibition of a microorganism and all the hurdles must be overcome in order for it to grow.

To ensure temperatures are kept at or below 4°C throughout the cold chain, refrigerated distribution is required with suitably designed vehicles, properly loaded to allow for air movement (Brackett, 1999). At supermarket level, LeBlanc et al. (1996) found 90 percent of produce items above 4°C in supermarket chill cabinets. Problems can also arise at consumer level, where products are held for extended periods in cars or experience elevated temperatures in domestic refrigerators. Time temperature indicators embedded in the packaging may have a significant role in ensuring that safe storage temperatures are used. Distributors, retailers, and consumers must be educated on the importance of low storage temperatures. Consumers also need to be educated on the nature of minimal processing technologies for fresh foods. In particular, they must realize that consumption of apparently fresh food beyond its use-by date is potentially hazardous.

General principles of good hygiene must apply throughout the distribution chain, particularly avoiding cross-contamination. For example, refrigerated trucks carrying vegetables on an outward journey may be used to "backhaul" animals or raw meats (Brackett, 1999). Truck use needs to be monitored and vehicles appropriately sanitized. In the food-service sector, training of operatives is important, because many of these are young and inexperienced, and may require food hygiene education (Beuchat and Ryu, 1997).

On meat products, an increase of CO_2 concentrations (50–100 percent) reduced growth of *Salmonella* (Silliker and Wolfe, 1980; Gray et al., 1984), but these concentrations would cause damage to most fresh-cut products. Most studies to date have been limited to meat products. In relation to the ability of *Salmonella* to grow in MAP on fresh-cut vegetables, more research is necessary. *Salmonella hadar* grows well on minimally processed cabbage, packaged in mono-oriented polypropylene bags. Population levels were not different from initial levels after 10 days storage at 4°C (Piagentini et al., 1997). However, on samples stored at 12°C and 20°C, *Salmonella* populations increased by 4 and 6 logs respectively after 10 days storage (Piagentini et al., 1997). Populations of *S. montevideo* on the surfaces of inoculated tomatoes stored at 10°C did not change significantly over 18 days. However, significant increases in population occurred within 7 days when tomatoes were stored at 20°C (Zhuang et al., 1995). Tamminga et al. (1978) showed that *S. typhimurium* developed well on peeled, steamed, and vacuum-packed potatoes at 5, 14, and 22°C.

6.5.2 Washing and Sanitizing Agents for Fresh-Cut Fruits and Vegetables

The washing or sanitizing step is one of the most important processes in the processing of fresh-cut products to remove dirt and microorganisms responsible for quality loss and decay. The product should be washed and rinsed with processing water of such quality that it does not contaminate the produce. It should be visually free of dust, dirt, and other debris and sanitized (Sapers, 2003). This step is also used to precool cut produce and remove cell exudates that adhere to product cut surfaces, supporting microbial growth or resulting in discoloration.

The relationship between human pathogens and the native microflora, including postharvest spoilage organisms, on produce has been described as of interest (Parish et al., 2003). It has been suggested that reducing/controlling the native microbial populations by washing and sanitizing can allow human pathogens to flourish on produce surfaces (Brackett, 1992). Concern has been expressed that reductions in surface populations reduces competition for space and nutrients, thereby providing growth potential for pathogenic contaminants (Parish et al., 2003). However, most foodborne illnesses are more likely to arise from contamination during food preparation and storage. Effective intervention strategies have been developed, but they cannot completely eliminate microbial food safety hazards associated with the consumption of uncooked produce. Prevention of microbial contamination at all steps, from production to consumption, is the most effective strategy to ensure that fresh-cut products are wholesome and safe for human consumption. The goal of washing is not necessarily to remove as many microorganisms as possible. Rather, the challenge is to ensure that the microorganisms present do not create human health risk and that if harmful microorganisms should, inadvertently, be present, that environmental conditions prevent their growth. Therefore, the best method to eliminate pathogens from produce is to prevent contamination in the first place. It is more difficult to decontaminate produce than it is to avoid contamination. However, this is not always possible and the need to wash and sanitize many types of produce remains of vital importance to prevent disease outbreaks.

"Sanitize" is defined by the 21 CFR 110 as "to adequately treat food-contact surfaces by a process that is effective in destroying vegetative cells of microorganisms of public health significance". The Food and Drug Administration's (FDA) Guide to Minimize Microbial Food Safety Hazards for Fresh Fruits and Vegetables (FDA, 1998) defines "sanitize" as "to treat clean produce by a process that is effective in destroying or substantially reducing the numbers of microorganisms of public health concern. The Guide also refers to reducing other undesirable microorganisms, without adversely affecting the quality of the product or its safety for the consumer." A variety of methods are used to reduce populations of microorganisms on fresh-cut produce. Each method has distinct advantages and limitations depending upon the type of produce, mitigation protocol, and other variables. In the last few years an important amount of information has been published concerning the efficacy of washing and sanitizing treatments in reducing microbial populations on fresh produce. A clear and well-documented comparison of different sanitation methods has been compiled in the Food Safety Guidelines for the Fresh-cut Produce Industry review (IFPA, 2002). This topic is covered in the review by Parish et al. (2003) and that of Sapers (2003). The efficiency of numerous physical, chemical, and biological methods for reducing the microbiological load of produce will be covered in this chapter, which includes the latest and most significant research findings. This chapter excludes other treatments with lethal effect on the plant tissue such as heat treatment, freezing, drying, fermentation, or the addition of acid dressings.

Traditional methods of reducing microbial populations on produce involve chemical and physical treatments. The control of contamination requires that these treatments be applied to equipment and facilities as well as to produce. Methods of cleaning and sanitizing produce surfaces usually involve the application of water, cleaning chemicals (e.g., detergent), and mechanical treatment of the surface by brush or spray washers, followed by rinsing with potable water. The rinse step may include a sanitizer treatment. Approved sanitizers suitable for equipment and food contact surfaces include chlorine-based sanitizers, iodine-based sanitizers, quaternary ammonium compounds, and acid-anionic sanitizers (21CFR173.315; 21CFR178.1010; IFPA, 2002). Researchers have pointed out that if washwater is not properly sanitized, it can become a source of microbial contamination (IFPA, 2002; Suslow, 2001). It is important to ensure that water used for washing and sanitizing purposes is clean so that it does not become a vehicle for contamination. However, the efficacy of the method used to reduce microbial populations is usually dependent upon the type of treatment, type and physiology of the target microorganisms, characteristics of produce surfaces (cracks, crevices, hydrophobic tendency and texture), exposure time, and concentration of cleaner/sanitizer, pH, and temperature (Parish et al., 2003).

The simple rinsing of produce in plain water reduces surface populations, although the reduction is usually less than 1 log (Nguyen-the and Carlin, 2000). It usually has no effect on microbial spoilage, and may even increase contamination or facilitate penetration of microorganisms for soft rot *Erwinia* spp. on potatoes or tomatoes (Bartz and Eckert, 1987). However, washing carrots in water at harvest reduced the incidence of microbial spoilage during storage by a factor of 3 to 5 (Lockhart and Delbridge, 1972). Because water washing only partially removes the microorganisms intrinsic to produce (Garg et al., 1990), microbial decay is still the major source of spoilage of fresh-cut produce (Brackett, 1994; Babic et al., 1996; Robbs et al., 1996). The most widely used sanitizing agents are highly effective in killing microorganisms suspended in water, in contrast to their limited efficacy against microorganisms attached to produce surfaces (Sapers, 2003). Conventional washing and sanitizing methods, even using some of the newer sanitizing agents such as chlorine dioxide, ozone, and peroxiacetic acid, were not capable of reducing microbial populations by more than 90 or 99 percent (Beuchat, 1998; Brackett, 1999; Sapers, 2003).

The condition and location of microorganisms on produce surfaces affect their resistance to the attachment by washing agents and to inactivation by antimicrobial agents (Sapers, 2003; Burnett et al., 2004, Lang et al., 2004; Ukuku and Fett, 2004). Sapers (2003) described that microbial resistance to sanitizing washes will depend, in part, on different factors such as whether the microbial populations have become firmly attached. Microbial resistance also depends on whether populations occupying inaccessible sites have penetrated into the interior of the commodity, or have become incorporated into biofilms. One of the characteristics of bacterial attachment to fruits and vegetables is the rapidity of attachment to the commodity surface. The effectiveness of washing will depend on the time interval between contamination and washing. If the time between contamination and washing increases, the effectiveness of microbial agents decreases (Sapers, 2003; Lang et al., 2004). Bacteria tend to concentrate where there are more binding sites. Attachment also might be in stomata (Seo and Frank, 1999), indentations, or other natural irregularities on the intact surface where bacteria could lodge. Bacteria also might attach at cut surfaces (Takeuchi and Frank, 2000; Liao and Cook, 2001) or in punctures or cracks in the external surface (Burnett et al., 2000).

A concern regarding washing system efficiency is the quality of washwater, especially if the water is recycled and not treated prior to reuse. The use of disinfectant chemicals in washwater provides a barrier to cross-contamination of produce. Economic considerations and wastewater discharge regulations make water recirculation a common practice in the fresh-cut industry (IFPA, 2002). In each country, the regulatory status of washwater disinfectants depends on some particular agencies. The definition of the product used to disinfect washwater depends on (1) the type of product to be washed and (2) in some cases, the location where the disinfectant is used (IFPA, 2002). In the United States, for instance, if the product to be washed is a fresh-cut produce, the washwater disinfectant is regulated by the FDA as a secondary direct food additive, unless it may be considered Generally Recognized As Safe (GRAS). If the product is a raw agricultural commodity that is washed in a food-processing facility, such as a fresh-cut facility, both the EPA and FDA have regulatory jurisdiction and the disinfectant product must be registered as pesticides with the EPA. A selected list of washwater disinfectants that have been approved by FDA (21 CFR173.315) is reported in the Food Safety Guidelines for the Fresh-cut Produce Industry (IFPA, 2002).

Chlorine Gas, Hypochlorite, Chloramines. Chlorine-based sanitizers have been used for sanitation purposes in food processing for several decades and are perhaps the most widely used sanitizers in the food industry (Ogawa et al., 1980; Walker and LaGrange, 1991; Cherry, 1999). Chlorine-based chemicals are often employed to sanitize produce and surfaces within produce-processing facilities, as well as to reduce microbial populations in water used during cleaning and packing operations.

Chlorine is obtained as a gas (Cl_2) or as liquid in the form of sodium (NaOCl) or calcium hypochlorite [$Ca(OCl)_2$]. Liquid chlorine and hypochlorites are generally used in the 50–200 ppm concentration range with a contact time of 1–2 min to sanitize produce surfaces and processing equipment. Higher concentrations have been investigated for use on seeds for sprout production. In water, these materials form the microbial biocide hypochlorous acid (HOCl), which is also called "available chlorine," or "free chlorine," and which has a high bactericidal activity against a broad range of microorganisms. Chlorination with hypochlorous acid remains a convenient and inexpensive sanitizer for use against many foodborne pathogens (White, 1992; Bryant et al., 1992). The factors that influence the effectiveness of chlorine include:

- pH of the solution;
- Concentration of available chlorine;
- Contact time;
- Temperature;
- Amount of organic material present in the water;
- Type of microorganism.

As sodium or calcium hypochlorite is added to water, the pH will increase as a result of the formation of sodium hydroxide (NaOH) (Reaction 1):

$$NaOCl + H_2O \longleftrightarrow NaOH + HOCl \quad (1)$$

$$HOCl \longleftrightarrow H^+ + OCl^- \quad (2)$$

Acids (H^+) such as muriatic or citric acid are added to lower the pH and increase the concentration of available chlorine, as shown in Reaction 2. These acids are "Generally Regarded As Safe" or (GRAS) by the FDA. In aqueous solutions, the equilibrium between hypochlorous acid (HOCl) and the hypochlorite ion (OCl^-) is pH-dependent, with the concentration of HOCl increasing as pH decreases. Typically, pH values between 6.0 and 7.5 are used in sanitizer solutions to minimize corrosion of equipment while yielding acceptable chlorine efficacy. At higher pH, the amount of available chlorine is dramatically reduced (Table 6.2). At lower pH, excessive loss of available chlorine may result from the formation of chlorine gas (Reaction 3):

$$HOCl + HCl \longleftrightarrow H_2O + Cl_2 \quad (3)$$

HOCl concentration is also significantly affected by temperature, the presence of organic matter, light, air, and metals. For example, increasing levels of organic matter decreases HOCl concentration and overall antimicrobial activity.

The concentration of available chlorine determines the oxidizing potential of the solution and its disinfecting power. Although minimum concentrations of available chlorine (1 ppm) are required to inactivate microorganism in clean water, higher concentrations are commonly used for most commodities. This offsets changes in the concentration of available chlorine resulting from the amount of crop treated and organic matter accumulated in the washwater during treatment of large amounts of fruits and vegetables (Adeskaveg, 1995).

TABLE 6.2 Effect of pH on Percent of Available Chlorine ($HOCl \rightarrow H^+ + OCl^-$)

	Percent Available Chlorine		
pH	0°C	20°C	30°C
4.0	—	—	100.0
5.0	99.8	99.7	99.6
6.0	98.5	97.7	96.9
7.0	87.0	79.3	75.9
8.0	40.2	27.3	23.9
9.0	6.3	3.7	3.1
10.0	0.7	0.4	0.3

Maximum solubility in water is observed near 4°C; however, it has been suggested that the temperature of processing water should be maintained at least 10°C higher than that of the produce items in order to reduce the possibility of microbial infiltration caused by a temperature-generated pressure differential. The opportunity for infiltration of microorganisms is also minimized when the sanitary condition of the water is maintained. At warmer temperatures, the disinfecting power of hypochlorous acid increases. In contrast, organic material in the washwater decreases available chlorine and could result in the formation of undesirable "combined chlorine". The amount of chlorine in wash tanks can be described as

$$\text{Total Chlorine} = \text{Available Chlorine} + \text{Combined Chlorine}$$

When a chlorine solution contacts cut produce that is to be washed, the sanitizer will react with the organic matter (such as vegetable tissue, cellular juices, soil particles) and available (free) chlorine will be depleted. Pirovani et al. (2001) showed that for fresh-cut spinach, the difference between available chlorine and total chlorine concentration was very small, probably because of the minimum amount of organic cellular compounds released by cutting the produce into relatively large pieces. Microorganisms are not very sensitive to combined chlorine; however, they react instantly or within seconds to available chlorine in clean water. In general, higher concentrations of available chlorine result in more effective inactivation of microbial populations in shorter contact time.

Another means of improving the efficacy of chlorine treatments is to monitor the oxidation–reduction potential (ORP) of the process water and to use this value as a means of controlling the treatment by addition of more hypoclorite or pH adjustment. The ORP or REDOX measured by electrical conductivity across a pair of electrodes is a method used to determine chlorine concentration, particularly applicable to water used in dump tanks or for cleaning or cooling purposes. There are readily available commercial systems for in-line monitoring and application of chlorine to maintain water cleanliness and to monitor periodically the concentration of hypochlorous acid (Adeskaveg, 1995). An ORP value of 650 mV is recommended. Colorimetric test kits are commonly used and are the most accurate for routine evaluations. For postharvest use, however, inaccurate measurements can result because of other salts dissolved in the water from organic and soil particles washed off the commodity. The ORP is often used as a guide to detect sudden changes in conductivity that may influence free chlorine concentration.

Electrolyzed water is a special case of chlorination where hypochlorous acid is generated when water containing a small amount of sodium chloride is subjected to electrolysis (Izumi, 1999). In this case, the concentration of available chlorine is between 10 and 100 ppm and it has a high oxidation potential of between 1000 and 1150 mV (Sapers, 2003). The results of electrolyzed water treatment are controversial. Izumi (1999) detected only a 1-log reduction in the microbial population on fresh-cut vegetables, whereas Horton et al. (1999) reported a 3.7- to 4.6-log reduction on apples. Other studies observed no significant differences between acidic electrolyzed water (AcEW) and acidified chlorine treatments on inoculated

E. coli and *L. monocytogenes* on lettuce leaves (Park et al., 2001). The disinfectant effect of AcEW, ozonated water, and sodium hypochlorite (NaOCl) solution on lettuce has been examined (Koseki et al., 2001). It was discovered that AcEW showed a significantly higher disinfectant effect than did ozonated water. It was confirmed that AcEW, ozonated water, and NaOCl solution removed aerobic bacteria, coliform bacteria, moulds, and yeasts from the surface of lettuce. It was also shown that the bacterial spores on the surface were not removed by any of the treatments in this study. However, it was also observed that the surface structure of lettuce was not damaged by any of the treatments. Thus, the use of AcEW for decontamination of fresh lettuce is suggested to be an effective means of controlling microorganisms (Koseki et al., 2001). The effectiveness of AcEW in killing *E. coli* O157:H7, *Salmonella enteritidis*, and *L. monocytogenes* on the surfaces of tomatoes has been reported to be useful in controlling pathogenic microorganisms on fresh products (Bari et al., 2003).

In spite of the efficiency of chlorine against microorganisms suspended in water, its effect on the number of microorganisms on vegetables is limited. Published data indicate that the most that can be expected at permitted concentrations ($<$200 ppm) is a 1- to 2-log population reduction (Brackett, 1987; Zhuang et al., 1995; Wei et al., 1995; Zhang and Farber, 1996; Beuchat et al., 1998; Sapers et al., 1999; Pirovani et al., 2001, 2004). Populations of *E. coli* inoculated onto lettuce leaves and broccoli florets were generally reduced $<$1 log CFU g^{-1} after a 5 min dip in 100 ppm free chlorine compared to a plain water dip (Behrsing et al., 2000). The results of Mazollier (1988) indicate that microbial reductions on leafy salad greens were essentially the same when treated with 50 or 200 ppm chlorine. Treatment of honeydew melons and cantaloupes with 200 ppm hypochlorite significantly reduced surface microbial populations compared to water-washed controls (Ayhan et al., 1998). The surface-active agent Tween 80 (80 to 100 mg/l) improved the activity of 100 mg/l chlorine slightly, but caused deterioration of lettuce leaves (Adams et al., 1989). Although a reduction of pH of the chlorine solution to between 4.5 and 5.0 increased lethality up to fourfold, longer wash times (from 5 to 30 min) did not result in increased removal of microorganisms.

Effects of chlorine on bacterial pathogens inoculated onto produce have been investigated with mixed results. Survival of *E. coli* O157:H7 on cut lettuce pieces after submersion for 90 s in a solution of 20 ppm chlorine at 20 or 50°C was not significantly different from the nonchlorine treatment (Li et al., 2001a, b). Ten-minute exposures of *Yersinia enterocolitica* on shredded lettuce to 100 and 300 ppm chlorine resulted in population reductions of roughly 2 to 3 logs (Escudero et al., 1999). Results at 4 and 22°C were not significantly different. In this same study, a combination of 100 ppm chlorine and 0.5 percent lactic acid inactivated *Y. enterocolitica* by more than 6 log. These results suggest that *Y. enterocolitica* may be more sensitive to chlorine than some other pathogens. Effects of sodium hypochlorite on *L. monocytogenes* have been widely studied *in vitro* and were shown to be very effective (Brackett, 1987; Tuncan, 1993). Zhang and Farber (1996) examined the effects of chlorine (200 ppm, 10 min) against *L. monocytogenes* on fresh-cut vegetables. The maximum observed log reductions of *L. monocytogenes* at

4 and 22°C, respectively, were 1.3 and 1.7 for lettuce and 0.9 and 1.2 for cabbage. Thus, the bactericidal effect on *L. monocytogenes* was higher at 22°C than at 4°C, and was higher on lettuce than on cabbage. However, reductions were only marginally greater when exposure time was increased from 1 to 10 min.

Brackett (1987) examined the effects of chlorine (200 mg/l) on brussel sprouts contaminated with *L. monocytogenes*. The counts of *L. monocytogenes* were reduced approximately 100-fold, 10-fold more than those treated with water. However, the elimination of *L. monocytogenes* from the surface of vegetables by disinfection is limited and unpredictable (Nguyen-the and Carlin, 1994). This ineffectiveness may be a result of a number of factors – an aqueous hypochlorite solution may not wet the hydrophobic surface of the waxy cuticle or penetrate into the crevices, creases, and natural openings of the vegetable (Adams et al., 1989). When inoculated into cracks of mature green tomatoes, *Salmonella montevideo* survived treatment with 100 ppm chlorine (Wei et al., 1995). Zhuang et al. (1995) examined the efficacy of chlorine treatment on inactivation of *S. montevideo* on and in tomatoes. Populations of *S. montevideo* on the surface and in the core tissues were substantially reduced by dipping tomatoes in solutions containing 60 or 110 ppm chlorine, respectively. The most useful effect of chlorine may be in inactivating vegetative cells in washing water and on equipment during processing as part of an HACCP system, thus avoiding buildup of bacteria and cross-contamination (Hurst, 2002).

Treatment of produce with higher concentrations of chlorine (>500 ppm) has also been studied, but in this case, a water rinse is necessary for fresh-cut produce. Beuchat et al. (1998) showed that the maximum reduction in human pathogen populations on apples, tomatoes, and lettuce was 2.3 log CFU cm^{-2} after dipping in solutions of 2000 ppm chlorine for 1 min. On fresh-cut cantaloupe cubes, 2000 ppm chlorine resulted in less than a 90 percent reduction in viable cells of several strains of *Salmonellae* (Beuchat and Ryu, 1997). Populations of *Salmonellae* or *E. coli* O157:H7 inoculated onto the surfaces of cantaloupes and honeydew melons were reduced between 2.6 and 3.8 log CFU (as compared to a water wash control) when treated for 3 min with 2000 ppm sodium hypochlorite or 1200 ppm acidified sodium chlorite (Park and Beuchat, 1999). Total microbial populations were reduced about 1000-fold when lettuce was dipped in water containing 300 ppm total chlorine, but no effect was seen against microbial populations on red cabbage or carrots (Garg et al., 1990). Coliform bacteria were reduced by 81 percent on parsley, 93 percent on lettuce, 98 percent on strawberries, and 85 percent on coriander after a 10-min contact time in a solution of 300 ppm chlorine (Lopez et al., 1988). However, the microbial population of cut potato strips was not effectively controlled by dips in 300 ppm hypochlorite (Gunes et al., 1997).

Recently, the greatest reductions in *Salmonella typhimurium* and *E. coli* counts were obtained after treatment with 0.2 and 0.3 percent cetylpyridinium chlorine (CPC) sprays (Yang et al., 2003). The CPC was applied in a laboratory-scale immersion spray system at spray pressures of 0.7 and 2.1 kg cm^{-2} with a spray time of 1.5 min. Increasing the spray pressure to 2.1 kg cm^{-2} led to further decreases in *S. typhimurium* and *E. coli* counts, and no color changes were observed in the lettuce after CPC treatment.

In relation to postdisinfection contamination, some authors have reported that disinfection of salad leaves or tomato slices with chlorine prior to inoculation with *L. monocytogenes* did not affect its subsequent growth during storage (Beuchat and Brackett, 1990, 1991). By contrast, Carlin et al. (1996) found that *L. monocytogenes* inoculated onto disinfected (10 percent hydrogen peroxide) endive leaves grew better than on water-rinsed produce, and gave higher counts after 7 days at 10°C than on leaves rinsed with water. Similarly, Francis and O'Beirne (1997) found that dipping shredded lettuce in chlorine (100 ppm) or citric acid (1 percent) solution followed by storage at 8°C, resulted in significant increases in *Listeria* populations compared with undipped samples.

Because chlorine reacts with organic matter, components leaching from tissues of cut produce surfaces may neutralize some of the chlorine before it reaches microbial cells, thereby reducing its effectiveness. As mentioned before, crevices, cracks, and small fissures in produce, along with the hydrophobic nature of the waxy cuticle on the surfaces of many fruits and vegetables, may prevent chlorine and other sanitizers from reaching the microorganisms. The inconsistencies of the results may be attributed also to experimental differences in the interval between contamination and treatment, leading to differences in degree and strength of bacterial attachment and the opportunity for biofilm formation (Sapers, 2003; Burnett et al., 2004; Lang et al., 2004; Ukuku and Fett, 2004).

Safety concerns about the reaction of chlorine with organic residues in the formation of potentially mutagenic or carcinogenic reaction products, such as trihalomethanes (Chang et al., 1988; Parish et al., 2003), and their impact on human and environmental safety have been raised in recent years. For this reason, its use for treatment of minimally processed vegetables is banned in several European countries, including Germany, The Netherlands, Switzerland, and Belgium. This is a cause of concern, because some restrictions in the use of chlorine might be implemented in other countries and therefore other alternatives to chlorine must be investigated.

Chlorine Dioxide and Acidified Sodium Chlorite. Chlorine dioxide is another antimicrobial agent for produce washing that has been evaluated as a postharvest sanitation treatment (Spotts and Peters, 1980; Roberts and Reymond, 1994). It is more effective against many classes of microorganisms at lower concentrations than free chlorine. The major advantages of chlorine dioxide (ClO_2) over HOCl include reduced reactivity with organic matter and greater activity at neutral pH. However, stability of chlorine dioxide may be a problem. It produces fewer potentially carcinogenic chlorinated reaction products than chlorine (Tsai et al., 1995; Rittman, 1997), although its oxidizing power is reported as 2.5 times that of chlorine (Apel, 1993; White, 1992). Unfortunately, there are several disadvantages to chlorine dioxide (White, 1992):

1. Chlorine dioxide cannot be transported and needs to be generated on site; thus, it is generally more expensive than chlorine.

2. Some generators produce free chlorine in addition to chlorine dioxide, which may react with organic material to form trihalomethanes (THM).
3. Although chlorine dioxide does not produce THM, the compound produces its own set of disinfection breakdown products (e.g., chlorite, chlorate), which may pose a direct threat to human health.
4. Simple assays for routine evaluations of concentration are not available.
5. Chlorine dioxide is toxic to humans and it commonly gives off gases from the washwater, causing noxious odors. Thus, it must be used in closed systems, outdoors, or in well-ventilated areas away from packing house workers.

Chlorine dioxide is approved for use at a maximum of 200 ppm for sanitizing of processing equipment and 3 ppm maximum for use on uncut produce (21CFR173.300). Only 1 ppm maximum is permitted for peeled potatoes. A potable water rinse or blanching, cooking, or canning must follow treatment of produce with chlorine dioxide.

There is less information about the effectiveness of ClO_2 than HOCl as a sanitizer for produce. As with HOCl, microbial susceptibility to ClO_2 differs with strain and environmental conditions of application. A population of *L. monocytogenes* inoculated onto shredded lettuce and cabbage leaves was reduced an additional 1.1 and 0.8 log at 4 and 22°C, respectively, after treatment with 5 ppm ClO_2 for 10 min when compared to washing in tap water (Zhang and Farber, 1996). Use of ClO_2 gas reduced the numbers of *E. coli* O157:H7 on injured green pepper surfaces (Han et al., 2000). Treatment of surface-injured green peppers with 0.6 and 1.2 ppm ClO_2 gas reduced populations of *E. coli* O157:H7 by 3.0 and 6.4 log cycles, respectively. These researchers noted that no significant growth of *E. coli* O157:H7 was observed on uninjured pepper surfaces, but significant growth occurred on injured pepper surfaces within 24 h at 37°C. Costilow et al. (1984) reported that 2.5 ppm ClO_2 was effective against microorganisms in washwater, but concentrations as high as 105 ppm did not reduce the microflora in or on cucumbers. Similar results were reported by Reina et al. (1995). Immersion of oranges in 100 ppm chlorine dioxide at 30°C for 8 min produced a 3-log reduction of nonpathogenic *E. coli* compared to about a 2-log reduction when immersed in deionized water only (Pao and Davis, 1999).

Acidified sodium chlorite has been approved for use on certain raw fruits and vegetables as either a spray or dip in the range 500–1200 ppm (21CFR173.325). Reactive intermediates of this compound are highly oxidative, with broad-spectrum germicidal activity. Park and Beuchat (1999) showed that acidified sodium chlorite achieves 3-log reduction of *E. coli* O157:H7 and *Salmonellae* inoculated onto cantaloupes, honeydew melons, and asparagus spears.

Ozone. Ozone is one of the several new sanitizing agents recently used in postharvest management systems (Graham, 1997; Xu, 1999). Ozone is effective in reducing bacterial populations in flume and washwater and may have some applications as a chlorine replacement in reducing microbial populations on produce (Sapers, 2003).

This compound is one of the strongest oxidizing agents commonly available (Bryant et al., 1992). Ozone is unstable at ambient temperatures and pressures, with a half-life of about 15 min, and it decomposes to oxygen. Therefore, ozone must be generated on site by passing air or oxygen through a corona discharge or UV light (Xu, 1999). One of the major advantages for ozone is that it does not directly form trihalomethanes (THM), although it may indirectly form them if halogens are present in the washwater. Levels of 0.5 to 4.0 μg/ml are recommended for washwater and 0.1 μg/ml for flume water (IFPA, 2002; Strasser, 1998). Ozone is generally not affected by pH within a range of 6–8, but its decomposition increases with high pH, especially above pH 8. Disinfection, however, may still occur at a high pH, because the biocidal activity of the compound is relatively rapid (White, 1992). The most serious drawbacks of ozone include:

1. It is lethal to humans with continuous exposure at high concentration (>4 ppm); thus, detectors are required for automatic shutdown of ozone generator (Tukey, 1993) and the room must be adequately vented to avoid worker exposure.
2. It has a corrosive nature to common materials, therefore often requiring stainless steel containers to be used.
3. Filtering of ozonated water is required to remove organic and particulate materials to be effective.

Ozone is an effective treatment for drinking water and it is widely used in bottled water purification system as an alternative to chlorine (Peeters et al., 1989; Korich et al., 1990; Finch and Fairbairn, 1991; Restaino et al., 1995). According to Restaino et al. (1995), bacterial pathogens such as *Salmonella typhimurium*, *Y. enterocolitica*, *S. aureus*, and *L. monocytogenes* are sensitive to treatment with 20 ppm ozone in water.

The use of ozone as an antimicrobial agent in food processing has been reviewed by Kim et al. (1999a) and Xu (1999). However, little has been reported about the inactivation of pathogens on produce. Treatment with ozonated water can extend the shelf-life of apples, grapes, oranges, pears, raspberries, and strawberries by reducing microbial populations and by oxidation of ethylene to retard ripening (Beuchat, 1998). Microbial populations on berries and oranges were reduced by treatment with 2–3 ppm and 40 ppm, respectively. Kim et al. (1999b) reported a 2 log g^{-1} reductions in total counts for shredded lettuce suspended in ozonated water with 1.3 mM ozone at a flow rate of 0.5 l/min.

The use of ozone gas could be another alternative to reduce microbial populations on the product prior to cutting. Scientists investigated the use of ozone gas during the storage of grapes and apples in an atmosphere containing ozone and both products showed a fungal growth reduction and increase in their shelf-life (Bazarova, 1982; Sarig et al., 1996). Fungal growth during storage of blackberries was also inhibited by 0.1–0.3 ppm ozone (Barth et al., 1995). The spoilage microorganisms of vegetables such as onions, potatoes, and sugar beets were reduced after storage in an ozone-containing atmosphere (Kim et al., 1999b). However, because of its strong oxidizing activity, ozone may also cause physiological injury of produce (Horvath et al., 1985). Bananas treated with ozone developed black spots after 8 days of exposure to 25–30 ppm gaseous ozone. Carrots exposed to ozone gas during storage had a lighter and less intense color than untreated carrots (Liew and Prange, 1994).

Recently, Garcia et al. (2003) reported the effects of treatment with ozone in combination with chlorine on the microbiological and sensory attributes of lettuce. They also reported the quality of the chlorine used for processing commercial lettuce salad. Chlorine, ozone, and chlorine–ozone mixtures reduced aerobic plate counts by up to 1.4, 1.1, and 2.5 log, respectively. As determined by appearance, commercial lettuce salads treated with chlorine, ozone, or a combination of both had a shelf-life of 16, 20, or 25 days, respectively. Water quality remained constant for longer periods and could, therefore, be reused for longer. Chlorine–ozone combinations may have beneficial effects on the shelf-life and quality of lettuce salads, as well as on the water used for rinsing or cleaning the lettuce (Garcia et al., 2003).

This report does not specifically address the antimicrobial effects of combinations of various mitigation strategies; however, it would be expected that combinations of sanitizers would have additive, synergistic, or antagonistic interactions (Parish and Davidson, 1993). For instance, the effect of prewashing procedures with chlorinated and ozonated water (Baur et al., 2004). The suitability of chlorinated and ozonated water for prewashing trimmed lettuce heads and washing shredded lettuce was assessed in terms of effects on sensory properties and microbiological quality. Initial counts of aerobic mesophiles and pseudomonas increased with storage time in all samples, although the lowest increases were obtained by washing samples in chlorinated water. Lettuce samples subjected to prewashing and/or washing showed slight edge but no surface browning. Washing in chlorinated water, whether before or after shredding, reduced browning and retarded the decline in overall visual quality relative to that observed in lettuce washed in tap water or left unwashed. Results using ozonated water were also inferior. Lettuce crispness, color and aroma were not markedly affected by washing procedure, although off odor washing in chlorinated water reduced development. Prewashing of lettuce heads with chlorinated, or to a lesser extent ozonated, water markedly reduced counts of *Enterobacteriaceae*, with virtually no resumption of growth upon storage.

Ozone is considered useful for treatment of process water, food contact surfaces, or whole produce, because it has an excellent ability to penetrate and does not leave a residue. An independent expert committee, sponsored by the Electric Power Research Institute, recommended that use of ozone as a disinfectant or food sanitizer be classified as "Generally Recognized As Safe (GRAS)" (Graham, 1997). Also considered are the potential for ozone to reduce pesticide levels in fruits and vegetables. However, the possible negative impact of ozone treatment on fruit and vegetable nutritional quality (Kong Fanchun and Shengmin Wang, 2003) warrants further study.

Halogen-Based Treatments. Little is known about the usefulness of halogen-based treatments (e.g., bromine, iodine) as sanitizers for produce. Some authors observed a synergistic antimicrobial relationship when bromine was added to chlorine solutions (Kristofferson, 1958; Shere et al., 1962). However, they are more costly, and, as with free chlorine, there are safety concerns about the production of brominated organic compounds and their impact on human and environmental safety.

The use of iodine-containing solutions as direct-contact sanitizers for produce is further limited because of a reaction between iodine and starch that results in a blue–purple color. Despite these limitations, iodine solutions such as iodophors (combinations of elemental iodine and nonionic surfactants or carriers) are used as equipment sanitizers by the meat industry. Iodophors are also used to sanitize utensils and as employee sanitizing dips. The dark color change indicates the sanitizer has been used up (Bartlett and Schmidt, 1957; Hays et al., 1967; Mosley et al., 1976; Lacey, 1979; Jilbert, 1988). Although iodophors are not approved for direct food contact, they might have some usefulness for treatment of produce items that are peeled before consumption. This type of use would require regulatory approval and a demonstration that produce treated by these compounds is safe for consumption.

Quaternary Ammonium-Based Sanitizers. Quaternary ammonium compounds or "quats," are cationic surfactants that are odorless, colorless, stable at high temperatures, noncorrosive to equipment, nonirritating to skin, and able to penetrate food contact surfaces more readily than other sanitizers (Walker and LaGrange, 1991). They are used primarily on floors, walls, drains, and on aluminum equipment. They are effective against *L. monocytogenes* and slightly more effective against molds than chlorine compounds. They have good penetrating ability because of their high surface-active capability and the mechanism of activity for quats possibly involves a breakdown of the cell membrane/wall complex. A unique feature is their ability to form a residual bacteriostatic film when applied to most hard surfaces. Quats are relatively stable to organic compounds. They are effective over a wide pH range (6–10). The major disadvantage of quats is that they have limited germicidal activity against coliforms. Although they are not approved for direct food contact, quats may have some usefulness with whole produce that must be peeled prior to consumption. As with iodine compounds, direct food contact would require regulatory approval and a demonstration that produce treated by quats is safe for consumption.

Acidic-Based Sanitizers. Acidification of low-acid products may act to prevent microbial proliferation, because many pathogens generally cannot grow at pH values below 4.5 (Parish et al., 2003). Most fruits naturally possess significant concentrations of organic acids such as acetic, benzoic, citric, malic, sorbic, succinic, and tartaric acids, which negatively affect the viability of contaminating bacteria. Other fruits, such as melons and papayas, as well as the majority of vegetables, contain lower concentrations of organic acids and pH values above 5.0 do not suppress the growth of pathogenic bacterial contaminants.

Lactic acid rinses are being recommended for decontamination of fruits and vegetables. Total plate count reductions of 1 to 1.5 log were reported on endive rinsed with 1.9 percent lactic acid for 1.5 min (Sapers, 2003). Total and fecal coliforms were reduced about 2 and 1 log g^{-1}, respectively, on mixed salad vegetables treated with 1 percent lactic acid (Torriani et al., 1997). Treatment with citric acid in the form of lemon juice has been shown to reduce populations of *Salmonella typhi* inoculated onto cubes of papaya and jicama (Fernandez Escartin et al., 1989). The antimicrobial activity of lemon juice against *Campylobacter jejuni* was investigated on cubes of watermelon and papaya (Castillo and Escartin, 1994). Six hours after treatment, populations of *C. jejuni* ranged from 0 to 14.3 percent of the original inoculum on cubes treated with lemon juice, and from 7.7 to 61.8 percent on cubes not treated with lemon juice. Acetic acid has been tested as an antimicrobial agent for apples (Sapers, 2003). Apples that had been inoculated with *E. coli* O157:H7 and air-dried for 30 min were treated with 5 percent acetic acid at 55°C for 25 min. Although the *E. coli* population was greatly reduced in the apple skin and stem areas, as many as 3 to 4 log survived in the calyx tissue (Delaquis et al., 2000a).

In another study, acetic acid was used to inactivate pathogenic bacteria on fresh parsley (Karapinar and Gonul, 1992). Populations of *Y. enterocolitica* inoculated onto parsley leaves were reduced >7 log cycles after washing for 15 min in solutions of 2 percent acetic acid or 40 percent vinegar. Treatment of whole parsley leaves for 5 min at 21°C with vinegar (7.6 percent acetic acid) reduced populations of *Shigella sonnei* more than 7 log per gram (Wu et al., 2000). Effects of organic acids as disinfectants on the growth of *L. monocytogenes* on produce, has received little attention. Lactic and acetic acid (1 percent) reduced numbers of *L. monocytogenes* on lettuce by only 0.5 and 0.2 log cycles, respectively, and were therefore not very effective (Zhang and Farber, 1996).

Various combinations of acetic acid, lactic acid, and chlorine were observed to reduce populations of *L. monocytogenes* on shredded lettuce (Zhang and Farber, 1996). Lactic or acetic acids in combination with 100 ppm chlorine were slightly more antagonistic toward *L. monocytogenes* than either acid or chlorine alone. A 2 min dip in 5 percent acetic acid at room temperature was the most effective treatment of several investigated for reducing populations of *E. coli* O157:H7 inoculated onto apple surfaces (Wright et al., 2000). The 5 percent acetic acid treatment reduced the population more than 3 log CFU/cm^2 as compared to less than a 3 log reduction by a commercial preparation with 80 ppm peroxyacetic acid. The effects of different disinfection treatments (6, 25, and 50 percent vinegar, 2 and 4 percent acetic acid, 80 ppm peracetic acid, 200 ppm sodium hypochlorite, and 200 ppm sodium dichloroisocyanurate) on the natural microbiota of lettuce were compared in terms of their effectiveness against the natural microbiota of lettuce (Nascimento et al., 2003). The results showed that the effectiveness levels for all sanitizing agents tested were equivalent to or higher than that for sodium hypochlorite at 200 ppm.

Organic acids, in particular vinegar and lemon juice, have potential as inexpensive, simple household sanitizers; however, it is not clear whether organic treatments would produce off-flavors in treated produce.

Alkaline-Based Sanitizers. Various high-pH cleaners containing sodium hydroxide, potassium hydroxide, sodium bicarbonate, and/or sodium orthophenylphenate (with or without surfactants) reduced populations of *E. coli* on orange surfaces (Pao et al., 2000). These same researchers determined that high-pH waxes used on fresh market citrus provided substantial inactivation of *E. coli* on orange fruit surfaces (Pao et al., 1999).

Trisodium phosphate (TSP) has been shown to be effective against *E. coli* and *Salmonella*. It has been used experimentally as a wash to decontaminate tomatoes (Zhuang and Beuchat, 1996) and apples (Sapers et al., 1999). Experimental applications for disinfection of fruits and vegetables have employed concentrations as high as 12 percent (Sapers, 2003). Treatment with 8 percent TSP decreased populations of *L. monocytogenes* only 1 log cycle. Resistance of *L. monocytogenes* to TSP was also reported by Zhang and Farber (1996). A 5-min treatment with 2 percent TSP produced 1 log reduction of *Salmonella chester* attached to the surface of apple disks (Liao and Sapers, 2000). *Salmonella* populations on the surface of tomatoes were reduced from 5.2 log CFU/cm^2 to nondetectable levels after 15 s in 15 percent TSP (Zhuang and Beuchat, 1996). A significant reduction in population was observed after 15 s in 1 percent TSP.

Trisodium phospate is classified as GRAS (21CFR182.1778) when used in accordance with GMPs (Sapers, 2003). The high pH of typical alkaline wash solutions (11–12) and concerns about environmental discharge of phosphates may be limiting factors for use of certain alkaline compounds on produce.

Peroxyacetic Acid. Peroxyacetic acid, or peracetic acid as it is sometimes called, is actually an equilibrium mixture of the peroxy compound, hydrogen peroxide and acetic acid (Ecolab, 1997; Ecolab, cited by Sapers, 2003). Peroxyacetic acid is approved for addition to washwater (21CFR173.315). It decomposes into acetic acid, water, and oxygen, all harmless residuals. It is a strong oxidizing agent and can be hazardous to handle at high concentrations, but not at strengths marketed to the produce industry. The superior antimicrobial properties of peroxyacetic acid are well known (Block, 1991). This agent is recommended for use in treating process water, but one of the major suppliers is also claiming substantial reductions in microbial populations on produce surface (Sapers, 2003). However, the only claim with cut corn was a 1 log reduction (Ecolab, 1997). The efficacy of peracetic acid against microorganisms on produce has not been reported extensively. When used at 40 and 80 ppm, a sanitizer that contains peracetic acid (Tsunami™ Ecolab, Mendota Heights, MN) significantly reduced *Salmonellae* and *E. coli* O157:H7 populations on cantaloupe and honeydew melon surfaces (Park and Beuchat, 1999). These treatments were less effective on asparagus spears. The brand of sanitizer used in this study is reported by the manufacturer to maintain its efficacy over a broader pH range and organic demand than hypochlorite, although it is more expensive. Nearly 100-fold reductions in total counts and fecal coliforms on cut-salad mixtures were observed after treatment with 90 ppm peroxyacetic (peracetic) acid or with 100 ppm chlorine (Masson, 1990). The subsequent inhibition of microbial growth during storage of salads was attributed to residual peracetic activity. Sanitizer formulation of peroxyacetic acid in combination with surfactants reduced populations of *Salmonella javiana*, *L. monocytogenes*, and *E. coli* O157:H7 by 96, 99.96, and 99.5 percent, respectively, compared with treatment in sterile water (Parish et al., 2003).

Hydrogen Peroxide. Hydrogen peroxide (H_2O_2) is a strong oxidizing agent against a wide range of bacteria, but is less active against fungi (Block, 1991). The properties are a result of its capacity to generate other cytotoxic oxidizing species such as hydroxyl radicals. Hydrogen peroxide is GRAS for some food applications (21CFR184.1366), but has not yet been approved as an antimicrobial wash agent for produce. It produces no residue because it is broken down to water and oxygen by catalase (Sapers, 2003). Juven and Pierson (1996) reviewed the antimicrobial activity of H_2O_2 and its use in the food industry. Use of H_2O_2 on whole and fresh-cut produce has been researched in recent years with an extensive investigation by Sapers (2003). Solutions of 5 percent hydrogen peroxide alone or combined with commercial surfactants can achieve substantially higher log reductions for inoculated apples than 200 ppm chlorine (Sapers et al., 1999). Hydrogen peroxide at concentration of 3 percent, alone or in combination with 2 or 5 percent acetic acid, sprayed onto green peppers, reduced *Shigella* populations approximately 5 log cycles, compared to less than a 1-log reduction by water alone (Peters, 1995).

In the same study, *Shigella* inoculated onto lettuce was reduced approximately 4 logs after dipping in H_2O_2 combined with either 2 or 5 percent acetic acid; however, obvious visual defects were noted on the treated lettuce. The same treatment gave similar results for *E. coli* O157:H7 inoculated onto broccoli florets or tomatoes with minimal visual defects. Treatment by dipping in H_2O_2 solution reduced microbial populations on fresh-cut bell peppers, cucumber, zucchini, cantaloupe, and honeydew melon, but did not alter sensory characteristics. Hydrogen peroxide has been investigated as a sanitizing method for fruits and vegetables infected with human pathogens to reduce the risk of produce-related foodborne illness. The efficacy of 1 percent hydrogen peroxide wash in decontaminating apples and cantaloupe melons containing *E. coli* was investigated (Sapers and Sites, 2003). The application of 1 percent H_2O_2 is an effective decontamination technique for *E. coli* infected apples. The same technique is less effective in cantaloupes.

Vapor treatments have been used to inhibit postharvest decay in some fruits and vegetables. Microbial populations on whole cantaloupes, grapes, prunes, raisins, walnuts, and pistachios were significantly reduced upon treatment with H_2O_2 vapor (Forney et al., 1991; Aharoni et al., 1994; Simmons et al., 1997; Sapers and Simmons, 1998; Ukuku and Sapers, 2001; Ukuku et al., 2001). However, hydrogen peroxide is phytotoxic to some commodities, causing browning in lettuce and mushrooms and anthocyanin bleaching in raspberries and strawberries

(Sapers and Simmons, 1998). Recently, a study on the consumer acceptance of fresh-cut iceberg lettuce treated with 2 percent hydrogen peroxide and mild heat has been published (McWatters et al., 2002). Surprisingly, the antibacterial treatment was more effective than the control treatment in maintaining sensory quality over 15 days of storage. This was so provided that the lettuce was initially intensely green. Of the participants, 75 percent indicated that they would be willing to buy precut packaged lettuce that had already been treated at the packing house or processing plant with an antibacterial solution (McWatters et al., 2002).

Irradiation. Ionizing radiation from ^{60}Co, ^{137}Cs, or machine-generated electron beams, alone or in combination with other treatments such as hot water, is used as a means of extending the shelf-life of produce (Diehl, 1995; Thayer et al., 1996). Irradiation with doses ranging from 0.5 to 2 kGy had been described to have no adverse effect on products stored a few days under refrigeration (Nguyen-the and Carlin, 2000). At 2 kGy, numbers of bacteria were usually reduced by 3 to 4 log cycles and yeast by 1 to 2 log cycles (Moy, 1983; Urbain, 1986; Farkas, 1997). Most medium- and high-level doses are not appropriate for produce because they can cause sensory defects (visual, texture, and flavor) and/or accelerated senescence due to irreparable damage to DNA and proteins (Thomas, 1986; Barkai-Golan, 1992). The shelf-life of raw vegetables is longer than that of a fresh-cut product, which may allow the development of damage induced by irradiation. In addition, produce treated by doses above the level of 1 kGy cannot use the term "fresh" (21CFR101.95).

Combinations of ionizing radiation with other treatments have been studied. A combination of 0.75 kGy irradiation with a 10 min dip in 50°C water provided much better control of postharvest spoilage organisms of papayas and mangoes than either treatment alone (Brodrick and van der Linde, 1981). A combination of UV and gamma radiation was not more effective than either treatment alone at preventing storage rot of peaches (Lu et al., 1993). Irradiation (0.43 kGy average dose) of segments from cut and peeled citrus fruits was not as effective as chemical preservatives at preventing spoilage during chilled storage (Hagenmaier and Baker, 1998a).

The shelf-life of packaged leaf vegetables stored at 10°C was extended by treatment with 1 kGy (Langerak, 1978). In this study, *Enterobacteriaceae* were eliminated on endive and the shelf-life was extended from 1 (for nonirradiated) to 5 days. Chervin and Boisseau (1994) reported that irradiation of shredded carrots was superior to chlorination and spin-drying. Microbial populations (measured as total plate counts) of shredded carrots treated with 0.5 kGy or chlorine and stored 9 days under refrigeration were 1300 and 87,000 CFU/g, respectively (Hagenmaier and Baker, 1998b). The same authors reported a similar reduction of microbial populations on cut iceberg lettuce treated with 0.19 kGy (Hagenmaier and Baker, 1997). A combination of hot water dips and 1.0 kGy irradiation doubled the shelf-life of mangoes from 25 to 50 days (El-Samahy et al., 2000). Effects of various doses of gamma irradiation on the sensory and physicochemical quality of fresh-cut iceberg lettuce were investigated in order to determine a suitable maximum dose (Xuetong and Sokorai, 2002). It is proposed that irradiation of fresh-cut lettuce in modified atmosphere packages at doses of 1 kGy, and perhaps 2 kGy, for safety and quality improvement, is feasible.

Warm water treatment in combination with modified atmosphere packaging reduces undesirable effects of irradiation on the quality of fresh-cut iceberg lettuce (Xuetong et al., 2003). These authors have observed that dipping cut lettuce at 47°C for 2 min prior to irradiation reduced accumulation of antioxidant compounds and phenols induced by irradiation. Lettuce treated with warm water and irradiated at 0.5 or 1 kGy retains good sensory quality with no marked loss of texture, vitamin C, or total antioxidant compounds (Xuetong et al., 2003).

Biocontrol. There are few published reports on the use of biocontrol agents to prevent growth of human pathogens on produce (Parish et al., 2003). In studies on onion, tomato, and cabbage, antagonists were isolated from the vegetables themselves, indicating that the epiphytic microflora of vegetables could be a natural barrier against microbial spoilage (Droby et al., 1991). The mode of action of antagonist has not been fully explained. It probably involves competition for nutrients, direct interaction with the pathogen (i.e., adhesion to the mycelium of the fungal pathogen and destruction), or induction of a defense reaction from the host tissues (Parish et al., 2003). The efficacy of microbial antagonist could be improved by the addition of nutrients, salts, and so on. Populations of *L. monocytogenes* inoculated onto endive leaves were inhibited by treatment with a mixed population of microorganisms originally isolated from endive (Carlin et al., 1996). Inoculation of *Lactobacillus casei* reduced multiplication of mesophilic bacteria (Vescovo et al., 1995). The introduction of *L. casei* caused a significant decrease in pH of the product, and, unfortunately, the authors gave no indication of the appearance of treated products. Strains of lactic acid bacteria were reported also to inhibit *A. hydrophila*, *L. monocytogenes*, *S. typhimurium*, and *S. aureus* on vegetable salads (Vescovo et al., 1996).

The application of microorganisms to prevent proliferation of postharvest spoilage organisms has been studied in its effect on extending shelf-life, more than to control human pathogens on produce surfaces (Liao, 1989; Smilanick and Denis-Arrue, 1992; Stanley, 1994; Janisiewicz and Bors, 1995; Korsten et al., 1995; Leibinger et al., 1997; Calvente et al., 1999; El-Ghaouth et al., 2000; Usall et al., 2000). These studies suggest that nonpathogenic microorganisms applied to produce surfaces might outcompete pathogens for physical space and nutrients, and/or may produce antagonistic compounds that negatively affect viability of pathogens. The use of bacteriophages to reduce population of *Salmonella* on fresh-cut fruit has been reported (Leverentz et al., 2001). Application of *Salmonella*-specific phages reduced populations about 3.5 log on honeydew melon slices (pH 5.8) stored at 5 or 10°C. *Salmonellae* were not reduced on apple slices, possibly because of the lower pH (4.2) of the fruit.

Plant-Derived Antimicrobial Compounds. Numerous plant-derived compounds with antimicrobial properties have been studied for use in food systems

(Cherry, 1999). Although their usefulness may be limited because of undesirable sensory effects, naturally derived food compounds and essences have shown antimicrobial activity against human pathogens in laboratory studies (Parish et al., 2003). Compounds such as various bacteriocins, cinnamaldehyde, diacetyl, benzaldehyde, pyruvic aldehyde, piperonal, basil methyl charvicol, vanillin, psoralens, jasmonates, allylisothiocyanate, lactoferricin, hop resins, and essences of garlic, clove, cinnamon, coriander, and mint have been studied for antimicrobial activity in various food systems (Isshiki et al., 1992; Tokuoka and Isshiki, 1994; Bowles et al., 1995; Delaquis and Mazza, 1995; Lis-Balchin et al., 1996; Cerrutti et al., 1997; Ulate-Rodriguez et al., 1997; Bowles and Juneja, 1998; Buta and Moline, 1998; Wan et al., 1998; Chantaysakorn and Richter, 2000; Fukao et al., 2000). The efficacy of chlorine dioxide, ozone, and thyme essential oil or a sequential washing on the destruction of *E. coli* O157:H7 on lettuce and baby carrots has been investigated (Singh et al., 2002). After the evaluation of the treatments applied singly and sequentially, scientists concluded sequential washing in thyme essential oil followed by aqueous chlorine dioxide and ozonated water was very effective in killing the pathogen (3 to 4 log reduction of initial load), but may have adverse effects on sensory properties of the product.

Alternative Technologies. Ultraviolet (UV) light is a nonchemical disinfectant system that uses an extremely rapid physical light energy of a specific wavelength to destroy microorganisms. The germicidal properties of UV irradiation (UVC 200–280 nm) are a result of DNA mutations induced by DNA absorption of the UV light. To achieve microbial inactivation, the UV radiant exposure must be at least 400 J/m^2 in all parts of the product (Sastry et al., 2000). The microbial inactivation is directly related to the UV energy dose received. The FDA has approved UV technology for use as a disinfectant to treat food as long as the proper wavelength of energy is maintained (IFPA, 2002). Ultraviolet light systems are available for water sterilization and they leave no chemical residues. Ultraviolet light systems are not affected by water chemistry, but efficacy as a washwater disinfectant is significant impacted by turbidity and so requires clear water to be effective. Potentially, UV irradiation may be utilized for disinfecting washwaters if particulates are removed and flow rates are slow enough to allow an adequate irradiation for disinfectation.

Water itself can remove microbial contamination, but this method requires filtration to remove the debris from the water. The bactericidal effect of UV irradiation has recently been investigated on the surface of Red Delicious apples, green leaf lettuce, and tomatoes inoculated with cultures of *Salmonella* spp. or *E. coli* O157:H7 (Yaun et al., 2004). It is proposed that UVC irradiation may be beneficial in ensuring the safety of fruits and vegetables, because apples inoculated with *E. coli* O157:H7 experienced the highest log reduction (approx. 3.3 log) after UVC irradiation at 24 mW/cm^2 (Yaun et al., 2004). Lower log reductions were observed on tomatoes inoculated with *Salmonella* spp. (2.19 log) and leaf lettuce inoculated with either *Salmonella* spp. and *E. coli* O157:H7 (2.65 and 2.79 log, respectively).

Ultraviolet light may be used in combination with other alternative process technologies, including various powerful oxidizing agents, such as ozone and hydrogen peroxide.

Other nonthermal alternative technologies, such as high-pressure, pulsed electric field, pulsed light, oscillating magnetic fields, and ultrasound treatments, have been investigated to reduce or eliminate microorganisms (FDA/CFSAN, 2000). The effect of moderate thermal and pulsed electric field treatments on textural properties of carrots, potatoes, and apples have been investigated (Lebovka et al., 2004), but there are no findings suggesting their use in controlling microorganism on fresh products.

The effectiveness of power ultrasound for microbial decontamination of minimally processed fruits and vegetables was studied by Seymour et al. (2002) in samples including iceberg lettuce, whole cucumber, cut baton carrots, capsicum pepper, white cabbage, spring onion, strawberries, curly leaf parsley, mint, and other herbs. The cleaning action of cavitation removed cells attached to the surface of fresh produce, rendering the pathogens more susceptible to the sanitizer. For large-scale trials, addition of chlorine to water in the tank gave a systematic difference in *E. coli* decontamination efficiency. However, the frequency of ultrasound treatment (25, 32–40, 62–70 kHz) had no significant effect on decontamination efficiency. With the high capital expenditure, together with the expensive process of optimization and water treatment, it is unlikely that the fresh produce industry would take up this technology. Furthermore, the additional 1 log reduction achieved by applying ultrasound to a chlorinated water wash does not eliminate the risk of pathogens on fresh produce completely (Seymour et al., 2002).

The process temperature during treatment cannot be as high for a produce labeled "fresh" as it is, in most cases, for liquid foods. There is, for that reason, little published research directly related to the impact of these technologies on the safety of fresh whole or cut produce (FDA, 2000).

Other treatments currently being evaluated more extensively by researchers include hot water or ethanol and hot water washes for defined time periods. Hot water treatments have been described to reduce the extent of microbial contamination and improve quality retention in fruits and vegetables (Lurie, 1998). The antimicrobial efficacy of these treatments has been demonstrated in garlic, green onions, sprouts, watercress (Park et al., 1998), mandarins (Gonzales-Aguilar et al., 1997), and nectarines (Obenland and Aung, 1997).

Several studies have indicated that heat stress can also maintain quality and reduce physiological breakdown upon subsequent storage. Browning, the major factor in loss in quality of fresh-cut lettuce, can be reduced by dipping in warm (50°C) water (Loaiza-Velarde et al., 1997). The microbial stability of fresh-cut iceberg lettuce was enhanced by a 3 min dip in chlorinated water at 47°C (Delaquis et al., 1999). However, the heat processing induced changes in flavor of the lettuce, and a chlorinaceous off-odor was detected by some panelists (Delaquis et al., 2000b). On papayas and mangoes, hot water has been used to suppress quiescent infections of anthracnose (Colletotrichum), whereas hot air has also been used to control brown rot decay of peaches during ripening. Populations of mesophilic aerobic microflora were significantly reduced (by approx. 1.73 to 1.96 log) by treatment with chlorinated water at 50 or 20°C and by water without chlorine at 50°C (Li et al., 2001a). Treatment with chlorinated water at 50°C was also observed to delay browning of lettuce and significantly reduces initial populations of yeasts and molds. Storage at 5°C prolonged lettuce shelf-life in terms of color and general appearance, although populations of mesophilic aerobic microflora, psychrotrophs, and *Enterobacteriaceae* in samples stored at 5 and 15°C increased with time, regardless of water treatment. Levels of fungi and lactic acid bacteria remained $<2\ \log_{10}$ CFU g^{-1} during storage at both temperatures, and mold populations were lower than those of mesophilic anaerobes by approximately 3 log (Li et al., 2001b). A serious disadvantage of these processes is the energy cost of heating and then subsequent cooling the commodities for cold storage. Ethanol alone at 30 percent concentration or in combination with fungicides has been effective in reducing decay of peaches in storage (Ogawa and Lyda, 1960; Feliciano et al., 1992).

6.5.3 Packaging: Controlled and Modified Atmospheres

Packaging technology has allowed food processors to provide consumers with safer and fresher foods. Microbial safety of fresh-cut fruits and vegetables depends on many factors, from inherent ones, such as indigenous microflora and physical and chemical properties, to environmental consequences, such as storage temperature and handling by processors, distributors, and consumers.

Studies show that although CO_2 has benefits reducing total microorganisms in fresh-cut produce because of its fungi- and bacteria-static characteristics (Bai et al., 2001), it has little impact on some foodborne pathogens, such as *C. botulinum*, *Y. enterocolitica*, *S. typhimurium*, *E. coli*, *S. sonnei*, *Staphylococcus aureus*, *L. monocytogenes*, and *Campylobacter* (Silliker and Wolfe, 1980; Hintlian and Hotchkiss, 1986; Farber, 1991; Nguyen-the and Carlin, 1994; Francis et al., 1999). Hao and Brackett (1993) illustrated the failure of 10 percent CO_2 to inhibit *E. coli* O157:H7 at refrigeration temperatures. A 3 percent O_2 + 10 percent O_2 atmosphere, although maintaining acceptable visual quality of shredded lettuce, did not appreciably affect microbial development (Barriga et al., 1991). However, different results showing that 75 percent CO_2 (balance N_2) was effective on *S. typhimurium* and *P. fragi* were reported by Lee and Cash (1998). Furthermore, 100 percent CO_2 has an inhibitory effect on the growth of psychrotrophic pathogens such as *A. hydrophila*, but at concentrations that would cause spoilage of many minimally processed fresh vegetables (Kader et al., 1989).

Juneja et al. (1996) also found that 2–50 percent CO_2/20 percent O_2/30–55 percent N_2 inhibited *Clostridium perfringens* at 4°C. More recent experiments on the surface growth of vegetable-associated microorganisms including spoilage and pathogens concluded that exposure to high O_2 alone did not inhibit microbial growth strongly, although a consistently strong inhibition was observed when high O_2 and 10–20 percent CO_2 were used in combination (Amanatidou et al., 1999). Other novel elements of MAP inc1ude the use of high oxygen and noble gas enriched atmospheres. Although atmospheres with 80 percent oxygen have been used in MAP of fresh meat for a few decades and appear safe, less is known about the effects of noble gases such as argon on microbial ecology. The microbial quality of the final packaged product should be monitored to ensure that it complies with international guidelines (Francis et al., 1999).

For fresh produce products, the proper gas mixture can significantly slow down respiration rates, but it is the permeability of the film that maintains the equilibrated package atmosphere (Yuan, 2003). Packaging materials must be carefully selected to ensure that their gas permeability properties match the respiration rates of the products being packaged. This is necessary in order to achieve package atmospheres within the technically useful range of 2–5 percent O_2 and 3–10 percent CO_2 (Barry-Ryan et al., 2000; Cliffe-Byrnes et al., 2003). Also, in order to attain the desired effect on a packaged product, the gas mixture headspace volume must be carefully considered. After packaging for distribution, it is difficult to determine if product safety will be maintained when it reaches the consumer. Within this time frame, an undetected leak in the package could have permitted the entry of air or temperature abuse could have occurred due to careless actions or ignorance.

Hurdle and Barrier Technology. In order to improve the effectiveness of the controlled and modified atmosphere reducing microbial development, hurdle and barrier technology are used, where multiple factors or technologies are employed to control microorganisms in foods effectively. Therefore, recent thoughts are to incorporate functional intrinsic and extrinsic elements to address the food-safety issue proactively and to minimize or eliminate food-safety risks. The extrinsic elements cover the processing environment that will impact the packaging. The intrinsic elements, such as intelligent or active packaging, are directly related to the packaging.

Extrinsic Factors. Several factors can be applied to food systems as hurdles, and more minimally processed foods are likely to embody this concept to ensure food safety. Some alternative technologies being studied with MAP include gamma irradiation (Grant and Patterson, 1991a, b; Thayer and Boyd, 2000), ozone (Rodgers et al., 2004), and high-pressure processing (Amanatidou et al., 2000). They were found to have strong synergies with reductions in test microorganisms and quality enhancements. Also, barriers like water activity, and composition of the medium and storage temperature (Ellis et al., 1994) can be effective in controlling and minimizing the growth of foodborne pathogens. For instance, *in vitro* experiments showed that the effect of anaerobiosis on *L. monocytogenes* may vary according to the composition and the availability of nutrients in the product. At pH 4.5, anaerobiosis reduced the growth rate and extended the lag time of *L. monocytogenes* grown in a tryptone soy bean medium with yeast extract, whereas the opposite effect was seen in a less rich medium (Buchanan and Klawitter, 1990).

Modified atmosphere packaging (MAP) is effective only when the temperature is maintained well below a predetermined appropriate level, because the effectiveness of MAP has a reciprocal relationship with temperature (Marshall et al., 1991;

Harris and Barakat, 1995; Mano et al., 1995; Özbas et al., 1997; Harrison et al., 2000). Consequently, maintaining the cold chain intact, from preparation to handling to distribution, is a prerequisite for the success of MAP. For instance, storage of products below 0°C under vacuum or below 2°C under CO_2 should be sufficient to inhibit *Yersinia* (Gill and Reichel, 1989). Moreover, Gibson et al. (2000) have reported that 100 percent CO_2 slows the growth rate of *C. botulinum*, and that this inhibitory effect is further enhanced with appropriate NaCl concentrations and chilled temperatures. MAP fresh-cut produce does not have one principle hurdle for controlling *C. botulinum* growth. It is the additive effect of several hurdles consisting of refrigeration, a competitive aerobic and facultative spoilage microflora, and the product's physiological processes that collectively will cause overt spoilage, at a temperature of 10°C or higher, prior to the presence of botulinal toxin. At 10°C, no toxin production from proteolytic strains is detected through 16 weeks of storage. The multiple hurdle scenario for produce is consistent with Sperber's (1982) observation that rarely is just a single preservative (hurdle) involved in the protection of food against the development of botulism.

Chilling rate affects the rate of bacterial growth and the rate of cell deterioration on MAP fresh-cut products, which then impact sensory properties of food products. Chilling extends the life of food products in good condition by retarding the rate of deterioration. Chilling cannot improve the quality of a poor product or stop the process of spoilage. The endemic microflora on fresh-cut produce play an important role in signalling the end of shelf-life and are also believed to suppress toxin production by *C. botulinum* (Larson and Johnson, 1999). However, some products, such as butternut squash and onions, have been demonstrated under temperature abuse conditions to have the potential of being acceptable after detection of botulinal toxin (Austin et al., 1998).

The fact that botulinal toxin was present prior to spoilage in the Solomon et al. (1990) study emphasizes the need for the industry to remain cognizant of the potential risk, regardless of its level, of botulism in MAP produce. MAP produce, with the exception of untreated raw potatoes, should always use oxygen-permeable packaging materials to avoid an anoxic package environment. Furthermore, the elimination or significant inhibition of the growth of spoilage bacteria should not be practiced. These bacteria serve as important indicators of temperature abuse and are expected to effectively compete against the growth of *C. botulinum*.

Another way of applying effective hurdles with MAP is to combine process and product design factors, such as adding antimicrobial compounds (Szabo and Cahill, 1998), salt dipping (Pastoriza et al., 1998), nisin (Fang and Lin, 1994), or organic acid (Zeitoun and Debevere, 1991). MAP produce (being a low oxygen application but supporting a healthy aerobic microflora along with a substantial lactobacillus population) also qualifies as an application that can rely on a competitive microflora for controlling the growth of *C. botulinum*. Larson and Johnson (1994) have demonstrated the effectiveness of this system using commercial supplies of MAP lettuce, cabbage, green beans, carrots, and broccoli.

Although it is generally accepted that *C. botulinum* cannot grow in food exposed to atmospheric oxygen levels, the redox potential (E_h) of most foods is usually sufficiently low to facilitate growth and hence toxigenesis (Kim and Foegeding, 1993; Phillips Daifas et al., 1999; Lawlor et al., 2000). Optimal growth is evident at E_h of -350 mV (Smoot and Pierson, 1979), but is possible in the range of $+30$ to $+250$ mV. The growth of the aerobic spoilage microflora rapidly decreases food E_h, improving conditions for the growth of *C. botulinum*. Hyperbaric CO_2 levels inhibit *C. botulinum*, while growth stimulation under atmospheric CO_2 levels has been described (Sperber, 1982). Atmospheres modified passively through the respiration of fresh mushrooms packed in polyvinylchloride potentially create conditions suitable for growth (Sugiyama and Yang, 1975). These findings prompted the USFDA to recommend the puncturing of packaging films applied to mushrooms. More recently, Larson et al. (1997) identified a probability of <1 in 10^5 for toxin production to occur in MAP vegetables.

Intrinsic Factors. Intelligent or active packaging technologies, collectively called "interactive packaging", have been introduced for use in conjunction with existing packaging techniques such as MAP (Rooney, 1995; Day, 2000). Interactive packaging concepts involve various technologies incorporated into the package or the packaging film to maintain or monitor the quality or safety of the product. These may interact with the food products, alter the packaging atmosphere, or respond to environmental changes and communicate to consumers. For example, active packaging prolongs shelf-life because the removal of oxygen inhibits the growth of aerobic spoilage microorganisms. With the aid of intelligent packaging, it is possible not only to extend the shelf-life further, but also to warn the consumer if a product has spoiled, thus allowing proper decision-making. Many factors can promote the growth of spoilage bacteria. Therefore, many interactive packaging applications, such as oxygen scavengers, ethylene scavengers, moisture absorbent pads, and antimicrobial films, balance out these factors.

Antimicrobial film technology looks promising; however, more research is necessary to validate the efficacy of embedded antimicrobials interacting with polymer films, gaseous headspace, and the food itself. Many potentially active compounds can be used as antimicrobials, such as silver-based compounds, bacteriocins such as nisin, enzymes such as lysozyme, and organic compounds such as triclosan (Yuan, 2003). As antimicrobials, they work directly on the bacterial cell wall by coming into contact and breaking down components leading to cell rupture. Two advantages of this technology are the decreased use of food additives and the combination of packaging with preservation.

For example, Microban®, by Microban Products Co. (Huntersville, NC), has incorporated triclosan into plastic to effectively add antibacterial action following migration to the surface of the food. The triclosan comes into contact with the cell wall and destroys the cell's ability to function, grow, and reproduce. Another example of antimicrobial films is the Microsphere® System created by Bemard Technologies, Inc. (Chicago, IL). When activated, it produces sodium dioxide from sodium chlorite, an oxidizer and well-known antimicrobial agent. Sodium dioxide inactivates the functional enzymes of the cell and thus kills the cell. Currently, commercially available antimicrobial films have not been widely accepted, mainly because of the limited features combined with their cost. Nevertheless, as mentioned earlier, antimicrobial film is a promising technology, and can be a major contributor to food safety.

In conclusion, MAP does not need to be limited to the gas mixture and film. From the center of packaging, it is beneficial to know which technology can better be integrated into packaging materials through gases or in polymers. Therefore, technical advice from researchers and packaging suppliers is essential. User-friendly software may be developed to assist industry in the future. Poor "package–product compatibility" will result in the creation of unintended atmospheres with uncertain microbiological implications (Bennik et al., 1998). Processors must know how to keep raw products clean and how other modern technologies act in synergy with packaging. Most of all, in order to make MAP safer, processors must know their products, the potential microbial hazard, know processes, apply effective hurdles, and validate the entire system through well-designed challenge tests (Yuan, 2003).

6.6 CONCLUSIONS

Reduction in populations of microflora on whole and fresh-cut produce is dependent upon the type of produce and the type of natural microflora present. It is important to use washing and sanitizing protocols that are efficient, because washing and sanitizing are unlikely to totally eliminate all pathogens after the produce is contaminated. There is still the need to find a sanitizer that maintains produce quality while enhancing safety by reducing populations of pathogenic microorganisms of public health significance. New treatments must be affordable and safe to carry out, have no adverse effect on quality, and be approved by the applicable regulatory agencies.

A variety of mitigation regimens and sanitizers are available to reduce microbial populations, depending upon the type of produce involved. Washing and sanitizing efficiencies depend on several factors, including characteristics of the produce surface, water quality, cleaner/sanitizer used, contact time, and presence and type of scrubbing action. However, further research is needed to determine additive, antagonistic, or synergistic effects of sanitation treatments when used in combination. Scientists must investigate the use of alternative technologies on the safety of whole and cut produce. It is crucial to determine the factors that limit the efficacy of washing in reducing microbial populations on produce and to devise means of overcoming such limitations.

REFERENCES

U.M. Abdul-Raouf, L.R. Beuchat, and M.S. Ammar, Survival and growth of *E. Coli* O157:H7 on salad vegetables. *Appl. Environ. Microbiol.*, 59, 1999–2006 (1993).

K. Abe and A.E. Watada, Ethylene absorbent to maintain quality of lightly processed fruits and vegetables. *J. Food Sci.*, 56, 1589–1592 (1991).

M.R. Adams, A.D. Hartley, and L.J. Cox, Factors affecting the efficacy of washing procedures used in the production of prepared salads. *Food Microbiol.*, 6, 69–77 (1989).

J.E. Adeskaveg, *Postharvest Sanitation to Reduce Decay of Perishable Commodities.* Perishables Handling Newsletter issue 82. University of California, Division of Agriculture and Natural Resources, Oakland, CA, 1995.

Y. Aharoni, A. Copel, and E. Fallik, The use of hydrogen peroxide to control postharvest decay on 'Galia' melons. *Ann. Appl. Biol.*, 125, 189–193 (1994).

R. Ahvenainen, *Novel Food Packaging Techniques.* CRC Press, Boca Raton, London, New York, Washington, DC, 2003, pp. 210–214.

A. Amanatidou, E.J. Smid, and L.G.M. Gorris, Effect of elevated oxygen and carbon dioxide on the surface growth of vegetable-associated microorganisms. *J. Appl. Microbiol.*, 86, 429–438 (1999).

A. Amanatidou, O. Schlüter, K. Lemkau, L.G.M. Gorris, E.J. Smid, and D. Knorr, Effect of combined application of high pressure treatment and modified atmospheres on the shelf life of fresh Atlantic salmon. *Innovative Food Sci. Emerg. Technol.*, 1, 87–98 (2000).

G. Apel, Chlorine dioxide. *Tree Fruit Postharvest J.*, 4, 12–13 (1993).

J.W. Austin, K.L. Dodds, B. Blanchfield, and J.M. Farber, Growth and toxin production by *Clostridium botulinum* on inoculated fresh-cut packaged vegetables. *J. Food Prot.*, 61, 324–328 (1998).

Z. Ayhan, G.W. Chism, and E.R. Richter, The shelf-life of minimally processed fresh cut melons. *J. Food Qual.*, 21, 29–40 (1998).

I. Babic, S. Roy, A.E. Watada, and W.P. Wergin, Changes in microbial populations in fresh-cut spinach. *Int. J. Food Microbiol.*, 31, 107–119 (1996).

A.S. Badawy, C.P. Gerba, and L.M. Kelley, Survival of rotavirus SA-11 on vegetables. *Food Microbiol.*, 2, 199–205 (1985).

G.A. Bagdasaryan, Survival of viruses of the enterovirus group (poliomyelitis, echo, coxsackie) in soil and on vegetables. *J. Hyg. Epidemiol. Microbiol. Immunol.*, 8, 497–505 (1964).

J.H. Bai, R.A. Saftner, A.E. Watada, and Y.S. Lee, Modified atmosphere maintains quality of fresh-cut cantaloupe (*Cucumis melon* L.). Sensory and nutritive qualities of food. *J. Food Sci.*, 66, 1207–1211 (2001).

M.L. Bari, Y. Sabina, S. Isobe, T. Uemura, and K. Isshiki, Effectiveness of electrolyzed acidic water in killing *Escherichia coli* O157:H7, *Salmonella enteritidis* and *Listeria monocytogenes* on the surfaces of tomatoes. *J. Food Prot.*, 66, 542–548 (2003).

R. Barkai-Golan, "Suppression of Postharvest Pathogens of Fresh Fruits and Vegetables by Ionizing Radiation," in I. Rosenthal, Ed., *Electromagnetic Radiations in Food Science.* Springer-Verlag, Berlin, 1992, pp. 155–194.

B.J. Barnett, M. Schwartze, D. Sweat, S. Lea, J. Taylor, B. Bibb, G. Pierce, and K. Hendricks, Outbreak of *Escherichia coli* O157:H713, Waco, Texas. Atlanta, GA, Epidemic Intelligence Service 44th Annual Conference, March, 1995, pp. 27–31.

M.I. Barriga, G. Trachy, C. Willemot, and R.E. Simard, Microbial changes in shredded iceberg lettuce stored under controlled atmospheres. *J. Food Sci.*, 56, 1586–1588 (1991).

C. Barry-Ryan, J.M. Pacussi, and K. O'Beirne, Quality of shredded carrots as affected by packaging film and storage temperature. *J. Food Sci.*, 65, 726–730 (2000).

M.M. Barth, C. Zhou, M. Mercier, and F.A. Payne, Ozone storage effects on anthocyanin content and fungal growth in blackberries. *J. Food Sci.*, 60, 1286–1287 (1995).

P.G. Bartlett and W. Schmidt, Surface-iodine complexes as germicides. *Appl. Microbiol.*, 5, 355–359 (1957).

J.A. Bartz, Washing fresh fruits and vegetables: lessons from treatment of tomatoes and potatoes with water. *Dairy Food Environ. Sanit.*, 19, 853–864 (1999).

J.A. Bartz and J.W. Eckert, "Bacterial Diseases of Vegetables Crops after Harvest," in J. Weichmann, Ed., *Postharvest Physiology of Vegetables*. Marcel Dekker, New York. 1987, pp. 351–376.

S. Baur, R. Klaiber, W.P. Hammes, and R. Carle, Sensory and microbiological quality of shredded, packaged iceberg lettuce as affected by pre-washing procedures with chlorinated and ozonated water. *Innov. Food Sci. Emerg. Technol.*, 5, 45–55 (2004).

V.I. Bazarova, Use of ozone in storage of apples. *Food Sci. Technol. Abstr.*, 14 (11), J1653 (1982).

J. Behrsing, M.S. Winkler, P. Franz, and R. Premier, Efficacy of chlorine for inactivation of *Escherichia coli* on vegetables. *Postharvest Biol. Technol.*, 19, 187–192 (2000).

M.H.J. Bennik, E.J. Smid, F.M. Rombouts, and L.G.M. Gorris, Growth of psychrotrophic foodborne pathogens in a solid surface model system under the influence of carbon dioxide and oxygen. *Food Microbiol.*, 12, 509–519 (1995).

M.H.J. Bennik, H.W. Peppelenbos, C. Nguyen-the, F. Carlin, E.J. Smith, and L.G.M. Gorris, Microbiology of minimally processed, modified atmosphere packaged chicory endive. *Postharvest Biol. Technol*, 9, 209–221 (1998).

M.E. Berrang, R.E. Brackett, and L.R. Beuchat, Growth of *Aeromonas hydrophila* on fresh vegetables stored under a controlled atmosphere. *Appl. Environ. Microbiol.*, 55, 2167–2171 (1989a).

M.E. Berrang, R.E. Brackett, and L.R. Beuchat, Growth of *Listeria monocytogenes* on fresh vegetables stored under controlled atmosphere. *J. Food Prot.*, 52, 702–705 (1989b).

L.R. Beuchat, Pathogenic microorganisms associated with fresh produce. *J. Food Prot.*, 59, 204–216 (1996).

L.R. Beuchat, Surface Decontamination of Fruits and Vegetables Eaten Raw: A Review. World Health Organization, Food Safety Unit WHO/FSF/FOS/98.2, 1998. Available at http://www.who.int/fsf/fos982~1.pdf.

L.R. Beuchat and R.E. Brackett, Survival and growth of *Listeria monocytogenes* on lettuce as influenced by shredding, chlorine treatment, modified atmosphere packaging and temperature. *J. Food Sci.*, 55, 755–758,870 (1990).

L.R. Beuchat and R.E. Brackett, Behaviour of *Listeria monocytogenes* inoculated into raw tomatoes and processed tomato products. *Appl. Environ. Microbiol.*, 57, 1367–1371 (1991).

L.R. Beuchat and J.H. Ryu, Produce handling and processing practices. *Emerg. Infect. Dis.*, 3, 459–465 (1997).

L.R. Beuchat, B.V. Nail, B.B. Adler, and M.R.S. Clavero, Efficacy of spray application of chlorinated water in killing pathogenic bacteria on raw apples, tomatoes and lettuce. *J. Food Prot.*, 61, 1305–1311 (1998).

S. Bidawid, J.M. Farber, and S.A. Sattar, Survival of hepatitis A virus on modified atmosphere packaged (MAP) lettuce. *Food Microbiol.*, 18, 95–102 (2001).

S.S. Block, "Peroxygen Compounds," in S.S. Block, Ed., *Disinfection, Sterilization and Preservation*, 4th edn. Lea and Febiger, Philadelphia, 1991, pp. 182–190.

J. Blostein, An outbreak of *Salmonella javiana* associated with consumption of watermelon. *J. Environ. Health*, 56, 29–31 (1993).

H.R. Bolin and C.C. Huxsoll, Effect of preparation procedures and storage parameters on quality retention of salad – cut lettuce. *J. Food Sci.*, 56, 60–62,67 (1991).

H.R. Bolin, A.E. Stafford, A.D. King, Jr., and C.C. Huxsoll, Factors affecting the storage stability of shredded lettuce. *J. Food Sci.*, 42, 1319–1321 (1977).

B.L. Bowles and V.K. Juneja, Inhibition of foodborne pathogens by naturally occurring food additives. *J. Food Safety*, 18, 101–112 (1998).

B.L. Bowles, S.K. Sackitey, and A.C. Williams, Inhibitory effects of flavor compounds on *Staphylococcus aureus* WRRC B124. *J. Food Safety*, 15, 337–347 (1995).

R.E. Brackett, Antimicrobial effect of chlorine on *Listeria monocytogenes. J. Food Prot.*, 50, 999–1003 (1987).

R.E. Brackett, Influence of modified atmosphere packaging on the microflora and quality of fresh bell peppers. *J. Food Prot.*, 53, 255–257 (1990).

R.E. Brackett, Shelf stability and safety of fresh produce as influenced by sanitation and disinfection. *J. Food Prot.*, 55, 808–814 (1992).

R.E. Brackett, "Microbiological Spoilage and Pathogens in Minimally Processed Refrigerated Fruits and Vegetables," in R.C. Wiley, Ed., *Minimally Processed Refrigerated Fruits and Vegetables*. Chapman and Hall, New York, 1994, pp. 269–312.

R.E. Brackett, Incidence, contributing factors and control of bacterial pathogens in produce. *Postharvest Biol. Technol.*, 15, 305–311 (1999).

G. Brandi, M. Sisti, G.F. Schiavano, L. Salvaggio, and A. Albano, Survival of *Aeromonas hydrophila, Aeromonas eaviae* and *Aeromonas sobria* in soil. *J. Appl. Bacteriol.*, 81, 439–444 (1996).

J.K. Brecht, Physiology of lightly processed fruits and vegetables. *Hort. Sci.*, 30, 18–22 (1995).

H.T. Brodrick and H.J. Van der Linde, *Technological Feasibility Studies on Combination Treatments for Subtropical Fruits. Combination Processes*. In Food Irradiation, Proceedings Series. International Atomic Energy Agency, Vienna, 1981, pp. 141–152.

E.A. Bryant, G.P. Fulton, and G.C. Budd, *Disinfection Alternatives for Safe Drinking Water*. VanNostrand Reinhold, New York, 1992, pp. 518.

R.L. Buchanan and L.A. Klawitter, Effects of temperature and oxygen on the growth of *Listeria monocytogenes* at pH 4.5. *J. Food Sci.*, 55, 1754 (1990).

R.K. Buick and A.P. Damoglou, The effect of vacuum packaging on the microbial spoilage and shelf-life of "ready-to-use" sliced carrots. *J. Sci. Food Agric.*, 38, 167–175 (1987).

S.L. Burnett, J. Chen, and L.R. Beuchat, Attachment of *Escherichia coli* O157:H7 to the surfaces and internal structures of apples as detected by confocal scanning laser microscopy. *Appl. Environ. Microbiol.*, 66, 4679–4687 (2000).

A.B. Burnett, M.H. Iturriaga, E.F. Escartin, C.A. Pettigrew, and L.R. Beuchat, Influence of variations in methodology on populations of *Listeria monocytogenes* recovered from lettuce treated with sanitizers. *J. Food Prot.*, 67, 742–750 (2004).

J.G. Buta and H.E. Moline, Methyl jasmonate extends shelf life and reduces microbial contamination of fresh-cut celery and peppers. *J. Agric. Food Chem.*, 46, 1253–1256 (1998).

V. Calvente, D. Benuzzi, and M.I.S. de Tosetti, Antagonistic action of siderophores from *Rhodotorula glutinis* upon the postharvest pathogen *Penicillium expansum. Int. Biodeter. Biodeg.*, 43, 167–172 (1999).

M. Cantwell, "Postharvest Handling Systems: Minimally Processed Fruits and Vegetables," in A.A. Kades, Ed., *Postharvest Technology of Horticultural Crops*, 2nd edn., Chapter 32 Publ. 3311. University of California, Division of Agriculture and Natural Resources, Oakland, CA, 1992.

M. Cantwell, Fresh-cut product biology and requirements. *Perishables Handling Newsletter*, 81, 4–9 (1995) University of California, Division of Agriculture and Natural Resources, Oakland, CA.

F. Carlin and C. Nguyen-the, Fate of *Listeria monocytogenes* on four types of minimally processed green salads. *Lett. Appl. Microbiol.*, 18, 222–226 (1994).

F. Carlin and M.W. Peck, Growth of and toxin production by non-proteolytic *C. botulinum* in cooked pured vegetables at refrigeration temperatures. *Appl. Environ. Microbiol.*, 62, 3069–3072 (1996).

F. Carlin, C. Nguyen-the, and A. Abreu Da Silva, Factors affecting the growth of *Listeria monocytogenes* on minimally processed fresh endive. *J. Appl. Bacteriol.*, 78, 636–646 (1995).

F. Carlin, C. Nguyen-the, and E.E. Morris, The influence of the background microflora on the fate of *Listeria monocytogenes* on minimally processed fresh broad leaved endive (*Cichorium endivia* var. latifolia). *J. Food Prot.*, 59, 698–703 (1996).

A. Castillo and E.F. Escartin, Survival of *Campylobacter jejuni* on sliced watermelon and papaya. *J. Food Prot.*, 57, 166–168 (1994).

Centers for Disease Control and Prevention (CDC), *Salmonella oranienburg* gastroenteritis associated with consumption of pre-cut watermelon. *MMWR*, 28, 522–523 (1979).

Centers for Disease Control and Prevention (CDC), Multistage outbreak of *Salmonella poona* infections – U.S. and Canada. *MMWR*, 40, 549–552 (1991).

Centers for Disease Control and Prevention (CDC), Multistage outbreak of *Salmonella* serotype montevideo infections, EPI-AID, 93–79 (1993).

Centers for Disease Control and Prevention (CDC), Outbreaks of *Escherichia coli* O157:H7 infection associated with eating alfalfa sprouts – Michigan and Virginia, June–July 1997. *MMWR*, 46, 741–745 (1997).

P. Cerrutti, S.M. Alzamora, and S.L. Vidales, Vanillin as an antimicrobial for producing shelf stable strawberry puree. *J. Food Sci.*, 62, 608–610 (1997).

21CFR101.95. Food Labelling: "Fresh," "freshly frozen," "fresh frozen," "frozen fresh," Code of Federal Regulations 21, Part 101.95."

21CFR110. Current Good Manufacturing Practice in Manufacturing, Packing, or Holding Human Food: Definitions, Code of Federal Regulations 21, Part 110.3(o).

21CFR173.300. Secondary Direct Food Additives Permitted in Food for Human Consumption: Chlorine Dioxide, Code of Federal Regulations 21, Part 173.300.

21CFR173.315. Chemicals Used in Washing or to Assist in the Peeling of Fruits and Vegetables. Code of Federal Regulations 21, Part 173, Section 173.315.

21CFR173.325. Secondary Direct. Food Additives Permitted in Food for Human Consumption: Acidified Sodium Chlorite Solutions, Code of Federal Regulations. Title 21, Part 173.325.

21CFR178.1010. Sanitizing Solutions. Code of Federal Regulations 21, Part 178, Section 178.1010.

21CFR182.1778. Sodium Phosphate. Code of Federal Regulations 21, Part 182, Section 182.1778.

21CFR184.1366. Hydrogen Peroxide. Code of Federal Regulations 21, Part 170–199, Section 184.1366, 463.

CFSAN, Draft assessment of the relative risk to public health from foodborne *Listeria monocytogenes* among selected categories of ready-to-eat foods. CFSAN-FDA-HHS/FSIS-USDA, Washington, DC, 2001.

T.L. Chang, R. Streicher, and H. Zimmer, The interaction of aqueous solutions of chlorine with malic acid, tartaric acid, and various fruit juices. A source of mutagens. *Anal. Lett.*, 21, 2049–2067 (1988).

P. Chantaysakorn and R.L. Richter, Antimicrobial properties of Pepsin-digested Lactoferrin added to carrot juice and filtrates of carrot juice. *J. Food Prot.*, 63, 376–380 (2000).

Y. Chen, W.H. Ross, V.N. Scott, and D.E. Gombas, *Listeria monocytogenes*: Low levels equal low risk. *J. Food Prot.*, 4, 570–577 (2003).

J.P. Cherry, Improving the safety of fresh produce with antimicrobials. *Food Technol.*, 53, 54–57 (1999).

C. Chervin and P. Boisseau, Quality maintenance of "ready-to-eat" shredded carrots by gamma-irradiation. *J. Food Sci.*, 59, 359–361 (1994).

V. Cliffe-Byrnes, C.P. McLaughlin, and D. O'Beirne, Effects of packaging film and storage temperature on the quality of modified atmosphere packaged dry coleslaw mix. *Int. J. Food Sci. Technol.*, 38, 187–189 (2003).

D.O. Cliver, "Foodborne Viruses," in M.P. Doyle, L.R. Beuchat, and T.J. Montville, Eds., *Food Microbiology – Fundamentals and Frontiers*. Washington, DC, ASM Press, 1997, pp. 437–446.

W.S. Conway, B. Leverentz, R.A. Saftner, W.J. Janisiewicz, C.E. Sams, and E. Leblanc, Survival and growth of *Listeria monocytogenes* on fresh-cut apple slices and its interaction with *Glomerella cingulata* and *Penicillium expansum*. *Plant Dis.*, 84, 177–181 (2000).

R.N. Costilow, M.A. Uebersax, and P.J. Ward, Use of chlorine dioxide for controlling microorganisms during handling and storage of fresh cucumbers. *J. Food Sci.*, 49, 396–401 (1984).

J.A. Daniels, R. Krishnamurthi, and S.S.H. Rizvi, A review of the effect of carbon dioxide on microbial growth and food quality. *J. Food Prot.*, 48, 532–537 (1985).

B. Day, Conference Proceedings, International Conference on Active and Intelligent Packaging, September 7–8, Gloucestershire, UK, Campden and Chorleywood Food and Drink Research Association, Chipping Campden, UK, 2000.

E. De Boer, Isolation of *Yersinia enterocolitica* and related species from foods. In Proceedings of the 2nd World Congress. Foodborne Infections and Intoxications. Institute of Veterinary Medicine, Robert Van Osterlag Institute, West Berlin, Germany, 1986, p. 481.

P.J. Delaquis and G. Mazza, Antimicrobial properties of isothiocyanates in food preservation. *Food Technol.*, 49, 73–84 (1995).

P.J. Delaquis, S. Stewart, P.M. Toivonen, and A.L. Moyls, Effect of warm chlorinated water on the microbial flora of shredded lettuce. *Food Res. Int.*, 32, 7–14 (1999).

P.J. Delaquis, S.M. Ward, and K. Stanich, Evaluation of pre-pressing sanitary treatments for the destruccion of *Escherichia coli* O157:H7 on apples destined for production of unpasterized apple juice. Technical report no. 9901, Agriculture and Agri-Food Canada, Pacific Agri-Food Research Centre, Summerland, BC, 2000a.

P.J. Delaquis, S. Stewart, M. Cliff, P.M. Toivonen, and A.L. Moyls, Sensory quality of ready-to-eat lettuce washed in warm, chlorinated water. *J. Food Qual.*, 23, 553–563 (2000b).

P. Delaquis, P. Toivonen, K. Walsh, K. Rivest, and K. Stanich, Chlorine depletion in sanitizing solutions used for apple slice disinfection. *Food Prot. Trends*, 24, 323–327 (2004).

C.L. Delmas and D.J.M. Vidon, Isolation of *Y. enterocolitica* and related species from food in France. *Appl. Environ. Microbiol.*, 50, 767–771 (1985).

F. Devlieghere, I. Lefevere, A. Magnin, and J. Debevere, Growth of *Aeromonas hydrophila* on modified atmosphere packaged cooked meat products. *Food Microbiol.*, 17, 185–196 (2000).

C. Diaz and J.H. Hotchkiss, Comparative growth of *E. coli* O157:H7, spoilage organisms and shelf life of shredded iceberg lettuce stored under modified atmospheres. *J. Sci. Food Agric.*, 70, 433–438 (1996).

J.F. Diehl, *Safety of Irradiated Foods*, 2nd edn. Marcel Dekker, Inc., New York, 1995.

K.L. Dodds and J.W. Austin, "*Clostridium botulinum*," in M.P. Doyle, L.R. Beuchat, and T.J. Montville, Eds., *Food Microbiology – Fundamentals and Frontiers*. ASM Press, Washington, DC, 1997, pp. 288–304.

M. Doganay, Listeriosis: Clinical presentation. *FEMS Immunol. Med. Microbiol.*, 35, 173–175 (2003).

A. Doherty, J.J. Sheridan, P. Allen, D.A. McDowell, I.S. Blair, and D. Harrington, Survival and growth of *Aeromonas hydrophila* on modified atmosphere packaged normal and high pH lamb. *Int. J. Food Microbiol.*, 28, 379–392 (1996).

L.K. Ng Doris and H.L. Seah, Isolation and identification of *Listeria monocytogenes* from a range of foods in Singapore. *Food Control*, 6, 171–173 (1995).

M.P. Doyle, Fruit and vegetable safety microbiological considerations. *Hort. Sci.*, 25, 1478–1481 (1990).

M.P. Doyle, Reducing foodborne disease: what are the priorities? *Nutrition*, 16, 647–694 (2000).

S. Droby, E. Chalutz, and C.L. Wilson, Antagonistic microorganisms as biological control agents of postharvest diseases of fruit and vegetables. *Postharvest News Inf.*, 2, 167–173 (1991).

Ecolab Inc., Catching the wave. *Food Qual.*, 4, 51–52 (1997).

A. El-Ghaouth, J.L. Smilanick, G.E. Brown, A. Ippolito, M. Wisniewski, and C.L. Wilson, Application of *Candida saitoana* and glycochitosan for the control of postharvest diseases of apple and citrus fruit under semi-commercial conditions. *Plant Dis.*, 84, 243–248 (2000).

W.O. Ellis, J.P. Smith, B.K. Simpson, H. Ramaswamy, and G.J. Doyon, Growth of and aflatoxin production by *Aspergillus flavus* in peanuts stored under modified atmosphere packaging (MAP) conditions. *Int. J. Food Microbiol.*, 22, 173–187 (1994).

S.K. El-Samahy, B.M. Youssef, A.A. Askar, and H.M.M. Swailam, Microbiological and chemical properties of irradiated mango. *J. Food Safety*, 20, 139–156 (2000).

M.E. Escudero, L. Velásquez, M.S. Di Genaro, and A.S. De Guzmán, Effectiveness of various disinfectants in the estimation of *Yersinia entero colitica* on fresh lettuce. *J. Food Prot.* 62, 665–669 (1999).

T.J. Fang and L.W. Lin, Inactivation of *Listeria monocytogenes* on raw pork treated with modified atmosphere packaging and nisin. *J. Food Drug Anal.*, 2, 189–200 (1994).

J.M. Farber, Microbiological aspects of modified-atmosphere packaging technology: A review. *J. Food Prot.*, 54, 58–70 (1991).

J.M. Farber, G.W. Sanders, and M.A. Johnson, A survey of various foods for the presence of Listeria species. *J. Food Prot.*, 52, 456–458 (1989).

J.M. Farber, S.L. Wang, Y. Cai, and S. Zhang, Changes in populations of *Listeria monocytogenes* inoculated on packaged fresh-cut vegetables. *J. Food Prot.*, 61, 192–195 (1998).

J. Farkas, "Physical Methods of Food Preservation," in M.P. Doyle, L.R. Beuchat, and T.J. Monteville, Eds., *Food Microbiology: Fundamentals and Frontiers*. American Society for Microbiology, Washington, DC, 1997, pp. 497–519.

FDA, Food and safety assurance program; development of hazards analysis critical control points; proposal rule. *Federal Register*, August 4, 1994.

FDA, Guidance for Industry. Guide to minimize microbial food safety hazards for fresh fruits and vegetables, Food Safety Initiative Staff, HFS-32. U.S. Food and Drug Administration, Center for Food Safety and Applied Nutrition, Washington, DC, 1998. Available at http://www.foodsafety.gov/~dms/prodguid.html. Accessed 02.03.06.

FDA/CFSAN, Center for Food Safety and Applied Nutrition, Kinetics of microbial inactivation for alternative food processing technologies. Food and Drug Administration, Center for Food Safety and Applied Nutrition, 2000. Available at http://vm.cfsan.fda.gov/~comm/ift-toc.html. Accessed 02.03.06.

A. Feliciano, A.J. Feliciano, J. Vendrusculo, J.E. Adaskaveg, and M.J. Ogawa, Efficacy of ethanol in postharvest benomyl-DCNA treatments for control of brown rot of peach. *Plant Dis.*, 76, 226–229 (1992).

D.R. Fenlon, J. Wilson, and W. Donachie, The incidence and level of *Listeria monocytogenes* contamination of food sources at primary production and initial processing. *J. Appl. Bacteriol.*, 81, 641–650 (1996).

E.F. Fernandez Escartin, A. Castillo Ayala, and J. Saldana Lozano, Survival and growth of *Salmonella* and *Shigella* on sliced fresh fruit. *J. Food Prot.*, 52, 471–472 (1989).

G.R. Finch and N. Fairbairn, Comparative inactivation of poliovirus type 3 and MS2 coliphage in demand-free phosphate buffer by using ozone. *Appl. Environ. Microbiol.*, 57, 3121–3126 (1991).

C.F. Forney, R.E. Rij, R. Denis-Arrue, and J.L. Smilanick, Vapor phase hydrogen peroxide inhibits postharvest decay of table grapes. *Hort. Sci.*, 26, 1512–1514 (1991).

G.A. Francis and D. O'Beirne, Effects of gas atmosphere, antimicrobial dip and temperature on the fate of *L. innocua* and *L. monocytogenes* on minimally processed lettuce. *Int. J. Food Sci. Technol.*, 32, 141–151 (1997).

G.A. Francis and D. O'Beirne, Effects of vegetable type, package atmosphere and storage temperature on growth and survival of *Escherichia coli* O157:H7 and *Listeria monocytogenes*. *J. Ind. Microbiol. Biot.*, 27, 111–116 (2001).

G.A. Francis, C. Thomas, and D. O'Beirne, The microbiological safety of minimally processed vegetables. *Int. J. Food Sci. Technol.*, 34, 1–22 (1999).

T. Fukao, H. Sawada, and Y. Ohta, Combined effect of hop resins and sodium hexametaphosphate against certain strains of *Escherichia coli*. *J. Food Prot.*, 63, 735–740 (2000).

A. Garcia, J.R. Mount, and P.M. Davidson, Ozone and chlorine treatment of minimally processed lettuce. *J. Food Sci.*, 68, 2747–2751 (2003).

G.D. García de Fernando, G.J.E. Nychas, M.W. Peck, and J.A. Ordóñez, Growth/survival of psychrotrophic pathogens on meat packaged under modified atmospheres. *Int. J. Food Microbiol.*, 28, 221–231 (1995).

R.M. García-Gimeno, M.D. Sanchez-Pozo, M.A. Amaro López, and G. Zurera-Cosano, Behaviour of *Aeromonas hydrophila* in vegetable salads stored under modified atmosphere at 4 and 15°C. *Food Microbiol.*, 13, 369–374 (1996).

N. Garg, J.J. Churey, and D.F. Splittstoesser, Effect of processing conditions on the microflora of fresh-cut vegetables. *J. Food Prot.*, 53, 701–703 (1990).

A.M. Gibson, R.C.L. Ellis-Brownlee, M.E. Cahill, E.A. Szabo, G.C. Fletcher, and P.J. Bremer, The effect of 100% CO_2 on the growth of non-proteolytic *Clostridium botulinum* at chill temperatures. *Int. J. Food Microbiol.*, 54, 39–48 (2000).

C.O. Gill and M.P. Reichel, Growth of cold tolerant pathogens *Y. enterocolitica*, *A. hydrophila*, and *L. monocytogenes* on high-pH beef packaged under vacuum or carbon dioxide. *Food Microbiol.*, 6, 223–230 (1989).

V.S. Gohil, M.A. Ahmed, R. Davies, and R.K. Robinson, Incidence of *Listeria* spp. in retail foods in the United Arab Emirates. *J. Food Prot.*, 58, 102–104 (1995).

D.A. Golden, M.J. Eyles, and L.R. Beuchat, Influence of modified-atmosphere storage on the growth of uninjured and heat-injured *Aeromonas hydrophila*. *Appl. Environ. Microbiol.*, 55, 3012–3015 (1989).

D.E. Gombas, Y. Chen, R.S. Clavero, and V.N. Scott, Survey of *Listeria monocytogenes* in ready-to-eat foods. *J. Food Prot.*, 66, 559–569 (2003).

G.A. Gonzalez-Aguilar, L. Zacarias, M. Mulas, and M.T. Lafuente, Temperature and duration of water dips influence chilling injury, decay and polyamine content in "Fortune" mandarins. *Postharv. Biol. Technol.*, 12, 61–69 (1997).

J.R. Gorny, "Fresh-Cut Product Preparation," in C.A. Davis, Ed., *Fresh-Cut Products: Maintaining Quality and Safety. Postharvest Horticulture Series No. 10*, Postharvest Outreach Program. Department of Pomology, University of California, 1996, pp. 7.2–7.7.

J.R. Gorny, *Packaging Design for Fresh-Cut Produce*, Chapter 4. IFPA, Alexandria, 2003, pp. 19–22.

J.R. Gorny, International Fresh-Cut Produce Association. Oral session presented at the IFT Annual Meeting, Las Vegas, July 2004.

J.R. Gorny and A.A. Kader, "Fresh-Cut Fruit Products," in C.A. Davis, Ed., *Fresh-Cut Products: Maintaining Quality and Safety. Postharvest Horticulture Series No. 10*, Postharvest Outreach Program. Department of Pomology, University of California, 1996.

D.M. Graham, Use of ozone for food processing. *Food Technol.*, 51, 72–75 (1997).

I.R. Grant and M.F. Patterson, Effect of irradiation and modified atmosphere packaging on the microbiological safety of minced pork stored under temperature abuse conditions. *Int. J. Food Sci. Technol.*, 26, 521–533 (1991a).

I.R. Grant and M.F. Patterson, Effect of irradiation and modified atmosphere packaging on the microbiological and sensory quality of pork stored at refrigeration temperatures. *Int. J. Food Sci. Technol.*, 26, 507–519 (1991b).

R. Gravani, "Incidence and Control of *Listeria* in Food-Processing Facilities," in E.T. Ryser, and E.H. Marth, Eds. *Listeria, Listeriosis and Food Safety*, 2nd ed. Marcel Dekker, New York, 1999, pp. 657–709.

R.J.H. Gray, P.H. Elliott, and R.I. Tomlins, Control of two major pathogens on fresh poultry using a combination potassium sorbate/CO_2 packaging treatment. *J. Food Sci.*, 49, 142–145 (1984).

P.M. Griffin and R.V. Tauxe, The epidemiology of infections caused by *E. coli* O157:H7, other enterohaemorrhagic *E. coli* and the associated Haemolytic Syndrome. *Epidemiol. Rev.*, 13, 60–98 (1991).

M.M. Guerra, J. McLauchlin, and F.A. Bernardo, *Listeria* in ready-to-eat and unprocessed foods produced in Portugal. *Food Microbiol.*, 18, 423–429 (2001).

G. Gunes, D.L. Splittstoesser, and C.Y. Lee, Microbial quality of fresh potatoes: effect of minimal processing. *J. Food Prot.*, 60, 863–866 (1997).

R.D. Hagenmaier and R.A. Baker, Low-dose irradiation of cut iceberg lettuce in modified atmosphere packaging. *J. Agric. Food Chem.*, 45, 2864–2868 (1997).

R.D. Hagenmaier and R.A. Baker, An evaluation of gamma irradiation for preservation of citrus salads in flexible packaging. *Proc. Florida State Hort. Soc.*, 110, 243–245 (1998a).

R.D. Hagenmaier and R.A. Baker, Microbial population of shredded carrot in modified atmosphere packaging as related to irradiation treatment. *J. Food Sci.*, 63, 162–164 (1998b).

Y.Y. Han, D.M. Sherman, R.H. Linton, S.S. Nielsen, and P.E. Nelson, The effects of washing and chlorine dioxide gas on survival and attachment of *Escherichia coli* O157:H7 to green pepper surfaces. *Food Microbiol.*, 17, 521–533 (2000).

Y.Y. Hao and R.E. Brackett, Influence of modified atmosphere on growth of vegetable spoilage bacteria in media. *J. Food Prot.*, 56, 223–228 (1993).

Y.Y. Hao, R.E. Brackett, L.R. Beuchat, and M.P. Doyle, Microbial quality and the inability of proteolytic *Clostridium botulinum* to produce toxin in film-packaged fresh-cut cabbage and lettuce. *J. Food Prot.*, 61, 1148–1153 (1998).

S.M. Harmon, D.A. Kautter, and H.M. Solomon, *Bacillus cereus* contamination of seeds and vegetable sprouts grown in a home sprouting kit. *J. Food Prot.*, 50, 62–65. (1987).

L.J. Harris and R.K. Barakat, Growth of *Listeria monocytogenes* and *Yersinia enterocolitica* on cooked poultry stored under modified atmosphere at 3.5, 6.5 and 10°C. *J. Food Prot.*, 58 (Suppl.), 38–42 (1995).

W.A. Harrison, A.C. Peters, and L.M. Fielding, Growth of *Listeria monocytogenes* and *Yersinia enterocolitica* colonies under modified atmosphere at 4 and 8°C using a model food system. *J. Appl. Microbiol.*, 88, 38–43 (2000).

A.H.W. Hauschild, B. Aris, and R. Hilsheimer, *C. botulinum* in marinated products. *Can. Inst. Food Sci. Technol. J.*, 8, 84–87 (1978).

A.H.W. Hauschild and K.L. Dodds, "*C. botulinum* and Other Clostridia that Produce Botulinum Neurotoxin," in *C. botulinum*: *Ecology and Control in Foods*. Marcel Dekker, New York, 1993; pp. 3–20.

H. Hays, P.R. Elliker, and W.E. Sandine, Microbial destruction by low concentrations of hypochlorite and iodophor germicides in alkaline and acidified water. *Appl. Microbiol.*, 15, 575–581 (1967).

C.B. Hintlian and J.H. Hotchkiss, The safety of modified atmosphere packaging: a review. *Food Technol.*, 40, 70–76 (1986).

T.W. Ho, B. Mishu, C.Y. Li, C.Y. Gao, D.R. Cornblath, J.W. Griffin, A.K. Asbury, M.J. Blaser, and G.M. Mckhann, Guillain barre syndrome in northern China: relationship to *Campylobacter jejuni* infection and glycolipid antibodies. *Brain*, 118, 597–605 (1995).

A.R. Horton, Y.C. Hung, K. Venkitanarayanan, G.O.I. Ezeike, and M.P. Doyle, Disinfection effects of electrolyzed water on *E. coli* O157:H7 and apple quality. Abstract 22D-6, Abstract of papers presented at 1999 IFT Annual Meeting, Chicago, IL, July 24–28, 1999.

M. Horvath, L. Bilitzky, and J. Huttner, *Ozone*. Elsevier, Amsterdam, 1985, pp. 68–74, 304–331.

J.H. Hotchkiss, M.J. Banco, F.F. Busta, A.C. Genigeorgis, R. Kociba, L. Rheaume, L.A. Smoot, J.D. Schuman, and H. Sugiyama, The relationship between botulinal toxin production and spoilage of fresh tomatoes held at 13 and 23°C under passively modified and controlled atmosphere and air. *J. Food Prot.*, 55, 522–527 (1992).

J.A. Hudson, J. Mott, and N. Penney, Growth of *Listeria monocytogenes*, *Aeromonas hydrophila* and *Yersinia enterocolitica* on vacuum and saturated carbon dioxide controlled atmosphere packaged sliced roast beef. *J. Food Prot.*, 57, 204–208 (1994).

W.C. Hurst, Sanitation of lightly processed fruits and vegetables. *Hort. Sci.*, 30, 22–24 (1995).

W.C. Hurst, "Safety Aspects of Fresh-Cut Fruits and Vegetables," in O. Lamikanra, Ed., *Fresh-Cut Fruits and Vegetables*. CRC Press, Boca Raton, London, New York, Washington, DC, 2002, pp. 45–90.

International Institute of Refrigeration (IIR). Informatory note on refrigeration in food. *Listeria monocytogenes* in refrigerated foods. January 2, 1998.

International Commission on Microbiological Specifications for Foods (ICMSF), *Microorganisms in Foods. 5. Microbiological Specifications of Food Pathogens*, T.A. Roberts, A.C. Baird-Parker, and R.B. Tompkin, Eds. Blackie Academic and Professional, London, 1996.

International Commission on Microbiological Specifications for Foods (ICMSF), *Microorganisms in Foods. 7. Microbiological Testing in Food Safety Management*. Kluwer Academic/Plenum Publishers, New York, 2002, pp. 285–312.

Institute of Food Science and Technology (IFST). *Guidelines for the Handling of Chilled Foods*, 2nd edn. IFST, London, ISBN 0905367073, 1990.

International Fresh-Cut Produce Association (IFPA), see http://www.fresh-cuts.org.

IFPA, *Food Safety Guidelines for the Fresh-Cut Produce Industry*, J.R. Gorny, Ed., 4th edn. IFPA, Alexandria, VA, 2002.

K. Isshiki, K. Tokuoka, R. Mori and S. Chiba, Preliminary examination of allyl isothiocyanate vapor for food preservation. *Biosci. Biotech. Biochem.*, 56, 1476–1477 (1992).

H. Izumi, Electrolyzed water as a disinfectant for fresh-cut vegetables. *J. Food Sci.*, 64, 536–539 (1999).

W.J. Janisiewicz and B. Bors, Development of a microbial community of bacterial and yeast antagonists to control wound-invading postharvest pathogens of fruit. *Appl. Environ. Microbiol.*, 61, 3261–3267 (1995).

W.R. Jilbert, "Quality Control and Sanitation Aspects of Fresh Squeezed Citrus Juice Processing," in R.F. Matthews, Ed., *Food Industry Short Course Proceedings*. Florida IFT and University of Florida Extension Service, Gainesville, 1988.

V.J. Jöckel, and W. Otto, Technologische und hygienische aspekte bei der herstellung und distribution von vorgeschnittenen salaten. *Arch. Lebensmittelhyg.*, 41, 129–132 (1990).

Y.K. Juneja, B.S. Marmer, and J.E. Call, Influence of modified atmosphere packaging on growth of *Clostridium perfringens* in cooked turkey. *J. Food Safety*, 16, 141–150 (1996).

B.J. Juven and M.D. Pierson, Antibacterial effects of hydrogen peroxide and methods for its detection and quantitation. *J. Food Prot.*, 59, 1233–1241 (1996).

F.K. Käferstein, The microflora of parsley. *J. Milk Food Technol*, 39, 837 (1976).

A.A. Kader, D. Zagory, and E.L. Kerbel, Modified atmosphere packaging of fruits and vegetables. *Crit. Rev. Food Sci. Nutr.*, 28, 1–30 (1989).

K.D. Kallender, A.D. Hitchins, G.A. Lancette, J.A. Schmieg, G.R. Garcia, H.M. Solomon, and J.N. Sofos, Fate of *Listeria monocytogenes* in shredded cabbage stored at 5 and 25°C under a modified atmosphere. *J. Food Prot.*, 54, 302–304 (1991).

M. Karapinar and S.A. Gonul, Removal of *Yersinia enterocolitica* from fresh parsley by washing with acetic acid or vinegar. *Int. J. Food Microbiol.*, 16, 261–264 (1992).

J.N. Karitsky, H.T. Osterholm, H.B. Greenberg, J.A. Keriath, J.R. Godes, C.W. Hedberg, J.C. Forfang, A.Z. Kapikian, J.C. McCullough, and K.E. White, Norwalk gastroenteritis: a community outbreak associated with bakery product consumption. *Ann. Intern. Med.*, 100, 519–521 (1984).

M.A. Karmali, B.T. Steele, M. Petric, and C. Lim, Sporadic cases of haemolytic uremic syndrome associated with faecal cytotoxin and cytotoxin-producing *E. coli*. *Lancet*, 1, 619–620 (1983).

J.M. Ketley, Pathogenesis of enteric infection by *Campylobacter*. *Microbiol.*, 143, 5–21 (1997).

J. Kim and P.M. Foegeding, "Principles of Control," in A.H.W Hauschild and K.L. Dodds, Eds., *C. botulinum: Ecology and Control in Foods*. Marcel Dekker, New York, 1993, pp. 121–176.

J.G. Kim, A.E. Yousef, and S. Dave, Application of ozone for enhancing the microbiological safety and quality of foods: a review. *J. Food Prot.*, 62, 1071–1087 (1999a).

J.G. Kim, A.E. Yousef, and G.W. Chism, Use of ozone to inactivate microorganisms on lettuce. *J. Food Safety*, 19, 17–34 (1999b).

A.D. King and H.R. Bolin, Physiological and microbiological storage stability of minimally processed fruits and vegetables. *Food Technol.*, 43, 132–135, 139 (1989).

B.P. Klein, Nutritional consequences of minimal processing of fruits and vegetables. *J. Food Qual.*, 10, 179–193 (1987).

S. Knochel and C. Jeppesen, Distribution and characteristics of *Aeromonas* in food and drinking water in Denmark. *Int. J. Food Microbiol.*, 10, 317–322 (1990).

L. Kong Fanchun and Qun Shengmin Wang, Application of ozone for pesticide degradation and preservation of fruits and vegetables. *Food Machinery*, 5, 24–26 (2003).

D.G. Korich, J.R. Mead, M.S. Madore, N.A. Sinclair, and C.R. Sterling, Effects of ozone, chlorine dioxide, chlorine and monochloramine on *Cryptosporidium parvum oocyst* viability. *Appl. Environ. Microbiol.*, 56, 1423–1428 (1990).

L. Korsten, E.S. De Jager, E.E. De Villiers, A. Lourens, J.M. Kotze, and F.C. Wehner, Evaluation of bacterial epiphytes isolated from avocado leaf and fruit surfaces for biocontrol of avocado postharvest disease. *Plant. Dis.*, 79, 1149–1156 (1995).

S. Koseki, K. Yoshida, S. Isobe, and K. Itoh, Decontamination of lettuce using acidic electrolyzed water. *J. Food Prot.*, 64, 652–658 (2001).

T. Kristofferson, Mode of action of hypochlorite sanitizers with and without sodium bromide. *J. Dairy Sci.*, 41, 942–949 (1958).

R.W. Lacey, Antibacterial activity of providone iodine towards non-sporing bacteria. *J. Appl. Bacteriol.*, 46, 443–449 (1979).

K. Lainé and J. Michard, Fréquence et abondance des *Listeria* dans des légumes frais découpés prets a lemploi. *Microbiol. Alim. Nutr.*, 6, 329–335 (1988).

M.M. Lang, L.J. Harris, and L.R. Beuchat, Evaluation of inoculation method and inoculum drying time for their effects on survival and efficiency of recovery of *Escherichia coli* O157:H7, *Salmonella* and *Listeria monocytogenes* inoculated on the surface of tomatoes. *J. Food Prot.*, 67, 732–741 (2004).

D.I. Langerak, The influence of irradiation and packaging on the quality of prepacked vegetables. *Ann. Nutr. Aliment.*, 32, 569–586 (1978).

A.E. Larson and E.A. Johnson, Challenge of vegetables in modified atmosphere packaging with *Clostridium botulinum.* Food Research Institute, University of Wisconsin, Madison, 1994 (unpublished).

A.E. Larson and E.A. Johnson, Evaluation of botulinal toxin production in packaged fresh-cut cantaloupe and honeydew melons. *J. Food Prot.*, 8, 948–952 (1999).

A.E. Larson, E.A. Johnson, C.R. Barmore, and M.D. Hughes, Evaluation of the botulism hazard from vegetables in modified atmosphere packaging. *J. Food Prot.*, 60, 1208–1214 (1997).

D. Law, Virulence factors of *Escherichia coli* O157:H7 and other Shiga toxin-producing *Escherichia coli*. *J. Appl. Microbiol.*, 88, 729–745 (2000).

K.A. Lawlor, M.D. Pierson, C.R. Hackney, J.R. Claus, and J.E. Marcy, Nonproteolytic *Clostridium botulinum* toxigenesis in cooked turkey stored under modified atmospheres. *J. Food Prot.*, 63, 1511–1516 (2000).

N.I. Lebovka, I. Praporscic, and E. Vorobiev, Effect of moderate thermal and pulsed electric field treatments on textural properties of carrots, potatoes and apples. *Inn. Food Sci. Emerg. Technol.*, 5, 9–16 (2004).

C.H. Lee and J.N. Cash, Comparative growth rates of bacteria on minimally processed, meat-vegetable product under modified atmospheres. *Food Sci. Biotechnol.*, 7, 6–12 (1998).

D.I. Leblanc, L. Start, B. Macneil, B. Goguen, and C. Beaulieu, Perishable food temperatures in retail stores. *Refrig. Sci. Technol. Proc.*, 6, 42–51 (1996).

W. Leibinger, B. Breuker, M. Hahn, and K. Mendgen, Control of postharvest pathogens and colonization of the apple surface by antagonistic microorganisms in the field. *Phytopathology*, 87, 1103–1110 (1997).

B. Leverentz, W.S. Conway, Z. Alavidze, W.J. Janisiewicz, Y. Fuchs, M.J. Camp, E. Chighladze, and A. Sulakvelidze, Examination of bacteriophage as a biocontrol method for *Salmonella* on fresh-cut fruit: a model study. *J. Food Prot.*, 64, 1116–1121 (2001).

Y. Li, R.E. Brackett, J. Chen, and L.R. Beuchat, Survival and growth of *Escherichia coli* O157:H7 inoculated onto cut lettuce before or after heating in chlorinated water, followed by storage at 5°C or 15°C. *J. Food Prot.*, 64, 305–309 (2001a).

Y. Li, R.E. Brackett, R.L. Shewfelt, and R.L. Beuchat, Changes in appearance and natural microflora on iceberg lettuce treated in warm, chlorinated water and then stored at refrigeration temperature. *Food Microbiol.*, 18, 299–308 (2001b).

C.H. Liao, Antagonism of *Pseudomonas putida* strain PP22 to phytopathogenic bacteria and its potential use as a biocontrol agent. *Plant. Dis.*, 73, 223–226 (1989).

C.H. Liao and G.M. Sapers, Attachment and growth of *Salmonella chester* on apple fruit and in vivo response of attached bacteria to sanitizer treatments. *J. Food Prot.*, 63, 876–883 (2000).

C.H. Liao and P.H. Cook, Response to trisodium phosphate treatment of *Salmonella* Chester attached to fresh-cut green pepper slices. *Can. J. Microbiol.*, 47, 25–32 (2001).

C.L. Liew and R.K. Prange, Effect of ozone and storage temperature on postharvest diseases and physiology of carrots (*Caucus carota* L.). *J. Am. Soc. Hortic. Sci.*, 119, 563–567 (1994).

T. Lilly, H.M. Soloman, and E.J. Rhodeharnel, Incidence of *C. botulinum* in vegetables packaged under vacuum or modified atmospheres. *J. Food Prot.*, 59, 59–61 (1996).

M. Lis-Balchin, S. Hart, S.G. Deans, and E. Eaglesham, Comparison of the pharmacological and antimicrobial action of commercial plant essential oils. *J. Herbs. Spices Medic. Plants*, 4, 69–86 (1996).

J.G. Loaiza-Velarde, F.A. Tomás-Barberán, and M.E. Saltveit, Effect of intensity and duration of heat-shock treatments on wound-induced phenolic metabolism in iceberg lettuce. *J. Am. Soc. Hort. Sci.*, 122, 873–877 (1997).

C.L. Lockhart and R.W. Delbridge, Control of storage diseases of carrots by washing, grading and postharvest fungicide treatments. *Can. Plant Dis. Surv. Dec.*, 52, 140–142 (1972).

L.V. Lopez, J.R. Romero, and J. Urbina, Eficiencia de desinfectantes en vegetales y frutas. *Alimentos*, 13, 25–30 (1988).

J.Y. Lu, S.M. Lukombo, C. Stevens, V.A. Khan, C.L. Wilson, P.L. Pusey, and E. Chaultz, Low dose UV and gamma radiation on storage rot and physiochemical changes in peaches. *J. Food Qual.*, 16, 301–309 (1993).

B.M. Lund and M.W. Peck, Heat resistance and recovery of non-proteolytic *Clostridium botulinum* in relation to refrigerated processed foods with an extended shelf-life. *J. Appl. Bacteriol. Symp. Suppl.*, 76, 115–128 (1994).

B.M. Lund and A.L. Snowdon, "Fresh and Processed Fruits," in B.M. Lund, T.C. Baird-Parker, G.M. Gould, Eds., *The Microbiological Safety and Quality of Food*. Aspen Publishers, Inc., Gaithersburg, Maryland, 2000, pp. 738–758.

S. Lurie, Postharvest heat treatments. *Postharv. Biol. Technol.*, 13, 257–269 (1998).

A.P. MacGowan, K. Bowker, J. McLauchlin, P.M. Bennett, and D.S. Reeves, The occurrence and seasonal changes in the isolation of *Listeria* spp. in shop bought food stuffs, human faeces, sewage and soil from urban sources. *Int. J. Food Microbiol.*, 21, 325–334 (1994).

J.A. Magnusson, A.D. King Jr, and T. Torok, Microflora of partially processed lettuce. *Appl. Environ. Microbiol.*, 56, 3851–3854 (1990).

S.B. Mano, J.A. Ordoñez, and G.D. García de Fernando, Growth survival of natural flora and *Listeria monocytogenes* on refrigerated uncooked pork and turkey packaged in modified atmospheres. *J. Food Safety*, 15, 305–319 (1995).

R. Marchetti, M.A. Casadei, and M.E. Guerzoni, Microbial population dynamics in ready-to-use vegetable salads. *Int. J. Food Sci.*, 4, 97–108 (1992).

M. Marcotte Irradiated strawberries enter the U.S. market. *Food Technol.*, 46, 80–86 (1992).

D.L. Marshall, P.L. Wiese-Lehigh, J.H. Wells, and A.J. Farr, Comparative growth of *Listeria monocytogenes* and *Pseudomonas fluorescens* on precooked chicken nuggets stored under modified atmospheres. *J. Food Prot.*, 54, 841–843 (1991).

M.L. Martin, D. Shipman, M.E. Potter, J.G. Wacbsmuth, K. Wells, K. Hedberg, and R.V. Tauxe, Isolation of *E. coli* O157:H7 from dairy cattle associated with two cases of haemolytic uremic syndrome. *Lancet*, 11, 1043 (1986).

R.B. Masson, Recherche de nouveax disinfectants pour les produits de 4eme gamme. In: Proc Congress Produits de 4eme Gamme et de 5eme Gamme, Brussels. C.E.R.I.A, 1990, pp. 101.

J.R. Mazollier, IVe gamme. Lavage-desinfection des salades. *Infros-Ctifl*, 41, 19 (1988).

K.H. McWatters, M.S. Chinnan, S.L. Walker, M.P. Doyle, and C.M. Lin, Consumer acceptance of fresh-cut iceberg lettuce treated with 2% hydrogen peroxide and mild heat. *J. Food Prot.*, 65, 1221–1226 (2002).

P.S. Mead, L. Slutsker, V. Dietz, L.F. McCaig, J.S. Bresee, C. Shapiro, P.M. Griffin, and R.V. Tauxe, Food-related illness and death in the United States. *Emerg. Infect. Dis.*, 5 (1999). Available at http://www.cdc.gov/ncidod/eid/vol5no5/mead.html. Accessed 02.03.06.

H. Michino, K. Araki, S. Minami, T. Nakayama, Y. Ejima, and K. Hiroe, "Recent Outbreaks of Infections Caused by *E. coli* O157:H7 in Japan," in J.P. Kaper and A.D. O'Brien, Eds., *E. coli O157:H7 and Other Shiga-Toxin Producing E. coli Strains.*" American Society for Microbiology, Washington, DC, 1998, pp. 73–81.

E.B. Mosley, P.R. Elliker, and H. Hays, Destruction of food spoilage indicator and pathogenic organisms by various germicides in solution and on a stainless steel surface. *J. Milk Food Technol.*, 39, 830–836 (1976).

J.H. Moy, "Radurization and Radication: Fruits and Vegetables," in E.S. Josephson and M.S. Peterson, Eds., *Preservation of Food by Ionizing Radiation.* CRC Press, Boca Raton, London, New York, Washington, DC, 1983, pp. 83–108.

M.S. Nascimento, N. Silva, M.P.L.M. Catanozi, and K.C. Silva, Effects of different disinfection treatments on the natural microbiota of lettuce. *J. Food Prot.*, 66, 1697–1700 (2003).

C. Nguyen-the, and F. Carlin, The microbiology of minimally processed fresh fruits and vegetables. *Crit. Rev. Food Sci. Nutr.*, 34, 371–401 (1994).

C. Nguyen-the and J.P. Prunier, Involvement of *Pseudomonas* in the deterioration of "ready-to-use" salads. *Int. J. Food Sci. Technol.*, 24, 47–58 (1989).

C. Nguyen-the, and F. Carlin, "Fresh and Processed Vegetables," in B.M. Lund, T.C. Baird-Parker, and G.M. Gould, Eds., *The Microbiological Safety and Quality of Food.* Aspen Publishers, Inc., Gaithersburg, Maryland, 2000, pp. 620–684.

S. Notermans, J. Dufrenne, and J.P.G. Gerrits, Natural occurrence of *C. botulinum* on fresh mushrooms (*Agaricus bisporus*). *J. Food Prot.*, 52, 733–736 (1989).

D.M. Obenland and L.H. Aung, Sodium chloride reduces damage to nectarines caused by hot water treatments. *Postharv. Biol. Technol.*, 12, 15–19 (1997).

J.M. Ogawa and S.D. Lyda, Effects of alcohol on spores of *Monilia fruticola* and other peach fruit rotting fungi of California. *Phytopathology*, 50, 790–792 (1960).

J.M. Ogawa, M.W. Hoy, B.T. Manji, and D.H. Hall, Proper use of chlorine for postharvest decay control of fresh market tomatoes. *California Tomato Rama, Informational Bulletin* no. 27, pp. 2 (1980).

Z.Y. Özbas, H. Vural, and S.A. Aytac, Effects of modified atmosphere and vacuum packaging on the growth of spoilage and inoculated pathogenic bacteria on fresh poultry. *Z. Lebensm. Unters. Forsch.*, 203, 326–332 (1996).

Z.Y. Özbas, H. Vural, and S.A. Aytac, Effects of modified atmosphere and vacuum packaging on the growth of spoilage and inoculated pathogenic bacteria on fresh poultry. *Fleischwirts chaft* 77, 1111–1116 (1997).

S. Pao and C.L. Davis, Enhancing microbiological safety of fresh orange juice by fruit immersion in hot water and chemical sanitizers. *J. Food Prot.*, 62, 756–760 (1999).

S. Pao, C.L. Davis, D.F. Kelsey, and P.D. Petracek, Sanitizing effects of fruit waxes at high pH and temperature on orange surfaces inoculated with *Escherichia coli*. *J. Food Sci.*, 64, 359–362 (1999).

S. Pao, C.L. Davis, and D.F. Kelsey, Efficacy of alkaline washing for the decontamination of orange fruit surfaces inoculated with *Escherichia coli*. *J. Food Prot.*, 63, 961–964 (2000).

M.E. Parish and P.M. Davidson, "Methods for Evaluation," in P.M. Davidson and A.L. Branen, Eds., *Antimicrobials in Foods*. Marcel Dekker, New York, 1993.

M.E. Parish, L.R. Beuchat, T.V. Suslow, L.J. Harris, E.H. Garret, J.N. Farber, and F.F. Busta, "Methods to Reduce/Eliminate Pathogens from Fresh and Fresh-Cut Produce," in *Comprehensive Reviews in Food Science and Food Safety*, Chapter 5. IFT/FDA, 2003.

C.M. Park and L.R. Beuchat, Evaluation of sanitizers for killing *Escherichia coli* O157:H7, *Salmonella* and naturally occurring microorganisms on cantaloupes, honeydew melons and asparagus. *Dairy Food Environ. Sanit.*, 19, 842–847 (1999).

W.P. Park, S.H. Cho, and D.S. Lee, Effect of minimal processing operations on the quality of garlic, green onion, soybean and watercress. *J. Sci. Food Agric.*, 77, 282–286 (1998).

C.-M., Park, Y.-C. Hung, M.P., Doyle, G.O.I. Ezeike, and C. Kim, Pathogen reduction and quality of lettuce treated with electrolyzed oxidizing and acidified chlorinated water. *J. Food Sci.*, 66, 1368–1372 (2001).

L. Pastoriza, G. Sampedro, J.J. Herrera, and M.L. Cabo, Influence of sodium chloride and modified atmosphere packaging on microbiological, chemical and sensorial properties in ice storage of slices of hake (*Merluccius merluccius*). *Food Chem.*, 61, 23–28 (1998).

J.E. Peeters, E. Ares Mazas, W.J. Masschelein, I. Villacorta Martinez de Maturana, and E. Debacker, Effect of disinfection of drinking water with ozone or chlorine dioxide on survival of *Cryposporidium parvum* oocysts. *Appl. Environ. Microbiol.*, 55, 1519–1522 (1989).

D.L. Peters, Control of enteric pathogenic bacteria on fresh produce. MSc, Lincoln, University of Nebraska Graduate College. 1995.

R.L. Petran, E.A. Zottola, and R.B. Gravani, Incidence of *Listeria monocytogenes* in market samples of fresh and frozen vegetables. *J. Food Sci.*, 53, 1238–1240 (1988).

M.E. Pirovani, D.R. Güemes, and A.M. Piagentini, Predictive models for available chlorine depletion and total microbial count reduction during washing of fresh-cut spinach. *J. Food Sci.*, 66, 860–864 (2001).

M. Pirovani, A. Piagentini, D. Güemes, and S. Arkwright, Reduction of chlorine concentration and microbial load during washing-disinfection of shredded lettuce. Int. *J. Food Sci. Technol.*, 39, 341–347 (2004).

D. Phillips Daifas, J.P. Smith, B. Blanchfield, and J.W. Austin, Effect of pH and CO_2 on growth and toxin production by *Clostridium botulinum* in English-style crumpets packaged under modified atmospheres, *J. Food Prot.*, 62, 1157–1161 (1999).

A.M. Piagentini, M.E. Pirovani, D.R. Güemes, J.H. Di Pentima, and M.A. Tessi, Survival and growth of *Salmonella hadar* on minimally processed cabbage as influenced by storage abuse conditions. *J. Food Sci.*, 62, 616–618 (1997).

M.D. Pierson and N.R. Reddy, *C. botulinum* status summary. *Food Technol.*, 42, 196–198 (1988).

L.D. Reina, H.P. Fleming, and E.G. Humphries, Microbiological control of cucumber hydrocooling water with chlorine dioxide. *J. Food Prot.*, 58, 541–546 (1995).

L. Restaino, E.W. Frampton, J.B. Hemphill, and P. Palnikar, Efficacy of ozonated water against various food-related microorganisms. *Appl. Environ. Microbiol.*, 61, 3471–3475 (1995).

E.J. Rhodehamel, FDA's concerns with *sous vide* processing. *Food Technol.*, 46, 73–76 (1992).

E.J. Rhodehamel, T. Lilly, J.H.M. Solomon, and D. Kautter, *Incidence of Clostridium botulinum in Modified Atmosphere Packaged Vegetables*. International Association of Milk, Food, Environ. Sanitarians, Atlanta, GA, (1993).

K.J. Richert, J.A. Albrecht, L.B. Bullerman, and S.S. Sunner, Survival and growth of *Escherichia coli* O157:H7 on broccoli, cucumber and green pepper. *Dairy Food Environ. Sanit.*, 20, 24–28 (2000).

P. Ringlé, J.P. Vincent, and M. Catteau, "Evolution de *Listeria* dans les produits de 4éme gamme," in C. Lahellec, Ed., *Les microorganismes contaminants dans les industries agroalimentaires: Colonisation, Détection*. Maitrise, 7éme Colloque de la Section Microbiologie Alimentaire, Société Francaise de Microbiologie, Paris, 1991, p. 324.

D.D. Rittman, "Can you have your cake and eat it too" with chlorine dioxide? *Water/Eng. Mag.*, April 1997.

P.G. Robbs, J.A. Bartz, G. McFie, and N.C. Hodge, Causes of decay of fresh-cut celery. *J. Food Sci.*, 61, 444–448 (1996).

A.J. Roberts and M. Wiedmann, Review. Pathogen, host and environmental factors contributing to the pathogenesis of listeriosis. *Cell. Mol. Life Sci.*, 60, 904–918 (2003).

R.G. Roberts and S.T. Reymond, Chlorine dioxide for reduction of postharvest pathogen inoculum during handling of tree fruits. *Appl. Environ. Microbiol.*, 60, 2864–2868 (1994).

R.M. Robins-Browne, "*Y. enterocolitica*," in M.P. Doyle, L.R. Beuchat, and T.J. Montville Eds., *Food Microbiology – Fundamentals and Frontiers*. American Society of Microbiology Press, Washington, DC, 1997, pp. 192–215.

A.M.C. Rodriguez, E.B. Alcala, R.M.G. Gimeno, and G.Z. Cosano, Growth modelling of *Listeria monocytogenes* in packaged fresh green asparagus. *Food Microbiol.*, 17, 421–482 (2000).

S.L. Rodgers, J.N. Cash, M. Siddiq, and E.T. Ryser, A comparison of different chemical sanitizers for inactivating *Escherichia coli* 157:H7 and *Listeria monocytogenes* in solution and on apples, lettuce, strawberries and cantaloupe. *J. Food Prot.*, 67, 721–731 (2004).

M.L. Rooney, *Active Food Packaging*. Blackie Academic and Professional, Glasgow, UK, 1995.

M.F. Saddik, M.R. El-Sherbeeny, and F.L. Bryan, Microbiological profiles of Egyptian raw vegetables and salads. *J. Food Prot.*, 48, 883–886 (1985).

Safety Alerts, Crunch pack apple packages recalled in 17 states, March 26, 2001a. Available at www.safetyalerts.com.

Safety Alerts, Boskovich Fresh-cut recalls fresh, halved and seeded red bell peppers, September 20, 2001b. Available at www.safetyalerts.com.

M.E. Saltveit, "Physical and Physiological Changes in Minimally Processed Fruits and Vegetables," in F.A. Tomás-Barberán and R.J. Robins, Eds., *Phytochemistry of Fruits and Vegetables*. Oxford University Press, London, 1997, pp. 205–220.

S. Sanz, M. Gimenez, and C. Olarte, Survival and growth of *Listeria monocytogenes* and enterohemorrhagic *Escherichia coli* O157:H7 in minimally processed artichokes. *J. Food Prot.*, 66, 2203–2209 (2003).

G.M. Sapers, "Washing and Sanitizing Raw Materials for Minimally Processed Fruit and Vegetable Products," in J.S. Novak, G.M. Sapers, and V.K. Juneja, Eds., *Microbial Safety of Minimally Processed Foods.* CRC Press, Boca Raton, London, New York, Washington, DC (2003) pp. 221–253.

G.M. Sapers and J.E. Sites, Efficacy of 1% hydrogen peroxide wash in decontaminating apples and cantaloupe melons. *J. Food Sci.*, 68, 1793–1797 (2003).

G.M. Sapers and G.F. Simmons, Hydrogen peroxide disinfection of minimally processed fruits and vegetables. *Food Technol.*, 52, 48–52 (1998).

G.M. Sapers, R.L. Miller, and A.M. Mattrazzo, Effectiveness of sanitizing agents in inactivating *Escherichia coli* in Golden Delicious apples. *J. Food Sci.*, 64, 734–737 (1999).

P. Sarig, T. Zahavi, Y. Zutkhi, S. Yannai, N. Lisker, and R. Ben-Arie, Ozone for control of postharvest decay of table grapes caused by *Rhizopus stolonifer. Physiol. Mol. Plant Pathol.*, 48, 403–415 (1996).

S.K. Sastry, A.K. Datta, and R.W. Worobo, "Ultraviolet Light," in *Kinetics of Microbial Inactivation for Alternative Food Processing Technologies.* Special supplement. *J. Food Sci.*, 65, 90–92 (2000).

S.A. Sattar, V.S. Springthorpe, and S.A. Ansari, "Rotavirus," in Y.H. Hui, J.R. Gorham, K.D. Murrell, and D.O. Cliver, Eds., *Foodborne Disease*, Handbook Vol. 2, Marcel Dekker, New York, 1994, pp. 81–11.

D.A. Schiemann, "*Y. enterocolitica* and *Y. pseudotuberculosis*," in M.P. Doyle, Ed., *Foodborne Bacterial Pathogens.* Marcel Dekker, New York, 1989, pp. 601–672.

D.V. Schlimme and M.L. Rooney, "Packaging of Minimally Processed Fruits and Vegetables," in R.C. Wiley, Ed., *Minimally Processed Refrigerated Fruits and Vegetables.* Chapman and Hall, New York, 1994, pp. 138–166.

V.N. Scott, Interaction of factors to control microbial spoilage of refrigerated foods. *J. Food Prot.*, 52, 431–435 (1989).

K.H. Seo and J.F. Frank, Attachment of *Escherichia coli* O157:H7 to lettuce leaf surface and bacterial viability in response to chlorine treatment as demonstrated by using confocal scanning laser microscopy. *J. Food Prot.*, 62, 3–9 (1999).

I.J. Seymour, D. Burfoot, R.L. Smith, L.A. Cox, and A. Lockwood, Ultrasound decontamination of minimally processed fruits and vegetables. *Int. J. Food Sci. Technol.*, 37, 547–557 (2002).

L. Shere, M.J. Kelley, and J.H. Richardson, Effect of bromide hypochlorite bactericides on microorganisms. *Appl. Microbiol.*, 10, 538–541 (1962).

R.L. Shewfelt, Quality of minimally processed fruits and vegetables. *J. Food Qual.*, 10, 143–156 (1987).

J.H. Silliker and S.K. Wolfe, Microbiological safety considerations in controlled atmosphere storage of meats. *Food Technol.*, 34, 59–63 (1980).

G.F. Simmons, J.L. Smilanick, S. John, and D.A. Margosan, Reduction of microbial populations on prunes by vapor-phase hydrogen peroxide. *J. Food Prot.*, 60, 188–191 (1997).

N. Singh, R.K. Singh, A.K. Bhunia, and R.L. Stroshine, Efficacy of chlorine dioxide, ozone, and thyme essential oil or a sequential washing in killing *Escherichia coli* O157:H7 on lettuce and baby carrots. *Lebensm. Wiss. Technol.*, 35, 720–729 (2002).

K. Sizmur and C.W. Walker, *Listeria* in prepacked salads. *Lancet*, 1, 1167 (1988).

J.L. Smilanick and R. Denis-Arrue, Control of green mold of lemons with *Pseudomonas* species. *Plant Dis.*, 76, 481–485 (1992).

C. Smith de Waal, K. Barlow, L. Alderton, and M.F. Jacobson, Outbreak alert!. Center for Science in the Public Interest, 2002. Available at http://www.cspinet.org/reports/outbreak_report.pdf.

L.A. Smoot and M.D. Pierson, Effect of oxidation reduction potential on the outgrowth and chemical inhibition of *C. botulinum* 10755A spores. *J. Food Sci.*, 44, 706–709 (1979).

H.M. Solomon, D.A. Kautter, T. Lilly, and E.J. Rhodehamel, Outgrowth of *C. botulinum* in shredded cabbage at room temperature under modified atmosphere. *J. Food Prot.*, 53, 831–833 (1990).

W.H. Sperber, Requirements of *C. botulinum* for growth and toxin production. *Food Technol.*, 36, 89–94 (1982).

R.A. Spotts and B.B. Peters, Chlorine and chlorine dioxide for control of "Anjou" pear decay. *Plant Dis.*, 64, 1095–1097 (1980).

D. Stanley, Yeasts and bacteria battle decay. *Agric. Res.*, 42, 8–9 (1994).

E.G. Steinbruegge, R.B. Maxcy, and M.B. Liewen, Fate of *L. monocytogenes* on ready to serve lettuce. *J. Food Prot.*, 51, 596–599 (1988).

J. Strasser, Ozone Applications in Apple Processing, tech application. Electric Power Research Institute, Inc., Palo Alto, CA, 1998.

H. Sugiyama and K.S. Rutledge, Failure of *Clostridium botulinum* to grow in fresh mushrooms packaged in plastic film over wraps with holes. *J. Food Prot.*, 41, 348–350 (1978).

H. Sugiyama and K.H. Yang, Growth potential of *C. botulinum* in fresh mushrooms packaged in semipermeable plastic film. *Appl. Microbiol.*, 30, 964–969 (1975).

T. Suslow, *Water Disinfection. A Practical Approach to Calculating Dose Values for Preharvest and Postharvest Applications*, Publication 7256. Oakland, CA: Division of Agriculture and Natural Resources, University of California, Oakland, CA, 2001, pp. 1–4.

E.A. Szabo and M.E. Cahill, The combined effects of modified atmosphere, temperature, nisin and ALTA™M2341 on the growth of *Listeria monocytogenes. Int. J. Food Microbiol.*, 43, 21–31 (1998).

K. Takeuchi and J.F. Frank, Penetration of *Escherichia coli* O157:H7 into lettuce tissues as affected by inoculum size and temperature and the effect of chlorine treatment on cell viability. *J. Food Prot.*, 63, 434–440 (2000).

S.K. Tamminga, R.R. Beumer, M.J.H. Keijbets, and E.H. Kampelmacher, Microbial spoilage and development of food poisoning bacteria in peeled, completely or partly cooked vacuum packed potatoes. *Archiv. Für. Lebensmittlelhygiene*, 29, 215–219 (1978).

R.V. Tauxe, *Salmonella*: a postmodern pathogen. *J. Food Prot.*, 54, 563–568 (1991).

D.W. Thayer and G. Boyd, Reduction of normal flora by irradiation and its effect on the ability of *Listeria monocytogenes* to multiply on ground turkey stored at 7°C when packaged under a modified atmosphere. *J. Food Prot.*, 63, 1702–1706 (2000).

D.W. Thayer, E.S. Josephson, A. Brynjolfsson, and G.G. Giddings, Radiation pasteurization of food. *Council Agri. Sci.Technol.*, Issue Paper No. 7(Apr), 1–12 (1996).

P. Thomas, Radiation preservation of foods of plant origin. V. Temperate fruits: pome fruits, stone fruits, and berries. *Crit. Rev. Fd. Sci. Technol.*, 24, 357–400 (1986).

K. Tokuoka and K. Isshiki, Possibility of application of allylisothiocyanate vapor for food preservation. *Nippon Shokuhin Kogyo Gakkaishi*, 41, 595–599 (1994).

S. Torriani, C. Orsi, and M. Vescovo, Potential of *Lactobacillus casei*, culture permeate, and lactic acid to control microorganisms in ready-to-use vegetables. *J. Food Prot.*, 60, 1564–1567 (1997).

L.H. Tsai, R. Higby, and J. Schade, Disinfection of poultry chiller water with chlorine dioxide: consumption and byproduct formation. *J. Agric. Food Chem.*, 43, 2768–2773 (1995).

B. Tukey, Overview of ozone use at Snokist Growers. *Tree Fruit Postharvest J.* 4, 14–15 (1993).

E.U. Tuncan, Effect of cold temperature on germicidal efficacy of quaternary ammonium compound, iodophor, and chlorine on *Listeria*. *J. Food Prot*, 56, 1029–1033 (1993).

D.O. Ukuku and W.F. Fett, Method of applying sanitizers and sample preparation affects recovery of native microflora and *Salmonella* on whole cantaloupe surfaces. *J. Food Prot.*, 67, 999–1004 (2004).

D.O. Ukuku and G.M. Sapers, Effect of sanitizer treatments on *Salmonella Stanley* attached to the surface of cantaloupe and cell transfer to fresh-cut tissues during cutting practices. *J. Food Prot.*, 64, 1286–1291 (2001).

D.O. Ukuku, V. Pilizota, and G.M. Sapers, Influence of washing treatment on native microflora and *Escherichia coli* 25922 populations of inoculated cantaloupes. *J. Food Safety*, 21, 31–47 (2001).

J. Ulate-Rodriguez, H.W. Schafer, E.A. Zottola, and P.M. Davidson, Inhibition of *Listeria monocytogenes, Escherichia coli* O157:H7, and *Micrococcus lutens* by linear furanocoumarins in a model food system. *J. Food Prot.*, 60, 1050–1054 (1997).

W.M. Urbain, "Fruits, Vegetables, and Nuts," in B.S. Schweigert, Ed., *Food Irradiation*. Academic Press, Orlando, FL, 1986, pp. 170–216.

J. Usall, N. Teixido, E. Fons, and I. Vinas, Biological control of blue mould on apple by a strain of *Candida sake* under several controlled atmosphere conditions. *Int. J. Food Microbiol.*, 58, 83–92 (2000).

A.H. Varnum and M.G. Evans, *Foodborne Pathogens* – An *Illustrated Text*. Wolfe Publishing, UK, England, 1991.

P. Varoquaux and J. Mazollier, "Overview of the European Fresh-Cut Produce Industry," in O. Lamikanra, Eds., *Fresh-Cut Fruits and Vegetables*. CRC Press, Boca Raton, London, New York, Washington, DC, 2002, pp. 21–43.

P. Varoquaux and R.C. Wiley, "Biological and Biochemical Changes in Minimally Processed Refrigerated Fruits and Vegetables," in R.C. Wiley, Ed., *Minimally Processed Refrigerated Fruits and Vegetables*. Chapman and Hall, New York, 1994, pp. 226–268.

S. Velani and D. Roberts, *Listeria monocytogenes* and other *Listeria* spp. in prepacked mixed salads and individual salad ingredients. *PHLS Microbiol. Digest*, 8, 21–22 (1991).

M. Vescovo, C. Orsi, G. Scolari, and S. Torriani, Inhibitory effect of selected lactic acid bacteria on microflora associated with ready-to-use vegetables. *Lett. Appl. Microbiol.*, 21, 121–125 (1995).

M. Vescovo, S. Torriani, C. Orsi, F. Macchiarolo, and G. Scolari, Application of antimicrobial-producing lactic acid bacteria to control pathogens in ready-to-use vegetables. *J. Appl. Bacteriol.*, 81, 113–119 (1996).

S.J. Walker and M.F. Stringer, Growth of *Listeria monocytogenes* and *Aeromonas hydrophila* at Chill Temperatures. Campden Food Preservation Research Association Technical Memorandum, CFPRA, Chipping Campden, UK, 1987, Vol. 462.

H.W. Walker and W.S. LaGrange, "Sanitation in Food Manufacturing Operations," in S.E. Block, Ed., *Disinfection, Sterilization, and Preservation* 4th Edn. Lea and Febiger, Philadelphia, PA, 1991.

J. Wan, A. Wilcock, and M.J. Coventry, The effect of essential oils of basil on the growth of *Aeromonas hydrophila* and *Pseudomonas fluorescens. J. Appl. Microbiol.*, 84, 152–158 (1998).

A.E. Watada, N.P. Ko, and D.A. Minott, Factors affecting quality of fresh-cut horticultural products. *Postharvest Biol. Technol.*, 9, 115–125 (1996).

C.I. Wei, T.S. Huang, J.M. Kim, W.F. Lin, M.L. Tamplin, and J.A. Bartz, Growth and survival of *Salmonella montevideo* on tomatoes and disinfection with chlorinated water. *J. Food Prot.*, 58, 829–836 (1995).

G.C. White, *Handbook of Chlorination and Alternative Disinfectants*, 3rd edn. Van Nostrand Reinhold, New York, 1992, p. 1308.

R.C. Wood, C. Hedberg, and K. White, A Multistate Outbreak of *Salmonella javiana* Infections Associated with Raw Tomatoes. CDC Epidemic Intelligence Service 40th Annual Conference, Abstracts, Centers for Disease Control, Atlanta, 1991, p. 69.

J.R. Wright, S.S. Sumner, C.R. Hackney, M.D. Pierson, and B.W. Zoecklein, Reduction of *Escherichia coli* O157:H7 on apples using wash and chemical sanitizer treatments. *Dairy Food Environ. Sanit.*, 20, 120–126 (2000).

F.M. Wu, M.P. Doyle, L.R. Beuchat, J.G. Wells, E.D. Mintz, and B. Swaminathan, Fate of *Shigella sonnei* on parsley and methods of disinfection. *J. Food Prot.*, 63, 568–572 (2000).

F.M. Wu, L.R. Beuchat, M.P. Doyles, V. Garret, J.G. Wells, and B. Swaminathan, Fate of *Escherichia coli* O157:H7 in coleslaw during storage. *J. Food Prot.*, 65, 845–847 (2002).

L. Xu, Use of ozone to improve the safety of fresh fruits and vegetables. *Food Technol.*, 53, 58–61, 3 (1999).

F. Xuetong and K.J.B. Sokorai, Sensorial and chemical quality of gamma-irradiated fresh-cut iceberg lettuce in modified atmosphere packages. *J. Food Prot.*, 65, 1760–1765 (2002).

F. Xuetong, P.M.A. Toivonen, K.T. Rajkowski, and K.J.B. Sokorai, Warm water treatment in combination with modified atmosphere packaging reduces undesirable effects of irradiation on the quality of fresh-cut iceberg lettuce. *J. Agric. Food Chem.*, 51, 1231–1236 (2003).

H. Yang, Y. Cheng, B.L. Swem, and Y. Li, Efficacy of cetylpyridinium chlorine on *Salmonella typhimurium* and *Escherichia coli* O157:H7 in immersion spray treatment of fresh-cut lettuce. *J. Food Sci.*, 68, 1008–1012 (2003).

F. Yildiz, "Initial Preparation, Handling, and Distribution of Minimally Processed Refrigerated Fruits and Vegetables," in R.C. Wiley, Ed., *Minimally Processed Refrigerated Fruits and Vegetables*. Chapman and Hall, New York, 1994, pp. 41–48.

J.T.C. Yuan, "Modified Atmosphere Packaging for Shelf-Life Extension," in J.S. Novak, G.M. Sapers, and V.K. Juneja, Eds., *Microbial Safety of Minimally Processed Foods*. CRC Press, Boca Raton, London, New York, Washington, DC, 2003, pp. 205–215.

B.R. Yaun, S.S. Sumner, J.D. Eifert, and J.E. Marcy, Inhibition of pathogens on fresh produce by ultraviolet energy. *Int. J. Food Microbiol.*, 90, 1–8 (2004).

A.A.M. Zeitoun and J.M. Debevere, Inhibition, survival and growth of *Listeria monocytogenes* on poultry as influenced by buffered lactic acid treatment and modified atmosphere packaging. *Int. J. Food Microbiol.*, 14, 161–169 (1991).

S. Zhang and J.M. Farber, The effects of various disinfectants against *Listeria monocytogenes* on fresh-cut vegetables. *Food Microbiol.*, 13, 311–321 (1996).

R.Y. Zhuang, L.R. Beuchat, and F.J. Angulo, Fate of *Salmonella montevideo* on and in raw tomatoes as affected by temperature and treatment with chlorine. *Appl. Environ. Microbiol.*, 61, 2127–2131 (1995).

R.Y. Zhuang and L.R. Beuchat, Effectiveness of trisodium phosphate for killing *Salmonella montevideo* on tomatoes. *Lett. Appl. Microbiol.*, 22, 97–100 (1996).

CHAPTER 7

Pathogen Survival on Fresh Fruit in Ocean Cargo and Warehouse Storage

LISE KORSTEN

Department of Microbiology and Plant Pathology and Forest and Agriculture Biotechnology Institute, University of Pretoria, South Africa

DEVON ZAGORY

Davis Fresh Technologies, LLC, Davis, CA, USA

Microbial Hazard Identification in Fresh Fruit and Vegetables, Edited by Jennylynd James

7.1 INTRODUCTION

During the past several years both the volume and diversity of refrigerated cargo being shipped in maritime containers has increased dramatically. In 2003, the production of refrigerated shipping containers increased by 15 percent over the previous year to 135,000 20 ft equivalents (World Cargo News, 2004). An estimated 61.9 million tons of marine refrigerated cargo were moved in 2001, which has since grown by 5 percent per annum (Tanner and Smale, 2005). With increased shipping of fresh fruits and vegetables come increased hazards. This is reflected in the increased number of disease outbreaks associated with the consumption of imported fresh fruits and vegetables (Tauxe, 1997; Centres for Disease Control and Prevention, 2003; Louisiana Food Safety and Information Clearinghouse, 2005). Although available data have rarely, if ever, implicated shipping and warehousing of fruits or vegetables in foodborne illness outbreaks, it would be prudent to review possible food-safety hazards associated with marine shipping and warehousing of fruit with an ultimate goal of developing programs to ensure that such hazards are minimized or do not materialize.

There are relatively few food-safety hazards that can be introduced or develop during ocean transport and warehousing of fresh fruits and vegetables. Nevertheless, these hazards must be considered and should be fully addressed to ensure food safety is not compromised within the food chain. The reality is that fresh fruits and vegetables can become contaminated with human pathogens at any point during production, harvesting, postharvest handling or subsequent transport, warehousing, and distribution. Once introduced, these pathogens can survive and in certain cases even increase on the contaminated product while it is being moved through the food chain. Once contaminated, foodborne pathogens can also not be reliably removed except through cooking. Pathogens that are therefore allowed to contaminate produce during ocean transport and warehousing will remain on those products despite the best efforts of subsequent handlers to prevent further contamination or to remove resident pathogens. This being the case, it is the responsibility of all members within the food-handling chain to prevent contamination with pathogens and not to pass a serious and undetected problem on to subsequent handlers and consumers.

Newly introduced quality management and food safety standards including Good Agricultural Practices (GAPs), Eurepgap, Good Manufacturing Practices (GMPs), Hazard Analysis Critical Control Point (HACCP), British Retail Consortium (BRC), and Nature's Choice have been instituted to address the various handling steps of produce production and distribution to try and reduce the probability of contamination. Food safety is a shared responsibility, which requires that all players in the food chain take responsibility and implement similar hygiene standards. However, at the present time, there is a notable lack of implementation of best practices or food-safety standards to ensure maintenance of product safety during ocean transport and warehousing. This chapter deals primarily with current hygiene standards applicable during fresh fruit and vegetable ocean distribution and warehousing.

7.2 FOOD SAFETY VS. FOOD QUALITY

The terms "food quality" and "food safety" are often interchangeably used. However, they are, and should be, differentiated as referring to separate issues. Food quality reflects the market desirability of the product, while food safety refers to the assurance that the food will not cause harm to the consumer when it is prepared and/or eaten according to its intended use (Jones, 1992). Quality may vary with season, weather conditions, market preparation, and market expectations. In short, quality is a variable depending on many factors. Safety ought not to be a variable as it is never acceptable that food should cause foodborne illnesses. A lack of food-safety assurance can result in a risk to the consumer, causing illness or even death. Ensuring and protecting the health of the nation is a governmental regulatory responsibility, while quality is "regulated" by market forces and minimum standards as dictated by supply and demand. Food quality may be assessed visually and may impact directly on the value of the product in the market, while foodborne pathogens may be introduced and remain undetected up to the consumer, where the impact can be severe.

In order to ensure delivery of top quality produce, several practices need to be integrated (Frenkel and Jen, 1989) including best production practices, optimum harvesting time, careful picking and packing practices, rapid cooling, maintenance of the cold chain, careful packing practices, selection of uniform products, freedom from phytosanitary risks or blemishes, and adherence to national and international quality standards. In order to retain product quality and prevent postharvest decay, an integrated approach needs to be followed further down the export chain (Brackett, 1993; Arul, 1994). Correct product handling and maintenance of the cold chain is therefore critical during distribution to ensure minimal physical injury, prevent decay development, and retain quality. From a food safety point of view it is important to consider the potential points where foodborne pathogens may be introduced into the food chain and how the environment can impact the pathogens' ability to survive and grow on the surfaces of contaminated fresh fruits and vegetables.

In developed countries, studies have shown a high annual prevalence ($\sim$10–15 percent in the population) of foodborne illness, which in the case of the United States has been estimated to be as high as 30 percent (Kaferstein, 2002). Contaminated food causes an estimated 76 million illnesses, 325,000 hospitalizations, and 5000 deaths each year in the United States alone (De Waal, 2002; World Health Organization, 2002). According to Schlundt (2002), this figure is most likely higher in developing countries, which often do not have disease alert and reporting structures in place. Furthermore, foodborne diseases most severely affect pregnant women, children, the elderly, and immunocompromised people (Jones, 1992; Roberts and Unnevehr, 1994; Buzby, 2001). In general, foodborne diseases impose a substantial burden on healthcare systems and markedly reduce economic productivity (Schlundt, 2002; Tauxe, 2002). In developing countries, inadequate food-safety systems negatively impact on general public health (World Bank, 2000). Internationally, food safety has become one of the most important requirements in global food trade (Schlundt, 2002; World Health Organization, 2002) and has, in certain cases, been regarded by developing countries as another form of technical barriers to trade.

Food safety refers to all hazards that may negatively affect the health of the consumer (Food and Agriculture Organization and World Health Organization, 2003). Currently, there are several recognized food-safety hazards, including pathogenic microorganisms, toxic chemicals, physical hazards, naturally occurring toxicants, food additives, tampering, and nutritional defects (Brackett et al., 1993). However, pathogenic microorganisms, toxic chemicals, and physical hazards are the main food safety hazards associated with fresh fruit and vegetables (Brackett et al., 1993; Schlundt, 2002). Of these, microbiological hazards are the most important and are globally considered the major cause of foodborne illnesses (Beuchat, 1995; Gast, 1997; Food and Agriculture Organization and World Health Organization, 2003). Traditionally, most food-safety concerns have dealt with processed food and high-risk products such as meat, chicken, and fish. In contrast, fresh fruits and vegetables were always regarded as wholesome and safe.

The global increase in trade, the growth in cooled shipping of fresh produce, and the national campaigns by governments, particularly in developed countries, to encourage healthful eating habits have resulted in increased levels of fresh fruit and vegetable consumption. This, together with the fact that fruits and vegetables were previously not regarded as natural hosts for foodborne pathogens and that contamination was mostly ascribed to accidental exposure, has resulted in a shift towards prevention of contamination as the basis of food-safety assurance within the fresh fruit and vegetable industries. Minimizing the risk of contamination has been driven by due diligence considerations by major retailers and has resulted in the development of voluntary standards for the produce industry. The focus of food-safety assurance systems has been mainly on unhygienic handling, and the use of untreated manure or contaminated irrigation water (Beuchat, 1999; World Bank, 2000). Very few studies have focused on the risks within the distribution chain.

7.3 MICROBIAL ASPECTS OF FRESH FRUITS AND VEGETABLES IN OCEAN CARGO AND WAREHOUSES

Microbial loads of between 10^2 and 10^6 colony-forming units (cfu) of yeasts and between 10^3 and 10^5 cfu of fungi per apple have been reported (Doores, 1983), while bacteria often occur in lower numbers on fruit (less than 5 percent of the total microbial community) (Teixidó et al., 1998). Kaneko et al. (1999) reported that total aerobic plate counts (APC) for intact vegetables could vary between 2.9 and 7.3 log cfu/g. Albrecht et al. (1995) reported 7.0 log cfu/g for ready-to-eat vegetables at the retail level. The difference in microbial loads on fresh fruits and vegetables often makes it difficult to compare data and set acceptable standards for wholesome produce. The Japanese local government, for instance, has defined foods with an APC <5.0 log cfu/g as being safe for human consumption. However, product quality is often still acceptable despite such high counts

(Nguyen-the and Carlin, 1994). There is little evidence suggesting that total microbial counts are a reliable predictor of the safety of the product (Zagory, 1999). Further, no realistic international standard exists for measuring natural microbial loads on fresh fruits and vegetables. This makes it difficult to assess when microbial loads exceed normal microbial profiles.

During typical postharvest treatments, microbial diversity and species richness may be reduced, ultimately resulting in open microbial niches that allow the introduction and survival of pathogenic microorganisms. Washing fruit can reduced microbial populations by 1–4 log (Brackett, 1992; Doores, 1983; Spotts and Cervantes, 1993). Carlin et al. (1996) reported that the background microflora of endive leaves was reduced by a chemical disinfectant, which permitted enhanced growth of *Listeria monocytogenes*. During cold storage most microbial populations will be adversely affected or eliminated, while psychrotrophic organisms will survive or proliferate. *Listeria monocytogenes* is a foodborne pathogen known to grow at refrigeration temperatures and is able to exploit this competitive advantage (Wells and Butterfield, 1997). It therefore represents an important risk factor in the food chain, because exported fresh fruits and vegetables can remain within the cold chain for extended periods of time, providing an opportunity for growth of Listeria.

If fruit and vegetables enter the food chain already contaminated by foodborne pathogens or become contaminated during ocean transport or warehousing, the pathogens may have the ability to survive the cold chain or proliferate. Such a product may also cross-contaminate other foods, particularly at retail where fresh fruits and vegetables are usually displayed together under variable temperature conditions. Le Blanc and co-workers found in 1996 that temperatures recorded in display cases for fruits and vegetables were on average 7.6–8.4°C (winter and summer, respectively), but the majority of these produce should have been stored below 4°C (Paull, 1999).

7.3.1 Foodborne Pathogens Associated with Fresh Fruits and Vegetables

A foodborne illness in its broadest definition encompasses "any illness that is contracted from the consumption of, or exposure to, food" (Acheson, 2000). Foodborne human pathogens known to contaminate produce include bacteria, viruses, and parasites such as protozoa (Jaykus, 1996; De Roever, 1999; Burnett and Beuchat, 2001; European Commission, 2002). Of these, bacteria are of greatest concern in terms of reported cases and seriousness of the illness (Brackett et al., 1993; Beuchat, 1995; Gast, 1997). Most foodborne pathogens, such as *Salmonella* spp., *Escherichia coli*, *Shigella* spp., *Campylobacter* spp., and hepatitis A, are enteric and their natural habitat is the intestinal tracts of animals and/or people (Table 7.1). Others include typical soil-borne pathogens such as *Clostridium* spp., *L. monocytogenes*, and *Bacillus cereus*, or waterborne pathogens, such as *Vibrio* spp., while others inhabit the skin and mucous membranes of humans, for example, *Staphylococcus aureus* (Iandolo, 2000). Beuchat (1998) listed several important foodborne

TABLE 7.1 Summary of Some Foodborne Pathogens Associated with Fresh Fruit and Vegetables that May Play a Role at the Ocean Cargo and Warehouse Stage of the Food Chain

Foodborne Pathogens	Reservoir	Route of Entry into the Food Chain	Most Likely Introduction at Ocean Cargo and Warehouse	Ability to Survive at Low Temperatures	Ability to Proliferate at Low Temperatures	Type of Crop/s from which Previously Isolated	Reported Disease Outbreaks: Vector/Pathogen/Source	Type of Pathogen Causing Disease	Minimum Infective Dose
Bacterial Foodborne Pathogens									
Aeromonas spp. e.g., *A. hydrophilia*	Drinking water, saline water, brackish water, sewage, fish	Untreated irrigation water	Polluted water contaminating product	Yes, but decline after 30 days at 5°C	Can grow rapidly in 2 weeks, can reach 10^6 cells/g at 4°C. Optimal growth temperature (OTG) = 28°C	Alfafa, asparagus, broccoli, cauliflower, endive, kale, lettuce, pepper, salad vegetables, spinach	Sporadic outbreaks on fresh produce	Questionable infections cytotoxic strains	Not known
Bacillus cereus	Soil	Soil	Dust	Yes	Spores survive OTG = 30–50°C	Green sprouts, mustard sprouts, soybean sprouts, sprouting seeds	Seed sprouts, vegetables and fruit	Gastrointestinal	$>10^5$/g.
Campylobacter spp., e.g., *C. jejuni*	Mainly poultry, animal, food and waterborne	Untreated irrigation and surface water fecal/oral route	Food handlers, birds, insects, horizontal transmission	Yes, two weeks at 4°C	No growth, require reduced oxygen for growth. OTG = 42°C	Cabbage, green onion, mushroooms, mustard sprouts, leafy vegetables, lettuce, parsley, potato, radish, salad vegetables, spinach	Cross-contamination with poultry	Infections, bacterial enteritis Campylobacteriosis	500 cells
Clostridium perfringers	Soil	Soil	MAP fish	Yes, spores survive 1–2 months at <0°C. Spores can germinate at 7°C	Vegetative growth at 12°C. OTG = 45°C	Vegetables	Coleslaw, modified packaging	Perfringens food poisoning	$>10^5$/g.

Escherichia coli *E. coli* (O157:H7)	Cattle primary reservoir/beef	Fecal/oral route	Contaminated hands	Yes, survive at <0°C for 9 months	*E. coli* O157:H7 can grow in apple cider at 8°C or on shredded lettuce kept at 12°C. OTG = 37°C	Alfalfa sprouts, broccoli, cabbage, cantaloupe, celery, cilantro, coriander, spinach, salads	Cantaloupe, unpasteurized apple cider, broccoli cross-contaminated with raw beef, garden salad (lettuce, peas, carrots)	Infections: Entero-hemorrhagic *E. coli* (O157:H7) Enterotoxigenic travellers diarrhea	10–1000
Listeria moncytogenes	Soil/water/mammal feces/grow naturally on plants	Raw product/fecal/oral route	Aerosols, dust, walls, floors, drains, hands	Yes, minimum −1.5°C, reportedly grow on produce kept at 3°C. Psychrotrophic	Yes particularly damp environments, grow on lettuce at 5°C, cauliflower at 4°C. OTG = 30–37°C	Bean sprouts, broccoli, cabbage, cucumber, leafy vegetables, leeks, peppers, potato, radish, salad vegetables, tomato	Coleslaw kept in a shed for a season, lettuce, raw celery/tomato, lettuce, mixed salads	Infections, human listeriosis	$>10^2$/g
Salmonella spp., e.g. *S. enterididis; S. heidelberg; S. muenchen; S. saintpaul*	Mammalian feces, natural reservoir in animals/birds	Fecal/oral route	Contaminated hands, dust, aerosols	Yes, minimum 5.5°C, but can survive between 2 and 4°C on lettuce for 42–56 days, on cabbage for 35–49 and on carrots for 28–49 days	No, OTG = 37°C	Alfalfa sprouts, almonds, artichokes, bean sprouts, beet leaves, cabbage, cantaloupe, cauliflower, celery, egg plant, endive, fennel, honeydew melon, lettuce, mango, mungbeen sprouts, mushrooms, mustard cress, nuts, potato, radish, parsley, peppers, spinach, strawberries, tomato	Raw tomatoes = *S. javana* Raw tomato = *S. montevideo* Raw bean sprouts = *S. saintpaul* Cantaloupe contamination between field and market place = *S. chester*	Salmonellosis	10–1000

(*Continued*)

TABLE 7.1 *Continued*

Foodborne Pathogens	Reservoir	Route of Entry into the Food Chain	Most Likely Introduction at Ocean Cargo and Warehouse	Ability to Survive at Low Temperatures	Ability to Proliferate at Low Temperatures	Type of Crop/s from which Previously Isolated	Reported Disease Outbreaks: Vector/Pathogen/Source	Type of Pathogen Causing Disease	Minimum Infective Dose
Shigella spp., e.g., *S. sonnei*	Human feces, contaminated food	Fecal/oral route, insects	Contaminated hands, flies, roaches	Yes, can survive 5°C for 3 days on lettuce	No OTG = 44°C	Honeydew melon, lettuce, parsley, salad vegetables	Contaminated raw vegetables, contaminated. lettuce, food handler	Shigellosis, gastroenteritis	10–1000
Staphylococcus aureus	Human nasal, skin, hands	Food handlers	Food handlers	Yes, minimum 6.5°C, survive <0°C for 10 weeks	Slow growth OTG = 30–37°C	Banana, carrot, lettuce, parsley, radish, salad vegetables, seed sprouts, sweet potato	Vegetable salad, food handler	Staphylococcal food poisoning	Vegetative cells killed by heat, heat stable, toxin min dose 200 ng
Yersinia spp., e.g., *Y. enterocolitica*	Swine natural reservoir	Manure, fecal	Unknown, possible food handlers	Yes minimum −1.5°C, can grow on produce at 4°C	Yes, commonly grow on produce during transport and storage, OTG = 28–29°C	Cabbage, carrot, cucumber, lettuce, potato, radish, salad vegetables	Raw vegetables as vehicles	Yersiniosis	Uncertain
Other Groups of Organisms of Importance in Food Safety									
Cryptosporidium spp. (Protozoa)	Mammalian feces	Fecal/oral, untreated irrigation, surface and drinking water	Contaminated food handler	Up to 40 days at 4°C	Oocycsts survive, obligate parasite	Apple cider, berries, vegetables	Drinking water, apple cider, berries, vegetables	Cryptosporidiosis	<30

Cyclospora spp. (Protozoa)	Human feces	Fecal/oral	Infected food handlers	Two months at 4°C	Oocysts survive	Basil, raspberries, lettuce	Water, berries, lettuce	Cyclosporiasis	Likely low
Hepatitis A, virus	Human feces and urine	Fecal/oral	Infected hands	Yes	No	Frozen strawberries, lettuce	Raspberry mouses, frozen strawberries, lettuce, tomato, picker/packer infected	Hepatitis infection	10–50
Norwalk virus	Man, survival of virus on soil	Fecal/oral	Infected hands	Yes, survive on soils, vegetables stored at 4°C for 30 days	No	Carrots, green salads, radish	Food handlers green salads	Viral gastroenteritis	100–1000
Fungi, e.g., *Aspergillus* spp. *Fusarium* spp. *Penicillium* spp.	Soil, plant debris	Infected fruit	Air	Yes	Yes, no toxin production under 10°C	Fruits, grains	Corn, wheat	Mycotoxins, mycotoxicosis	Depend cn type of toxin and host

Source: Harris, 1997; Food Safety and Information Clearinghouse, Food and Drug Administration and Center for Food Safety and Applied Nutrition, 2003; Pestnet, Snyder, 2003).

pathogens associated with fresh fruits and vegetables and expressed their relevant importance as percent frequency in outbreaks of known etiology. For instance, *E. coli* O157:H7 and *Salmonella* were responsible in >8 percent, *L. monocytogenes* varied between 4 and 8 percent, and *Campylobacter* 3 percent of reported cases.

The two most important foodborne pathogens associated with fresh fruits and vegetables have been *Salmonella* spp. and *L. monocytogenes* (Beuchat, 1999). These pathogens were most commonly isolated from a range of different fruits and vegetables (Table 7.1). Foodborne pathogens are widespread and international. These pathogens were isolated in 20 different countries (Beuchat, 1999). The most were reported from the United States (25), followed by Spain (14), Canada (13), and Saudi Arabia (12). The more regular testing in the United States, and access to more diagnostic laboratories might explain the higher number in the developed countries. However, increased international trade in fresh fruits and vegetables ensures that pathogens, together with their food vectors, will continue to be widely distributed.

7.3.2 Foodborne Pathogen Survival on Fruit and Vegetable Surfaces

Although viruses do not multiply on plant surfaces, they can survive once introduced (Table 7.1). Several viral pathogens have been detected on green onions (CDC, 2003). Hepatitis A can be transmitted via infected workers (Louisiana Food Safety and Information Clearinghouse, 2005) at any point where product handling takes place. Survival of foodborne pathogens on the surfaces of fresh produce becomes important in situations where the initial inoculation dose is high enough to exceed the minimal infective dose (MID) of the pathogen. Several foodborne pathogens, such as *Salmonella* spp., *Shigella* spp., *E. coli*, and hepatitis A virus, have relative low MID levels (Table 7.1) and may therefore pose a potentially higher food safety risk for the end user.

7.3.3 Impact of Temperature on Foodborne Pathogens Within Fresh Fruit and Vegetable Chains

Several foodborne pathogens have the ability to survive under different cold temperature conditions depending on the type of crop and length of storage time (Table 7.1). The most important psychrotrophic organism from a food safety point of view is *L. monocytogenes*. Several reports describe *Listeria* surviving or growing on numerous fresh fruit and vegetable surfaces when kept at temperatures as low as 3°C (Beuchat, 1995, 1999; Food and Drug Administration and Centre for Food Safety and Applied Nutrition, 2001). De Roever (1999) described several studies where foodborne pathogens such as *Yersinia enteritidis* were shown to grow at low temperatures (4°C). Berrang et al. (1989) showed that *Aeromonas hydrophila* reached populations exceeding 10^6 cells/g on fresh asparagus, broccoli, and cauliflower when kept at 4°C for two weeks. *S. aureus* can grow slowly on food products kept at 8°C (Beuchat, 1999). *Salmonella* can increase rapidly on cut tomato surfaces stored for 2 weeks at 10°C (Zhuang et al., 2005). In the case of citrus, *Salmonella* typhimurium could still be isolated from spiked fruit held at 7°C after 3 weeks, although the total number of viable cells decreased rapidly over time (Fig. 7.1) (Korsten et al., unpublished data). *Escherichia coli* O157:H7 grew in apple cider when kept at 8°C (Zhao et al., 1993).

The longer fresh produce is held before consumption the longer psychrotrophic organisms have to grow. The National Advisory Committee on Microbiological Criteria for Foods reviewed the microbiological safety of fresh produce and found that longer food chains could permit amplification of pathogens and thereby create a health risk for the consumer (De Roever, 1999). Viruses are also known to survive longer on fresh produce kept cold and some can survive longer than the shelf-life of the product (Beuchat, 1995). For instance, Badawy et al. (1985) showed that rotavirus could survive on lettuce, radish, and carrots for 25–30 days when kept at 4°C but only 5–25 days at ambient temperatures.

Increased foodborne illnesses and product recalls associated with imported fresh fruits and vegetables have recently led to a greater awareness of temperature and to effective relative humidity management during shipping and storage (Paull, 1999). High condensate on produce is often noted when containers are opened after shipment. Moisture in storage and shipment facilities and on produce may favor microbial growth and, combined with low temperatures, is particularly favorable for growth of the foodborne pathogen *L. monocytogenes* (Beuchat, 1999).

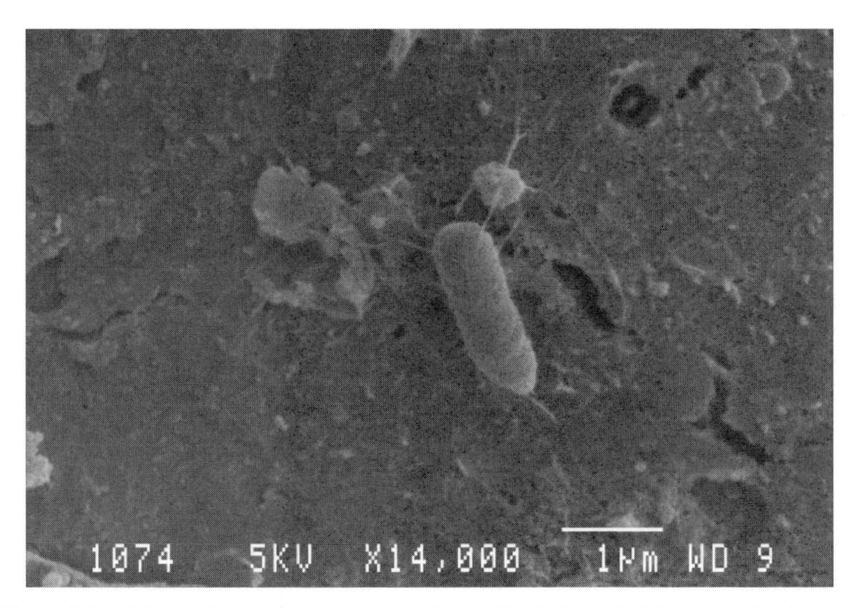

Figure 7.1 *Salmonella typhimurium* survival on spiked citrus fruit in a simulated export chain maintained at 7°C (unpublished data, courtesy L. Jooste).

7.4 MICROBIAL CONTAMINATION POINTS DURING OCEAN TRANSPORT AND WAREHOUSING

The potential risk factors can be grouped two categories: (1) pathogens already introduced onto the fruit or vegetable surfaces before entering the ocean cargo stage of the food chain, and (2) pathogens introduced onto the product at the transport, shipping, and warehouse/coldroom stage.

Incidental introductions at the beginning of the food chain are mostly associated with fecal contamination in the environment through contaminated water or soil, product exposure to untreated manure, animal, insect, or bird droppings, or contamination through handling of produce by infected people. These aspects have been dealt with more extensively in previous chapters. Of the various potential sources of postharvest contamination listed by Burnett and Beuchat (2001), inadequate facility sanitation, unhygienic transport containers, improper unhygienic handling, and dust are applicable during ocean transport and warehousing. For the most part, ocean cargos transporting fresh fruit and vegetables are packed, closed, palletized, and strapped before being loaded into marine containers or in conventional reefers. On arrival at the destination port these pallets may be transferred to temporary holding facilities such as warehouses or cold stores. The fruits and vegetables are thus protected from most general or cross-contamination. Nevertheless, there are several points that can present a potential risk of introducing foodborne pathogens.

7.4.1 Impact of Microbial Air Quality

Numerous microbial reservoirs are present in indoor environments, contributing to immigrating microorganisms and thus creating the potential for introducing foreign organisms onto surfaces (Stetzenbach et al., 2004). Airflow patterns in facilities are capable of supporting and spreading particulates from 0.001 to 1000 microns in size (Ecolab, 2001). Microorganisms are known to fall within this range (viruses 0.01–0.3 μm, bacteria 0.5–2 μm, and fungal spores 2–200 μm), and microbial flow patterns are therefore critical in hygiene programs in the healthcare and food sectors (Stetzenbach et al., 2004). The concentration of airborne microflora depends on the physical design of the facility, airflow patterns, human activity, general hygiene, and climatic and ecological conditions (Den Aantrekker et al., 2003; Medrela-Kuder, 2003). In addition, the external climatic and ecological conditions surrounding the facility may also impact on airflow patterns within the facility. This is particularly true for cold stores or warehouses. Within containers, ocean cargo and warehouse facilities the impact of contaminated air on product contamination with foodborne pathogens has not been conclusively illustrated.

Once a product has been exposed to airborne microorganisms it can become contaminated, resulting in product infestation and product decay (Den Aantrekker et al., 2003). Airborne pathogens can contaminate fresh produce if the product is exposed to dust, or remains uncovered for extended periods of time (Den Aantrekker et al.,

2003). According to Northcutt et al. (2004), airborne bacteria originate from contaminated surfaces or areas in processing facilities where cleaning and disinfestation is minimal. Such microorganisms can move through the air while adhering to dust or skin particles and will eventually settle on surfaces as a result of gravitation (Den Aantrekker et al., 2003).

Air conditioning systems, and particularly evaporator cooling coils in coldrooms, produce cold and wet environments inside the apparatus (Evans et al., 2004). In chilled rooms, evaporator fans draw large quantities of air over evaporator coils and distribute it through the room. Any contaminants in the air will be drawn over the cooling coils and some will be deposited. Microbes can multiply in the presence of moisture and can form biofilms. In one study, *S. aureus* and *B. cereus* were found in the air environment of most facilities studied (Evans et al., 2004). These organisms were circulated via the ventilation systems. Extended environmental persistence of *S. enteriditis* may have allowed widespread dispersal of the pathogen (Gast et al., 2004). *Listeria monocytogenes* is able to grow in moist conditions at refrigerated temperatures as low as 3°C (Beuchat, 1995, 1999; Food and Drug Administration and Center for Food Safety and Applied Nutrition, 2003). Thus, refrigeration units are an ideal environment for colonization and growth of *Listeria* and ought to be included in programs of regular cleaning and sanitation. In general, cleaning procedures in most facilities only focus on occasional cleaning of the evaporators and it is not part of a routine cleaning schedule. The evaporator coils can therefore represent a contamination source responsible for effective continuous circulation pathogens within the food-handling environment.

7.4.2 Food-Safety Risks in Maritime Containers

Containers that have previously been used to transport animals or animal products and have since not been thoroughly cleaned and sanitized could potentially carry contaminants that could be transferred to pallets, boxes, or bins carrying produce (Geldreich and Bordner, 1970). Containers that have been used to transport animal or animal products are not to be used for the transport of plant and plant products. However, traceability systems are not always in place to ensure that containers used for transportation of fresh fruits and vegetables are only used for plant-based foods (J. Botha, Maersk, personal communication).

Currently, containers are cleaned at holding ports prior to dispatch and are inspected for damage and taint (PPECB, 2005). However, disinfestation of containers is currently not part of the cleaning schedule. Furthermore, hygiene levels in terms of microbial contamination are not being monitored prior to dispatch and no records exist that can reflect the levels of sanitation or contamination. Containers are normally cleaned with high-pressure steam (70°C) (PPECB, 2005). In our studies, this cleaning approach resulted in higher bacterial counts than on uncleaned containers (Korsten, unpublished data). This may be because, during spray cleaning, microorganisms can be transferred to the air through aerosol formation (Den Aantrekker et al., 2003).

7.4.3 Food-Safety Risks in Warehouse and Coldroom Facilities

General levels of facility sanitation may influence microbial populations and the ultimate presence of foodborne pathogens on fresh fruit and vegetable surfaces. According to Madden (1992), Martin and co-workers reported in 1986 a large outbreak of gastroenteritis, caused by *Shigella sonnei* contaminating lettuce in a warehouse. In 1970, Geldreich and Bordner described the likelihood of crop contamination during shipping in dirty freight cars, cargo ships, boats, or trucks. They also described several studies where higher levels of foodborne pathogen contamination were recorded in products sampled at the market compared to those sampled in the field. Their study highlighted the importance of effective facility sanitation. In contrast to general sanitation standards that are currently being implemented in the fields through implementation of GAPs and through GMP, Eurepgap, HACCP, and British Retail Consortium (BRC) standards in packinghouses, few similar standards have been implemented in the transport, warehouse, and coldroom sectors of the food chain.

7.4.4 Impact of Personal Hygiene of Warehouse and Cooling Facility Workers

In a recent study of the apple, citrus, and litchi export chains (Korsten et al., unpublished results), *S. aureus* was the only human pathogen isolated from port terminal and coldroom environments and hands of workers and consumers. Some strains of *S. aureus* are capable of producing a highly heat-stable protein toxin that can cause illness in humans (FDA, 2003). To what extent *S. aureus* poses a potential risk for consumers is not certain, but it has thus far not been reported from fresh fruits. *Staphylococcus aureus* is ubiquitous in the environment and eradication of reservoirs is therefore important (Roberson et al., 1994). The presence of *S. aureus* in or on foods commonly indicates contamination that has been directly introduced by workers. This can happen through direct handling during repacking or inspections or by sneezing or coughing onto the product (Sandel and McKillip, 2004; Jaykus 1996). According to Sandel and McKillip (2004), approximately 50 percent of people carry *S. aureus* as commensals, with the predominant site of colonization being the throat and nose. In his study, Roberson et al. (1994) found that if a worker's nose was colonized by *S. aureus*, their hands would also be colonized.

7.4.5 Pest Control and Domestic Animals

Insect and rodent control at the warehouse is important, because enteric pathogens can be transmitted via these routes. *Shigella* and *Listeria* have been described as being insect transmittable, and animals and birds can introduce *Salmonella* and roaches transmit *Shigella* spp. (Louisiana Food Safety and Information Clearinghouse, 2005). The recently introduced Business Anti-Smuggling Coalition (BASC) standard (BASC, 2005) to ensure drug-free exports from Latin American countries into the United States utilizes sniffer dogs to inspect fresh produce consignments. During the physical inspection process the dogs sniff the pallets at the bottom, mid, and top section on all four sides for possible hidden drugs. In the case of packed palletized fresh asparagus, the first author observed that the dogs came in direct contact with the asparagus through the open display sections of the boxes. The possibility exists that paws or noses could touch the product and thereby transmit a foodborne pathogen. Taking into consideration that produce in this situation will remain in transit for several days or weeks, some of these organisms may survive (*Salmonella* and *E. coli*) or even increase (*Listeria*).

7.5 FOOD-SAFETY MANAGEMENT WITHIN THE FOOD CHAIN

Pathogens can be introduced into warehouses through various mechanisms and vectors and, once established, can contaminate the fruits and vegetables being stored or handled within these facilities. Pathogens can enter on fruits or vegetables that were contaminated in the field during production or harvesting, on bins, trays, boxes, or pallets that have been stored under unsanitary conditions, through cross-contamination from foods of animal origin transported or stored together with produce, or on the hands of workers who are ill and practice poor personal hygiene. Prevention of pathogen contamination in warehouses rests on comprehensive programs that address and reduce the risk from each of the sources mentioned above. Facility cleaning and sanitation programs should ensure that the entire facility is maintained in a clean and sanitary condition. Of special importance is that drains, cooling coils, drip pans, ice machines, and other areas that are routinely cold and wet are regularly cleaned, sanitized, and swabbed to prevent the survival and reproduction of *L. monocytogenes*. For instance, quaternary ammonium sanitizers are effective in excluding *Listeria* from containers, coldrooms, or warehouses.

According to Beuchat and Ryu (1997), temperature control is critical at every stage of postharvest handling if any success is to be achieved in minimizing the growth of bacterial pathogens. Produce should be stored so as to allow proper circulation of cold air. However, both *L. monocytogenes* and *Bacillus* spp. are known to grow at refrigerator temperatures. Viruses do not grow at all on fruits and vegetables. It is thus clear that maintaining proper cold temperatures is not a reliable guarantor of safety. Because maintenance of proper temperatures cannot be assured at retail or in the home, prevention of contamination and not temperature control should be the primary guarantor of safety. Standard pest control procedures should be followed to monitor and trap insects and rodents that can potentially transmit foodborne pathogens, particularly in warehouses or distribution centers where produce may stay for periods of time. The "first-in first-out" principle (FIFO) should also be practiced to prevent loss of quality and minimize time available for pathogen growth. Effective stock control forms the basis of any effective quality management system.

Workers can transmit foodborne pathogens via handling and when not following adequate personal hygiene practices. Workers in the infectious stages of hepatitis A, for instance, should be excluded from contact with food products and should not be involved in inspecting or repacking fruits or vegetables. Appropriate guidelines and proper training of workers to inform supervisors when they are ill, and training of supervisors to look for and recognize the symptoms of these illnesses, can help prevent such contamination. All workers should be regularly trained in proper personal hygiene, basic food safety, and careful handling of fresh produce. Through these programs, the probability of contamination within the warehouse can be reduced. Nevertheless, it is the responsibility of everybody in the food industry to know where their products come from, how they were handled, and to confirm that previous handlers practiced appropriate food-safety precautions and had the appropriate programs to identify, address, and reduce food-safety hazards. Although these perspectives may be new to some sectors of the produce industry, it is rapidly becoming requisite due diligence and should be incorporated into everyone's food-safety programs.

7.6 CONTAINER AND FACILITY SANITATION PRACTICES

Facilities should be so designed as to allow thorough cleaning and to minimize cross-contamination (Brackett, 1992; Hurst and Schuler, 1992). The general level of sanitation will directly influence microbial populations and the likelihood of the presence of pathogens on food products. Pathogenic microorganisms may be found in food-handling premises on floors, in drains, and on food contact surfaces (Food and Drug Administration, 1998). Places that are often neglected in the cleaning process, such as ceilings and walls, have been reported to be potential sources of *L. monocytogenes* and other pathogenic microorganisms (Brackett, 1992). *Listeria* has been described as one pathogen that all food-handling facilities should be aware of (International Specialty Supply, 2005). *Listeria monocytogenes* is common in most food environments, can grow at low temperatures in moist conditions, and may contaminate produce if condensation from refrigeration units or ceilings drips onto produce, particularly uncovered pallets.

Known reservoirs for this pathogen include floors, drains, ceilings, refrigeration units, forklifts, wooden pallets, maintenance tools, overhead rails, and overhead pipes (International Specialty Supply, 2005). "Hidden" areas include rusted material, equipment frames, air filters, shelving, door seals, cracked walls, and inadequately sealed surface panels. Typical "reservoirs and hidden areas" are common in most warehouses or cold stores. Cold-storage facilities, and particular refrigeration coils, refrigeration drip pans, forced air-cooling fans, and drain tiles should therefore be cleaned and sanitized on a regular basis (Gorny and Zagory, 2003). Sanitation programs should be so designed and validated as to ensure that all reservoirs and hidden areas are effectively cleaned to prevent the establishment of human pathogens. Quaternary ammonium sanitizers are particularly effective against Gram-positive bacteria, such as *Listeria* (Beuchat, 1999).

Facility design is critical to ensure easy access for regular and adequate cleaning, because poorly designed facilities can provide niches for microbial growth. In ocean cargo, ozone is sometimes pumped through the various levels of reefers once fresh fruit consignments have been cleared. This effectively sanitizes conventional reefers (personal discussions of the first author with port authorities at Antwerp port). However, it is important to point out that sanitizers are only effective on clean surfaces. Port terminals have different cleaning programs that may not include a deep clean program. Floors are often dirty because of the movement of forklifts and bulk produce. However, deep cleaning of these facilities can only happen in off seasons when lower volumes are being handled (personal communication with quality managers at Antwerp, Rotterdam, and Hamburg port terminals). Any large facility that holds fruit or vegetables while they are being moved through the export chain should be regularly cleaned and sanitized. Removal of debris, particularly between pallet stacking structures and inspection bins, is an essential part of an effective sanitation program. Regular cleaning of floors, drains, cooling coils, and drip pans in air-conditioning systems is crucial to maintain adequate hygiene levels. Unhygienic hot spots or reservoirs of filth should be determined for all facilities and should be cleaned more regularly. Monitoring cleaning efficiency should also be part of any cleaning program. Taking environmental microbial swabs from walls and floors, as well as careful observations to identify problem areas, could be of importance to determine cleaning intervals and to assess the effectiveness of the cleaning and sanitation program.

Managing container hygiene is an essential part of food-safety assurance. Traditionally, maritime container cleaning has been outsourced and they are usually inspected at ports for basic cleanliness. Containers should be thoroughly cleaned, kept in good repair, and used only for the transport of uncontaminated food items (Geldreich and Bordner, 1970). Inspection of containers is currently focused on odors, checking for damage to containers and inspecting cleanliness prior to sealing. However, the current inspection process does not include hygiene monitoring. In order to effectively manage sanitation levels in the export chain, maritime containers should be sanitized between loads. Further, a traceability system should be introduced to ensure that containers are marked and used only for transporting fresh fruits and vegetables.

7.7 IMPACT OF TRAINING ON SUSTAINING FOOD-SAFETY ASSURANCE

In contrast to training programs provided for primary agricultural workers, which are focused on careful picking, packing, and handling practices and basic personal hygiene, workers in port terminals, distribution, and repack centers are often not trained in similar basic food-handling and sanitation practices and personnel hygiene principles. A recent study by Walker et al. (2003) highlighted the lack of basic knowledge of hygiene amongst 104 small food business operators. This was described as a barrier to the effective implementation of HACCP. In a recent study of apple and citrus distribution (Korsten, 2005, unpublished data) careful handling (offloading from reefers and placement of pallets in stacking racks in coldrooms/warehouses) was found to be seriously lacking. Dockworkers operating forklifts work against a tight time schedule to offload vessels (containers and/or pallets). Moving pallets into the warehouse or coldrooms and placing them on racks or in allocated spaces happens quickly, and forklift drivers at times damage boxes as they rapidly move pallets into position. Dockworkers can be required to move both food and nonfood products and are, in some cases, temporary workers ill-trained in safe product handling (personal observations and communications with dock authorities at three ports in the EU). Once a product is damaged it provides a suitable environment for multiplication of microbial organisms. Safe handling of the pallets, boxes (during restacking), or individual fruit (during repacking) remains critical to ensure product integrity and ultimate safety.

Another serious lack of knowledge in terms of product handling and basic hygiene was found to be at the point of repacking of fruit at warehouses and distribution centers. The import agent will often arrange for individual boxes within a pallet or consignment to be repacked in the warehouse or distribution center. This will usually happen in cases where product decay within a pallet or box was observed or after inspections, or in cases where boxes were damaged due to poor handling. In these cases, training in personnel hygiene and careful product handling is often not required prior to contracting temporary repackers (personal observations and communications with importers and warehouse managers by the first author within EU facilities). In general, worker knowledge of safe product handling and personal hygiene was found to be seriously lacking from the point of transport to the ports, loading, containerizing, offloading, warehousing, and distribution. This lack of basic knowledge of hygiene and handling practices often resulted in increased wounding of produce and potential contamination. As with prefarm gate requirements for training, a similar responsibility resides with the appropriate authorities and role players further down the food chain to ensure that food safety of fresh fruits and vegetables is not compromised.

7.8 CONCLUSIONS

The accidental introduction of foodborne pathogens to fruit or vegetable surfaces via exposure to contaminated unhygienic facilities or through human handling may happen at any point within the food chain. Once introduced, foodborne pathogens can attach, survive, and reproduce, resulting in a potential threat to the end consumer. Various extrinsic and intrinsic factors, such as environmental conditions, fruit condition, shelf-life, and other competing residential microorganisms may influence the "in transit" growth or survival of these foodborne pathogens. This review identifies potential points of introduction and discusses the likelihood of foodborne pathogen survival in this section of the food chain. Produce may arrive already contaminated or may be accidentally contaminated via contaminated air, unhygienic human handling during inspection, repacking, or by exposure to unsanitary contact surfaces or environments.

The impact of basic facility sanitation and personal hygiene practices at the ocean cargo and warehouse/coldroom should not be underestimated. Marine container and warehouse or coldroom sanitation levels are often not up to the same standard as in the rest of the food chain. A lack of adequate worker training and the lack of awareness of the importance of personal hygiene and careful product handling and how it may impact on product safety is a further matter of concern. Developing an effective sanitation program and hygiene system that is audited to ensure compliance should be an essential component of a comprehensive food-safety system.

For some time, maintaining quality has been the main challenge faced by the global fresh produce industries, and systems have been developed to ensure its maintenance. Compliance with newer food-safety expectations has become a major entry requirement into export markets or for retaining market access. However, sharing the responsibility for food safety further down the chain has not yet been fully realized and implemented. Although known foodborne illness outbreaks have not been directly attributed to contamination during ocean transport and warehousing, this may be more because the transport sector has received less scrutiny than the farm and processing sectors. The sources of most outbreaks of foodborne illnesses are never identified, yet millions of people become sick annually. This militates a conservative approach to the safety of all perishable foods, particularly during long-distance transport or extended handling and holding of produce. Until we can validate that food-handling practices within this sector of the chain are not introducing hazards, we must assume that there is potential for the theoretical hazards discussed in this chapter to become a real threat and we should treat them appropriately.

The impact of a lack of effective food-safety legislative and market pressure on international carriers remains a point of concern. Several studies have shown that it is difficult for developing export countries to comply with up-front quality regulations and food-safety requirements from developed countries. This is mainly because of the fragmented nature of supply chains, lack of trust and accountability, limited technical investments, and a poor communication network in developing countries. In addition, a clear economic benefit to implementing food-safety practices and programs is not evident to many players in the fruit export, logistics, and marketing sectors. In order to buy at a lower price, procurement managers are at times willing to overlook a lack of comprehensive effectively implemented food-safety programs. This continues to undercut the goodwill and honest efforts of other players in the food chain. In order to consistently deliver healthful and wholesome products, all players within the food chain should be committed to ensuring product safety. Although food safety is a shared responsibility, it is currently not adequately reflected within the middle sector of the food chain.

REFERENCES

D.W.K. Acheson, "Food-Borne Illnesses," in J. Lederberg, Ed., *Encyclopedia of Microbiology*, Vol. 2. Academic Press, London, 2000, pp. 390–411.

J.A. Albrecht, F.L. Hamouz, S.S. Sumner, and V. Melch, Microbial evaluation of vegetable ingredients in salad bars. *J. Food Prot.*, 58, 683–685 (1995).

J. Arul, "Emerging Technologies for the Control of Postharvest Diseases of Fresh Fruits and Vegetables," in C.L. Wilson and M.E. Wisniewski, Eds., *Biological Control of Postharvest Disease Theory and Practice*. CRC Press Inc., Boca Raton, FL, 1994, pp. 1–6.

A.S. Badawy, C.P. Gerba, and L.M. Kelly, Survival of rotavirus SA-11 on vegetables. *Food Microbiol.*, 2, 199–205 (1985).

BASC, Business Anti-Smuggling Coalition. What is BASC? 2005. Available at http://www.wbasco.org/english/basc.htm. Accessed 4/10/2005.

M. Berrang, R. Bracket, and L.R. Beuchat, Growth of *Aeromonas hydrophila* on fresh vegetables stored under controlled atmosphere. *Appl. Environ. Microbiol.*, 55, 2167–2171 (1989).

L.R. Beuchat, Pathogenic microorganisms associated with fresh produce. *J. Food Prot.*, 59, 204–216 (1995).

L.R. Beuchat. *Surface Decontamination of Fruits and Vegetables Eaten Raw: A Review.* World Health Organization: Switzerland, 1998.

L.R. Beuchat, Surface decontamination of fruits and vegetables eaten raw: A review. Pesticide.Net. WHO Food Safety Programme Document (November 1999). Available at http://www/pestlaw.com/x/international/WHO-19991100A.html. Accessed 30/9/2005.

L.R. Beuchat and J. Ryu, Produce handling and processing practices. *Emerg. Infect. Dis.*, 3, 459–469 (1997).

R.E. Brackett, Shelf stability and safety of fresh produce as influenced by sanitation and disinfection. *J. Food Prot.*, 55, 808–814 (1992).

R.E. Brackett, "Microbial Quality," in R.L. Swefelt and S.E. Prussia, Eds., *Postharvest Handling, a Systems Approach.* Academic Press Inc., New York, 1993, pp. 126–145.

R.E. Brackett, D.M. Smallwood, S.M. Fletcher, and D.L. Horton, "Food Safety: Critical Points Within the Production and Distribution System," in R.L. Swefelt and S.E. Prussia, Eds., *Postharvest Handling, a Systems Approach.* Academic Press Inc., New York, 1993, pp. 302–323.

S.L. Burnett and L.R. Beuchat, Foodborne pathogens: human pathogens associated with raw produce and unpasteurised juice and difficulties in decontamination. *J. Indust. Microbiol.*, 27, 104–110 (2001).

J.C. Buzby, Children and microbial foodborne illness. *Food Review, USDA's Economic Research Service*, 19, 32–37 (2001).

F. Carlin, C. Nguyen-the, and C.E. Morris, Influence of background microflora on *Listeria monocytogenes* on minimally processed fresh broad-leaved endive (*Cichorium endivia* var. *latifolia*). *J. Food Prot.*, 59, 698–703 (1996).

Centres for Disease Control and Prevention (CDC), Foodborne Outbreaks Due to Bacterial Etiologies, Atlanta, USA, 2003. Available at http://www.cdc.gov/foodborneoutbreaks/us_outb/fbo2003/bacterial03.pdf. Accessed 27/9/2005.

CDC, Foodborne Disease Outbreaks Due to Viral Etiologies, 2003. Centre for Disease Control and Prevention (CDC), Atlanta, USA. Available at http://www.cdc.gov/foodborneoutbreaks/us_outb/fbo2003/viral03.pdf. Accessed 27/9/2005.

E.D. Den Aantrekker, R.R. Beumer, S.J.C. Van Gerwen, M.H. Zwietering, M. Van Schothorst, and R.M. Boom, Estimating the probability of recontamination via the air using Monte Carlo simulations. *Int. J. Food Microbiol.*, 87, 1–15 (2003).

C. De Roever, Microbial safety evaluations and recommendations on fresh produce. *J. Food Control*, 10, 117–143 (1999).

C.S. De Waal, Safe food from a consumer perspective. *Food Control*, 14, 75–79 (2002).

S. Doores, The microbiology of apple and apple products. *Crit. Rev. Food Sci.*, 19, 133–149 (1983).

Ecolab, Protecting Your Brand from Airborne Contamination, 2001. Available at http://www.ecolab.com/Initiatives/foodsafety/Images/Protecting_Your_Brand_AirQuality.pdf. Accessed 26/8/2005.

European Commission, *Health & Consumer Protection Directorate-General. Scientific Committee on Food. Risk Profile on the Microbiological Contamination of Fruits and Vegetables Eaten Raw*, 29 April 2002. FU, Brussels.

J.A. Evans, S.L. Russell, C. James, and J.E.L. Corry, Microbial contamination of food refrigeration equipment. *J. Food Eng.*, 62, 225–232 (2004).

Food and Agriculture Organization (FAO) and World Health Organization (WHO), Assuring Food Safety and Quality. Guidelines for Strengthening National Food Control Systems. FAO Food and Nutrition Paper 76, Rome, Italy, 2003.

Food and Drug Administration (FDA), Guide to Minimize Microbial Food Safety Hazards for Fresh Fruits and Vegetables. United States Department of Agriculture, Centre for Disease Control and Prevention, 1998. Available at http://www.foodsafety.gov/~dms/prodguid.html. Accessed 26/11/2004.

Food and Drug Administration/Center for Food Safety and Applied Nutrition (FDA/CFSAN), *Analysis and Evaluation of Preventive Control Measures for the Control and Reduction/Elimination of Microbial Hazards on Fresh and Fresh-Cut Produce*, 2001. Available at http://www.cfsan.fda.gov/%7Ecomm/ift3-toc.html. Accessed 13/10/2005.

C. Frenkel and J.J. Jen, "Tomatoes," in N.A.M. Eskin, Ed., *Quality and Preservation of Vegetables*. CRC Press, Inc., Boca Raton, FL, 1989, p. 73.

K. Gast, *Food Safety for Farmers Markets*. Kansas State University, Agricultural Experiment Station and Cooperative Extension Services, USA, 1997.

R.K. Gast, W. Mitchell, and P.S. Holt, Detection of airborne *Salmonella enteritidis* in the environment of experimentally infected laying hens by an electrostatic sampling device. *Avian Dis.*, 48, 148–154 (2004).

E.E. Geldreich and R.H. Bordner, Fecal contamination of fruits and vegetables during cultivation and processing for market. A review. *J. Milk Food Technol.*, 34, 184–195 (1970).

J. Gorny and D. Zagory, "Produce Food Safety," in *The Commercial Storage of Fruits, Vegetables and Florist and Nursery Stocks*. USDA-ARS, Agriculture Handbook Number 66, Revised 2003.

L. Harris, *Food Safety for Fresh Produce: Fact Sheet*. Department of Food Science and Technology: University of California at Davis, 95616–8631 (1997).

W.C. Hurst and G.A. Schuler, Fresh produce processing – an industry perspective. *J. Food Prot.*, 55, 824–827 (1992).

J.J. Iandolo, "*Staphylococcus*," in J. Lederberg, Ed., *Encyclopedia of Microbiology*, Vol. 4. Academic Press, London, 2000, pp. 387–393.

International Specialty Supply, *Listeria*, 2005. Available at www.sproutnet.com/listeria.htm, accessed August 2005.

L. Jaykus, The application of quantitative risk assessment to microbial food safety risks. *Crit. Rev. Microbiol.*, 22, 279–293 (1996).

J.M. Jones, *Food Safety*. Eagan Press, St. Paul, Minnesota, USA, 1992.

F.K. Kaferstein, Actions to reverse the upward curve of food illness. *Food Control*, 14, 101–109 (2002).

K.-I. Kaneko, H. Hayashidani, Y. Ohtomo, J. Kosuge, M. Kato, K. Takahashi, Y. Shiraki, and M. Ogawa, Bacterial contamination of ready-to-eat foods and fresh products in retail shops and food factories. *J. Food Prot.*, 62, 644–649 (1999).

Louisiana Food Safety and Information Clearinghouse's, Foodborne Pathogens, 2005. Available at http://www.gnofn.org/fsaic/foodborne.htm. Accessed 13/10/2005.

J.M. Madden, Microbial pathogens in fresh produce – the regulatory perspective. *J. Food Prot.*, 55, 821–823 (1992).

E. Medrela-Kuder, Seasonal variations in the occurrence of culturable airborne fungi in outdoor and indoor air in Cracow. *Int. Biodeteri. Biodegrad.*, 52, 203–205 (2003).

C. Nguyen-the and F. Carlin, The Microbiology of Minimally Processed Fresh Fruits and Vegetables. *Crit. Rev. Food Sci. Nutr.*, 34, 371–401 (1994).

J.K. Northcutt, D.R. Jones, K.D. Ingram, A. Hinton, Jr. and M.T. Musgrove, Airborne microorganisms in commercial shell egg processing facilities. *Int. J. Poultry Sci.*, 3(3), 195–200 (2004).

R.E. Paull, Effect of temperature and relative humidity on fresh commodity quality. *Postharvest Biol. Technol.*, 15, 263–277 (1999).

PPECB, *Export Directory*, 3rd edn. Makulu-Malachite Publishing, Helderberg, 2005.

J.R. Roberson, L.K. Fox, D.D. Hankock, and J.M. Gay, Ecology of *Staphylococcus aureus* isolated from various sites on dairy farms. *J. Dairy Sci.*, 77, 3354–3364, (1994).

T. Roberts and L. Unnevehr, New approaches to regulating food safety. *Food Rev.*, 42, 3–8 (1994).

M.K. Sandel and J.L. McKillip, Virulence and recovery of *Staphylococcus aureus* relevant to the food industry using improvements on traditional approaches. *Food Control* 15, 5–10, (2004).

J. Schlundt, New directions in foodborne disease prevention. *Int. J. Food Microbiol.*, 78, 3–17 (2002).

R.A. Spotts and L.A. Cervantes, Use of filtration for removal of conidia of *Penicillium expansum* from water in pome fruit packinghouses. *Plant Dis.*, 77, 828–830 (1993).

L.D. Stetzenbach, M.P. Buttner, and P. Cruz, Detection and enumeration of airborne biocontaminants. *Curr. Opin. Biotechnol.*, 15, 170–174 (2004).

D. Tanner and N. Smale, Sea transportation of fruits and vegetables: an update. *Stewart Postharvest Rev.*, 1, 1 (2005).

R.V. Tauxe, Emerging foodborne diseases: An evolving public health challenge. *Emerg. Infect. Dis.*, 3, 425–434 (1997).

R.V. Tauxe, Emerging foodborne pathogens. *Int. J. Food Microbiol.*, 78, 31–41 (2002). Available at http://www.elsevier.com/gej-ng/10/19/57/107/27/show/toc.htt. Accessed 15/9/2005.

N. Teixidó, I. Viñas, J. Usall, and N. Magan, Control of blue mold of apples by preharvest application of *Candida sake* grown in media with different water activity. *Phytopathology*, 88, 960–964 (1998).

E. Walker, C. Pritchard, and S. Forsythe, Food handlers' hygiene knowledge in small food businesses. *Food Control*, 14, 339–343 (2003).

J.M. Wells and J.E. Butterfield, *Salmonella* contamination associated with bacterial soft rot of fresh fruits and vegetables in the marketplace. *Plant Dis.*, 81, 867–872 (1997).

World Bank, Food Safety Issues in the Developing World. Technical Paper No. 469, 2000.

World Cargo News, Reefer market stays hot. Press Release, 2004, pp. 27–30.

World Health Organization (WHO), *Food Safety Issues*. WHO Global Strategy for Food Safety, Safer Food for Better Health. WHO, Geneva, Switzerland, 2002.

D. Zagory, Effects of post-processing handling and packaging on microbial populations. Proceedings of the Beltsville Agricultural Symposium, May 3–6, 1998. *Postharvest Biol. Technol.*, 15, 313–321 (1999).

T. Zhao, M.P. Doyle, and R.E. Beser, Fate of enterohemorrhagic *Escherichia coli* O157:H7 in apple cider with and without preservatives. *Appl. Environ. Microbiol.*, 59, 2526–2530 (1993).

R.Y. Zhuang, L.R. Beuchat, and F.J. Angulo, Fate of *Salmonella montevideo* on and in raw tomatoes as affected by temperature and treatment with chlorine. *Appl. Environ. Microbiol.*, 61, 2127–2131 (1995).

CHAPTER 8

Fresh Produce Safety in Retail Operations

TONI HOFER and BRETT GARDNER
Raley's Supermarkets, Sacramento, CA, USA

TOM FORD
Ecolab, Greensboro, NC, USA

Microbial Hazard Identification in Fresh Fruit and Vegetables, Edited by Jennylynd James

8.1 INTRODUCTION

8.1.1 Historical Perspective on Safe Food-Handling Practices for Produce

Retailers are concerned about the increase of reported foodborne illnesses associated with produce. At the same time, we see an upward trend in the consumption of fresh produce in the United States (CDC, 2001; Sivapalasingam et al., 2004). Many of the outbreaks have been from life-threatening bacteria, viruses, and parasites. The U.S. population in general has an ever-growing percentage of immunocompromised individuals (CDC, 2001). In 1997, the FDA estimated cost of a foodborne illness outbreak at $6.5 billion to $34.9 billion. Regulators and legislators are compelled to initiate programs that provide safe food. The FDA issued science-based guidance to the industry in 1998 (FDA, 1998) after analyzing increases in the number of foodborne illnesses associated with both domestic and imported fresh fruits and vegetables. After the September 11, 2003 attack on the United States, the Department of Homeland Security announced that one of the biggest threats to security was the food supply. Bioterrorism is now a key concern for grocery retail operators in their responsibility for providing safe food to customers.

The "farm to table" food chain is a tangled web of commerce – grower/shippers, co-packers, distributors, and wholesale/terminal markets. All of these can change from day to day, commodity to commodity, and season to season. The growing demand for "fresh" produce, year round, has prompted changes to the distribution patterns, the products, and postharvest/processing technologies (Calvin, 2003). At the same time, known and emerging pathogens in produce have increased, and so has the virulence of the pathogens themselves (Griffin et al., 2002).

The average consumer assumes that produce misting systems are sufficient for washing fruits and vegetables offered on the retail sales floor. He assumes therefore that produce is grab-and-go, ready-to-eat. Consumers do not always wash fruits and vegetables thoroughly before consuming. It should be noted that with many commodities, a thorough washing can affect the quality and taste of the product.

The general trend in the United States is to cook vegetables less. This practice, while helping to maintain the vitamins and nutrients, eliminates a kill step for the pathogens. As legislators and regulators pass laws to ensure the food supply is safe, retailers are asking growers to provide indemnification and assurances that the food purchased is safe.

Few consumers have a food-safety program for their home refrigerators, kitchens, and pantries. One suspects that, as a result, there are no reliable statistics demonstrating a link between foodborne illnesses and cross-contamination because of improper storage practices in a home refrigerator. However, many of the foodborne outbreaks associated with fresh produce are traced back to food-service establishments. A well-documented *E. coli* O157:H7 outbreak involving cut watermelons at a steak house in Idaho resulted from cross-contamination at point of preparation in the kitchen (Food Safety Network, 2000). Graphs on the FDA website providing data on foodborne illnesses only list the commodities involved and not how the contamination occurred.

Retailers, grocers, and food-service providers must be diligent, not only with their internal food-safety programs, but also in educating their consumers. Retailers must educate customers about how to store and prepare produce to avoid foodborne illnesses. Because food safety and home economics courses are not mandatory in schools, students are not generally aware of the basic principles of food safety and so do not bring these into the home or eventually the workplace (Taege, 2002).

In most cases, the cause of foodborne illness is difficult to determine. The most common reasons for foodborne illnesses are identified as the following (USHHS Public Health Service, 2000).

1. Emerging pathogens.
2. Improper food preparation and storage practices among consumers.
3. Insufficient training of food workers.
4. An increasingly global food supply.
5. An aging population and an increase in immunocompromised population. These high-risk groups represent approximately 20 percent of the U.S. population. Individuals in these groups may become ill from smaller doses of microorganisms and may be more likely to die of foodborne illness than the average person (Smith, 1997).
6. An increase in demand for processing/packaging/transportation of fresh fruits and vegetables. Advances in agronomy, processing, preservation, and packaging technologies, shipping, and marketing have made produce available 365 days of the year. New procedures and new sources often introduce new pathogens, viruses, and protozoa/parasites (opportunistic microorganisms).

8.1.2 Farm-to-Table Food Chain

The farm-to-table food chain is ever changing. Retailers are responsible for offering safe food for sale to consumers. This implies that they know exactly where it is grown, processed, and under what type of condition it is transported to each retail location. Retailers source produce from individual growers, co-op wholesalers/distributors, and terminal markets. Many times the produce is packed and repacked prior to the retailer taking possession of the product. Tomatoes from one farm look the same as tomatoes from another farm. Sometimes they are repacked together under a co-operative label. Traceback stops there. The practicality of labeling each piece of fruit is mind-bending, and so far, impossible.

Food safety is a responsibility for retailers and food-service operators. We know that when a retailer makes a mistake, it hurts the entire industry. Companies that have survived a foodborne outbreak will be the first to state that marginally meeting food-safety specifications is not enough anymore. Companies should meet or exceed existing food-safety standards.

Consumers expect the food they buy to be safe. Health-conscious American consumers are turning to the produce industry to provide a year-round supply of ready-to-eat, quick, healthy, and often exotic meals and snacks. The United States is bringing in more and more fresh fruit and vegetables from other countries. Imports rose from 13.8 billion pounds in 1993 to 22.7 billion in 2002.

Food safety, as it relates to produce, involves several players: the grower, processor, distributor, retailer, and consumer. Each has an important role in ensuring and maintaining the safety of the food. The grower has the initial responsibility to assure the product enters the system in a safe and wholesome state. The conditions and methods used to grow and harvest the food play an integral role in the safety of the food. Improper agricultural practices can lead to a foodborne outbreak miles away, and months in the future. Processing areas offer many opportunities for contamination. Equipment, water, and food handlers can all increase or reduce the quality and safety of produce. The distribution chain can also allow for introduction or replication of harmful agents, because sanitary conditions and temperature impact quality. Conditions at retail are critical to keeping the level of quality and safety at their highest. Finally, the consumer must know and practice safe food handling to ensure that the product remains wholesome as it is handled, stored, and prepared. The retailer can and must exhibit as much influence as possible at each of these points in the food chain for the produce he acquires and offers for sale to the customers.

8.2 IDENTIFYING FOOD-SAFETY RISKS IN RETAIL OF FRUITS AND VEGETABLES

8.2.1 Identifying Variable Risk Levels Associated with Produce

The first step in delivering safe produce to the customer is identifying the risks, specific to each of the produce items a retailer offers for sale. Much risk for serious outbreaks of foodborne illnesses is evident. However, the risks are ever changing and new threats have evolved in recent years. Items once thought to be safe, because of internal and external unique properties, have been found to sometimes allow for growth of human pathogens. For example, once believed to be safe because of their inherent low pH, both orange juice and tomatoes have been involved in foodborne outbreaks (MMWR 1999, 2005).

In addition to identifying the risks associated with each of the produce items for sale, the retailer must also have interventions in place at the most effective point in the product's life cycle to provide the highest possible safety. First, retailers must have knowledge of growing practices and Good Agricultural Practices (GAPs) to evaluate any vendor's program adequately. Once the product has been introduced into the retailer's custody, a risk analysis using the Hazard Analysis Critical Control Point (HACCP) process may be considered.

Hazards associated with produce include a full spectrum of chemical, biological, and physical threats. Retailers should work with their suppliers to identify how existing and new pathogens may be eliminated from the path the produce takes to the customer. Hazard identification in different types of produce should be a part of the retailer's buying process. Outbreaks caused by pathogens such as *Salmonella*, *Shigella*, *E. coli*, and the virus causing hepatitis A have been associated with specific groups of produce and most have had a recognized epidemiological pathway. The retailer should work within these known associations to eliminate the risk for customers.

After identifying the potential hazard, the retailer should also play a role in assuring the most effective intervention process has been applied at critical points of the process. In some cases, the interventions may be placed most effectively in the growing phase of the produce life cycle. For example, bean sprouts should be grown in an environment where they are protected from potential contamination. Other products such as green leafy vegetables would benefit from an intervention provided at the retail level. All intervention attempts should include education of the consumer in the safe handling of the products they purchase and consume.

8.2.2 Identifying Control Points in the Fresh Produce Distribution Chain

Retailers must work with suppliers to identify and limit the risk associated with produce entering the system. Control points should be identified at any point contamination may occur, from the delivery of produce until it is consumed. Table 8.1 reviews the phases in produce production and distribution, identifying some potential risks and control points.

8.2.3 Regulatory Guidelines for High-Risk Products

In response to outbreaks associated with produce, a growing number of regulations and guidance documents have been put in place to reduce the risk to the public. These regulations and guidance documents have been enforced everywhere along the path to the marketplace. Risk has been addressed at the retail level as well. The requirement for washing produce has been a portion of the Food Code for several years (FDA, 2005). A 5-log reduction has been a requirement for juiced items at retail since January 19, 2001 (FDA, 2001). Use of legislation regulating food safety will continue as key causative aspects of outbreaks related to foodborne illnesses emerge and steps can be taken to prevent them.

TABLE 8.1 Identification of Control Points and Potential Risks in the Fresh Produce Chain

Control Point	Risk
Growing/Harvest	Introduction of pesticide/fertilizer
	Viral/bacteriological contamination by worker
Processor	Water contamination with chemical/biological agents
	Viral/bacteriological contamination by worker
	Equipment contamination of product with chemical/biological agents
	Temperature abuse allowing proliferation of bacteria
Distributor	Equipment contamination of product with chemical/biological agents
	Temperature abuse allowing proliferation of bacteria
Retailer	Water contamination with chemical/biological agents
	Viral/bacteriological/chemical/physical contamination by worker
	Equipment contamination of product with chemical/biological agents
	Temperature abuse allowing proliferation of bacteria
Consumer	Cross-contamination
	Temperature abuse allowing proliferation of bacteria

8.3 ADDRESSING THE RISKS IN RETAIL OF FRUITS AND VEGETABLES

8.3.1 Establishing Procurement Standards

The delivery of exceptional produce to the customer, the goal of every retailer, is achieved only by beginning with exceptional produce being shipped from the supplier. Even though produce quality relies heavily on growing conditions, a retailer must have a system in place that allows for the highest quality and safest produce to reach its customers. A retailer should have buying specifications and procedures to assess that product safety and quality are met or exceeded.

8.3.2 Defining Product Specifications and Buying Expectations

"Excellence in–Excellence out" should be the goal for receiving and distributing produce. Quality Assurance must play an integral role in establishing, monitoring, verifying, and addressing product quality and – ideally – integrating it into the buying process. The first step is the development of buying specifications.

Retailers must develop clearly defined product quality and safety specifications and communicate these to their suppliers. Standards should include requirements that all growers have documented GAPs, Food Security Plans, Good Manufacturing Practices (GMPs), and HACCP plans in place. In addition, retailers should define all expected product attributes and incorporate these into the specifications. Specifications may include size, shape, color, ripeness, quality, Brix, texture, defect levels, other quality attributes, packaging, and transportation standards. The retailer should document standards and practices, and have assessment methods for each standard.

8.3.3 Assessing Supplier Performance for Incorporation into the Buying Equation

An inspection program for measuring against set specifications is an important component of quality assurance and food-safety programs. Retailers should not expect produce to arrive at the distribution center exceeding the communicated specification. Even if the product left the suppliers' doors above the desired specification, conditions in transit may have a serious impact on product quality. A Quality Assurance check at the distribution center assures the product has met specifications. Because many quality attributes of food also affect food safety, this process should have a positive impact on food safety. Thus the buying team, working along with the quality assurance team, can improve quality for the customer.

A retail Quality Assurance program can have two philosophies for product specifications and receiving: the bull's eye approach and the hurdle approach. A bull's eye approach would assess whether or not a product possesses the most desirable qualities. For example, a product would have in its specification a most desirable Brix, size, color, ripeness, and so on, and the Quality Assurance team receiving the produce would score the product against these ideals. A hurdle approach to Quality Assurance specifications would accept any product that exceeds the minimum standard described by that specification. An ideal situation would be to take the bull's eye approach and tie it to the buying process. Suppliers with the best product (suppliers who hit the bull's eye) would be rewarded with more business. This would result in the highest quality product reaching the store shelves and the consumer.

8.3.4 Retail Store Risks

In the store, a retailer should ensure an intervention process is in place to address risk. The process should include several steps to limit contamination or decrease the growth of bacteria:

1. Retailers should have receiving practices to identify and accept only fresh produce that meets their specifications. Temperature, condition, and quality should be evaluated.
2. In storage, temperatures should be monitored; product should be culled and held under ideal relative humidity until being brought to the sales floor.
3. Before produce is processed, produce should be trimmed, crisped, or otherwise prepared for conditions on the sales floor.
4. On the sales floor, produce racks and wet racks should hold food at the ideal temperature and moisture conditions. Employees should clean the racks to assure quality is maintained at its highest level.
5. At the checkout line, baggers should be trained to bag produce separately from raw meat like chicken, beef, and pork.

8.4 ESTABLISHING SAFE FOOD-HANDLING STANDARDS FOR PROCESSING AND DISPLAYING FRUITS AND VEGETABLES

In many ways, a retail store must function like a processing facility. The essential steps in producing safe food include all the good manufacturing practices from food processing. Like food processors, food retailers receive, store, and process foods, adhering to safe food-handling practices every day. However, assuming the dual roles of food processor and retail facility is not easy. Unlike food processors, retailers have thousands of customers walking through their facility on a weekly basis. Food-processing operations are designed with easy-to-clean floors, walls, ceilings, plumbing, and equipment. Food stores, on the other hand, are designed for ease of shopping and for displaying food.

Grocers must design perishables departments to service customers, and prepare, process, and hold food. Equipment must be small enough to fit in the department, but also must be easy to clean. A perishables department must have the same sanitation systems in place as food-processing facilities. Retailers must ensure cleaning removes harmful agents effectively from equipment and utensils. A retail operation must have food-safety systems in place, as it manufactures and serves food. Customers expect and deserve safe food, and it takes an intricate system to ensure food safety is a promise delivered daily.

8.4.1 Specification

A buying specification is important in ensuring the supplier understands and is able to provide the food safety needs a retailer requires and expects. A buying specification must be an active document at the store level. Without this, food safety is a wish, not a reality. The store team must have knowledge and direction about what is expected in product quality and safety. Ideally, the store should have a method to communicate to the distribution center whether the food received has met the standard.

8.4.2 Receiving and Storage

The first step in delivering food safely to the customer happens when the food enters the store. The receiving step is critical. Ideally, it ensures only that products exceeding the desired specification are allowed into the store. The keys to the receiving process are product specifications, inspection methodologies, and empowerment of Quality Assurance staff. The importance of these three factors may be described as follows:

1. *Specification*: Knowing what to monitor and monitoring what is important;
2. *Inspection Method*: Having an effective method of monitoring what is received; and
3. *Empowerment*: Having an empowered employee who can take the appropriate action to ensure unacceptable products do not enter the store.

8.4.3 Inspection Method

The store must have an effective method to determine whether the food entering the depot meets the standard. All food must be evaluated when received. Trained personnel should be available to evaluate product 24 hours a day, if necessary. The person must be trained in what to look for and how to look for it. He or she must have the correct tools, such as thermometers, scales, and even ultraviolet lights, to make a decision.

8.4.4 Empowerment

The person responsible for receiving food must have the authority to reject a substandard shipment. Thus, essential retail food safety entails having the right specifications, the right method, and persons with authority trained to evaluate the shipment. The purpose of having a receiving program is to stop poor-quality food from entering the store. Too often, however, stores' focus is on loss prevention, counting cases, and verifying the number of cases received, as opposed to monitoring the safety and quality of the food entering.

8.5 ESTABLISHING SAFE FOOD-HANDLING STANDARDS AT STORE LEVEL

8.5.1 Processing

A produce department in a grocery store accomplishes a myriad of processing steps every day, from crisping vegetables, trimming and culling, to more complex operations. The more risky operations from a food-safety perspective are activities such as coring pineapples, dicing melons, juicing oranges, and making dip. These complex operational activities can carry significant food-safety risks if not accomplished correctly. A produce department is unique in that it prepares ready-to-eat foods using methods conducted at store level. These ready-to-eat foods are potentially hazardous. Retailers have initiated 5-log reduction programs for retail-level juicing programs, an FDA requirement since 2001. All retailers must manage sanitation, hygiene, and temperature to assure the processing conducted reduces the potential development of a food-safety issue.

8.5.2 Display

Once a food has been processed into a ready-to-eat, potentially hazardous food, the retailer must protect it. The most important step in this phase is maintaining the food at a temperature that reduces bacterial growth. This can be a difficult task, made more complicated by facilities not designed to display ready-to-eat foods. Some older produce racks may operate at levels in the 45–55°F range – well into the danger zone for growth of human pathogens. These display cases, which were originally designed to maintain quality and yield of whole pieces of produce, may allow bacterial growth in cups of diced fruit. It is very important for retailers to have sanitation standards in place for juice, cut fruit, melons, and vegetables that reduce or eliminate the risk associated with the commodity. Foodborne outbreaks associated with these commodities over the last decade showed the produce items with highest potential for contamination (FDA/CFSAN, 2001). These Standard Sanitation Operating Procedures should include:

1. Processing
   (a) Washing all fruit, melons and vegetables thoroughly prior to prechilling and processing.
   (b) Prechilling all fruit, melons, and vegetables prior to processing.
   (c) Processing all cut fruit, melons, and vegetables in the produce back room on clean and sanitary preparation surfaces with clean utensils.
2. Personal Hygiene
   (a) Employees following best practices for food preparation including proper dress code, hair restraints, and hand washing.
3. Time/Temperature Controls and Documentation
   (a) Processing any cut fruit, melons, and vegetables within a two-hour period.
   (b) Prechilling any cut fruit, melons, and vegetables to below 41°F prior to display; any cut produce shall be displayed and stored at 41°F or below.
   (c) Employees monitoring and documenting temperatures of cut produce every two hours, taking effective and thorough corrective action for any out-of-compliance product identified.
4. Labeling
   (a) Labeling and dating each package of cut produce with the following:
      - Production date and time;
      - Sell by date;
      - Prominent "Keep Refrigerated" labels to ensure that customers know to keep the product refrigerated after purchasing.

Retailers should also be aware that merchandising schematics requiring plastic racking and facing of products may hinder airflow and increase the temperature of the product. Other merchandising practices, such as using ice to refrigerate products, can also compromise product temperatures, making it difficult for the store team to maintain the proper temperature. Temperature control in the Produce Department depends on the equipment used and the merchandising techniques chosen. However, the retailer must meet the challenges daily.

8.5.3 Auditing Against the Standard

The customer expects and deserves safe food. The only way a retailer can ensure it lives up to this expectation is to verify that the systems it has in place are working well. This requires a verification process that might include the use of start-up checklists to ensure equipment has been effectively cleaned and sanitized prior to start of production. Production logs and other operational documents can be used to verify that procedures have been followed and document that every process has been carried out correctly. Training programs that document who was trained, and competency in the desired task, may also be used to verify the process. Finally, a retailer can use an unbiased, third-party auditor to verify the effectiveness of the food-safety system.

One can argue that a produce department is the department within a grocery that operates under the greatest risk as it relates to food safety. It is the one department charged with processing foods into ready-to-eat forms. By operating like a food processor within a grocery store, a produce department can limit its potential for risk and increase its level of food safety. By initiating a firm process of control through receiving, processing, and display, and by verifying these steps, a store can be a leader in food safety.

8.6 TAKING THE FOOD SAFETY MESSAGE TO THE CONSUMER

8.6.1 Serving the Consumer and Food Industry as an Advocate for Public Health

The role of the retailer has to extend beyond the grocery store. Food safety should not stop once the food goes into the shopping cart. Imagine all the food-safety information surrounding a particular product being automatically attached to the food. Any time a customer selected food and placed it in a cart, everything that consumer needed to know about handling, cooking, and storing that food would be there. The concept is innovative. Companies can already use checkout receipts to carry product-specific information to the customer. Other chains have taken advantage of frequent shopper programs to track customers and inform them when purchased items have been involved in a recall. These are examples of how retailers use technology to fill the gap in consumer education and extend the reach of food safety. Retailers should fill the role of educator and resource in the realm of food safety for the consumer. How this role is filled is limited only by the creativity of the retailer.

When a consumer stands in front of a produce rack, deciding which bunch of green onions to buy, her mind is focused on color, size, freshness, quality, and value. Her mind is not on hepatitis A, tainted irrigation water, *E. coli* or pesticides. She has entered into a relationship of trust with the retailer. Her belief is that the retailer has minimized risks associated with a particular product, and that allows her to select her product on factors that are important to her. This trust can be reinforced and built. The grocer should be the best source of food-safety information for the consumer. In addition, it is to a retailer's advantage to educate customers about products and food safety. Educating a customer about food safety may retain a customer who shops with confidence and total trust. The retailer can be a community resource.

8.6.2 Use of the Media and Internet

The retailer can use the media – newspapers, radio, television, and the Internet – to inform consumers about products and topical information. In addition, the retailer can reach customers before they begin their trip to the grocery store. As they grab a cart in the parking lot, signs on the shopping cart handle highlight food-safety topics as well as the products within the store.

8.6.3 Smart Cards and Other Customer Membership Benefits

Smart cards – computer screens that alert customers to products they consume or that are on sale – are in use around the world. Point-of-sale signage and information are everywhere throughout the store, on shelves, on cases, on the walls, hanging from the ceilings, and even on the floor. As the customer checks out, frequent shopper information is gathered. Special product information is linked to the information printed on their receipts and coupons. Bag stuffers and even the outside of the shopping bags are also used to keep the customer informed and aware. Imagine having available all these points of contact with the customer to communicate important food-safety information about any product they have purchased. The Future can be the Here and Now if retailers use all of the tools available to communicate with consumers. Food-safety knowledge can strengthen consumer trust and loyalty. Knowledge is power!

8.6.4 Mechanics of Education by Retailers

The customer learns about produce food safety from a variety of sources. They visit a grocer's website. They listen to a consumer affairs person discuss recipes with a local chef on a radio program as they drive to stores. They read about organic produce on signs attached to shopping cart handles as they walk into stores. Signs on shelves provide product information such as country of origin, nutritional information, and ingredients used in processing. Signage indicates whether produce has been tested for pesticides or been validated by a third-party auditor. Videos playing in the store suggest preparations. Demo stations allow for sampling. At the checkout, information can be inserted in the bag or printed on the shopping bag to communicate key food-safety practices. In these examples, the grocer is informing, educating, and assisting the consumer through every step of the process.

8.6.5 Information Programs: Regulatory and Industry Sources

Although the grocer is a close, trusted source of information for the customer, he is only one source of information. The consumer must be surrounded by sources of information if food safety is to be successfully managed. The Government is a major source of reliable, timely, and appropriate information for the consumer. Federal governmental agencies (the FDA, USDA, and CDC) are charged with setting and monitoring food-safety standards, as well as communicating key food-safety information to the consumer. Local government has also been very successful at educating and informing consumers in the areas of food safety. "Fight Bac!" has been a very successful program and can serve as a model for how food-safety knowledge can be carried to the consumer level (www.fightbac.org).

It takes a "community" effort to encircle the customer with food-safety information. Only a coordinated and concerted effort by a team of producers, suppliers, government agencies, and retailers will ensure trust with the customer.

8.7 FUTURE DEVELOPMENTS OF FOOD SAFETY IN RETAIL OF FRUITS AND VEGETABLES

8.7.1 Producing Safer Food

Future improvements in produce food-safety will mean producing safer food, tracking that food, and evaluating its condition. It will mean having products and technology that improve the safety of produce at retail, and verifying that the system works.

In the future, producers will have to be innovative and effective in improving the safety of the produce they deliver. Growing, harvesting, packing, and shipping must have food-safety interventions at each step. Interventions are being implemented in the following areas: hygiene of field workers, groundwater quality, and maintenance of irrigation water quality. Improved integrated pest management, processing food-safety systems, surface treatments for produce to reduce bacteria loads, and packaging technology to increase safety and shelf-life are all needed.

8.7.2 Enhanced Traceback and Epidemiological Capabilities

The future is now in product traceability. It is now possible to trace products via radio frequency identification technology (RFID) at all points of its travel to the store. Once products arrive at the store, it is possible to track the performance of display cases to assure the produce is being held in perfect conditions to maximize quality and safety. With frequent shopper technology, retailers have contacted customers who purchased products that were part of a recall to notify them and to reclaim the product. Another technology known as genetic fingerprinting has the ability to identify sources of foodborne outbreaks rapidly and can genetically match multiple outbreaks across the nation. This technique can also be used to trace the pathogenic agent back to its source – all the way to the field where it was grown.

8.7.3 Chemicals to Reduce Bacteria on Produce at Retail

A product can become contaminated at any point in its path from the farm to the customer. The closer one can place an intervention to the point of consumption, the safer the product will be. Products are now available to reduce the level of pathogens safely and easily at store level. This brings the power to sell safe food right into the store.

8.7.4 Third-Party Verification

Retailers are using third-party entities to verify that the suppliers, distributors, and their own stores are operating correctly. It is now possible for a retailer to communicate product specifications to suppliers and verify that the products meet those standards at the production location. The third-party auditor also verifies that conditions and people are contributing to the quality and safety of the product being produced at that location. They can also verify that the operator has systems in place to identify and make corrections when practices drift from their desired level. Third-party organizations are also verifying systems and standards in transit and at distribution centers. Perhaps, however, the most effective use of third-party auditing occurs at the store level. This is the front line of the food-safety battle. This is the closest point in the process to the customer. The store level is where key food-handling steps occur, sanitation is accomplished, and opportunities for improved personal hygiene are most abundant. Third-party auditing groups can provide feedback on compliance. More importantly, they can coach proper behavior, reinforce procedures, and train to desired competency for each task. Produce food safety must occur at store level and third-party auditors can make sure a retailer is executing food-safety practices properly.

8.8 SUMMARY

In retail, the Produce Department processes ready-to-eat potentially hazardous foods. In doing so, they manage the greatest food-safety risk in retail. The way the Produce Department manages the risk is critical. It must

1. Work with suppliers to establish and communicate product specifications;
2. Verify that the produce maintains its high level of food safety throughout all phases from the field to the store;
3. Train store associates in procedures and practices and verify that they execute them correctly; and
4. Take time to educate and inform the customer.

Produce food safety at retail is the last step along the path from field to the customer. It is the basis of the relationship between the retailer and the customer, a relationship built on trust. This trust is maintained by a retailer's ability to deliver on the promise to provide safe food.

REFERENCES

L. Calvin, "Produce, Food Safety, and International Trade. Response to U.S. Foodborne Illness Outbreaks Associated with Imported Produce," in J.C. Buzby, Ed., *International Trade and Food Safety*, Chapter 5. Economic Research Service/USDA, Washington, D.C., 2003, pp. 74–96.

CDC, *Outbreaks Associated with Fresh and Fresh-Cut Produce. Incidence, Growth, and Survival of Pathogens in Fresh and Fresh-Cut Produce.* U.S. Food and Drug Administration – Center for Food Safety and Applied Nutrition (September), 2001.

FDA/CFSAN, Food safety from farm to table: A national food safety initiative. Report to the President, May 1997. Available at http://www.cdc.gov/ncidod/foodsafe/report.htm, accessed October 2005.

FDA/CFSAN, FDA Survey of Imported Fresh Produce, FY 1999 Field assignment, 2001. Available at http://www.cfsan.fda.gov/~dms/prodsurb.html. Accessed April 20, 2004.

Food and Drug Administration (FDA), Guide to Minimize Microbial Food Safety Hazards for Fresh Fruits and Vegetables, 1998. Available at http://vm.cfsan.fda.gov/~acrobat/prodguid.pdf, accessed September 2005.

Food and Drug Administration (FDA), Hazard analysis and critical control point (HACCP); procedures for the safe and sanitary processing and importing of juice; final rule. *Fed. Regist.*, 66, 6138–6202 (2001).

Food and Drug Administration (FDA), Code of Federal Regulations. [Title 21, Volume 2] [Revised as of April 1, 2005]. U.S. Government Printing Office via GPO Access. CITE: 21CFR120, (4) 2001 Food Code 3-202.110, p. 46.

Food Safety Network, Milwaukee *E. coli* tied to outside source. Reuters/AP, August 1, 2000. Available at http://archives.foodsafetynetwork.ca/fsnet/2000/8-2000/fs-08-01-00-01.txt, accessed October 2005.

P.M. Griffin, P.S. Mead, and S. Sivapalasingam, "*Escherichia coli* O157:H7 and other enterohemorrhagic *E. coli*," in M.J. Blaser, P.D. Smith, J.I. Ravdin, et al., Eds., *Infections of the Gastrointestinal Tract*, 2nd edn. Lippincott Williams & Wilkins, Philadelphia, PA, 2002, pp. 627–642.

MMWR, Outbreak of *Salmonella* serotype Muenchen infections associated with unpasteurized orange juice – United States and Canada, June 1999. *MMWR*, 48(27), 582–585 (1999).

MMWR, Diagnosis and management of foodborne illness. Centers for Disease Control and Prevention. *MMWR*, 50(RR02), 1–69 (2000).

MMWR, Outbreaks of *Salmonella* infections associated with eating roma tomatoes – United States and Canada, 2004. *MMWR*, 54(13), 325–328 (2005).

S. Sivapalasingam, C.R. Friedman, L. Cohen, and R.V. Tauxe, Fresh produce: a growing cause of outbreaks of foodborne illness in the United Sates, 1973 through 1997. *J. Food Prot.*, 67(10), 2342–2353 (2004).

J.L. Smith, Long-term consequences of foodborne toxoplasmosis: Effects on the unborn, the immuno-compromised, the elderly, and the immunocompetent. *J. Food Prot.*, 60, 1595–1611 (1997). Available at www.ucfoodsafety.ucdavis.edu/Consumer_Advice/Foodborne_Illness_At_Risk_Populations.htm, accessed October 2005.

A.J. Taege, The new American diet and the changing face of foodborne illness. Department of Infectious Diseases. *Cleveland Clin. J. Med.*, 69(5), 419–424 (2002). Available at http://www.ccjm.org/pdffiles/taege502.pdf, accessed October 2005.

USHHS Public Health Service, HHS Initiatives to Reduce Foodborne Illness. US Department of Health and Human Services, HHS Fact Sheet, 2000. Available at http://www.cfsan.fda.gov/~lrd/hhsfsi2.html, accessed October 2005.

CHAPTER 9

Consumer Handling of Fresh Produce from Supermarket to Table

CHRISTINE BRUHN
Center for Consumer Research, University of California, Davis, CA, USA

9.1 INTRODUCTION

Fruit and vegetable consumption is viewed as health enhancing, because produce has been shown to be protective against several chronic diseases. In the last decade, 50 percent or more people say they are increasing their consumption of

Microbial Hazard Identification in Fresh Fruit and Vegetables, Edited by Jennylynd James

fruits and vegetables. Improved handling and distribution systems, as well as exposure to locally grown products, have increased consumer exposure to and demand for quality produce. High-quality produce is second only to a clean, neat store as top factors in selecting a supermarket. From 1992 to 2004, 90 percent of consumers or more rate high-quality produce as very important in supermarket selection (Research International, 2004). High-quality fruits and vegetables rated very important to all income and geographic groups, but is especially valued among the highest income households, where 97 percent rated it very important in 2000 (Research International, 2000). Although consumption of fruit and vegetables enhances health, these products may also increase risk for foodborne illness if the products are not handled properly from production through to the consumer's table.

9.2 PRODUCE SELECTION

The Food Marketing Institute's annual survey of 1000 households consistently indicates that good taste is the most important factor influencing purchase (Research International, 2000). Perceptions of health do have an influence, however. Produce is viewed as a healthy food choice. In each of the last 10 years, 70 percent or more consumers responding to the Food Marketing Institute annual survey say that they have increased produce consumption for a healthier diet (Research International, 2004). Similarly, Americans responding to Parade's annual study of the nation's shopping practices say they are eating more complex carbohydrates, including vegetables (50 percent), salads (49 percent), and fruits (47 percent) (Hales, 2004). Consumers view fruits and vegetables as good sources of vitamins, minerals, and fiber, helpful in calorie control, and a possible cancer preventative. In 1997, Californians said they were eating more fruits and vegetables because they were "trying to eat healthier" (30 percent), "liking the taste" (30 percent), "lowering disease risk" (6 percent), and "weight reduction" (5 percent) (Foerster et al., 1998). Apples, bananas, carrots, bagged salads, and broccoli are the top five fruits and vegetables consumers report eating more of in 2004 (Nelson, 2004). Produce continues to be viewed as health enhancing. Focus groups conducted over the Internet by Vance Research Services found consumers were aware of the link between produce and cancer prevention (Nelson, 2004). When asked what health-related reasons made them eat more produce, people responded that they were cutting back on calories, reducing cholesterol, following a diet, or following suggestions from a health professional. More households with children said they were increasing their produce consumption compared to households without children. Marketplace disappearance data confirm an increase in fruit and vegetable consumption, with the greatest increase in the fresh compared to frozen or canned category (Economic Research Service, 2005).

Consumers associate good taste and optimal nutritional value with appearance. Ripeness and freshness are rated as most important in initial purchase (Packer, 2001a). Appropriate color, ripeness, shape, and size are also important quality criteria (Gaovindasamy et al., 1997). Although color varies by produce and variety, red blush is preferred in some products, such as in peaches and nectarines (Bruhn, 1995). A characteristic odor is desirable as it indicates ripeness and reflects eating quality. Generally, larger sized products are priced at a premium; however, some prefer medium or smaller sizes depending on intended use. Scars, scratches, and other marks lower quality rating (Kader, 2002); however, some consumers will purchase lower grades if the price is sufficiently low and other factors indicate good eating quality.

Attitude studies indicate consumers tend to prefer locally grown produce, both as a result of perceptions of higher quality and to support the local economy. Many people do not know what produce is grown locally, however (Bruhn, 1992; Lockeretz, 1986).

At this time, branding does not appear to be a major factor related to consumer perceptions of quality. Almost 90 percent of consumers believe branded and non-branded items are about the same in nutritional value, and about 80 percent consider them comparable in storage life and taste. In regard to safety, 75 percent of consumers consider branded and nonbranded items comparable (Packer, 2000).

When asked how produce can be promoted to encourage purchase, consumers suggest quality be the focus, with spoiled produce kept out of the display, products held at proper temperature, recipes and preparation tips offered, and tasting provided so consumers can verify quality (Packer, 1996a; Packer, 1996b).

9.3 CONSUMER PERCEPTION OF PRODUCE SAFETY

Most consumers are confident in the safety of the food supply. However, perception varies over time, dependent on news coverage of food-safety issues. The Food Market Institute annual survey indicated that, in 2004, 82 percent of consumers were completely or mostly confident that food in the supermarket is safe. Confidence peaked at 84 percent in 1996, up from 72 percent in 1992 (Research International, 2004).

Modifications by biotechnology and treatment by food irradiation are newer technologies that generate concern among some consumers. When potential food-safety problems were identified in a 2004 survey, 56 percent considered pesticide residues a serious hazard, compared to 73 percent for bacterial contamination, 31 percent for food irradiation, and 34 percent for biotechnology (Research International, 2004).

Concern about pesticide residue is the most frequently volunteered food-safety concern associated with produce, with 30 percent of consumers concerned about residues on fresh fruits, compared to 22 percent concerned about residues on fresh vegetables (Packer, 2001b). Disease and contamination was the primary food-safety concern associated with vegetables, specified by 36 percent of consumers. In 2001, 62 percent of consumers mentioned fruits as being prone to food-safety concerns, and 41 percent mentioned vegetables. Apples, grapes, and strawberries are more frequently volunteered as being prone to food-safety concerns than other fruits, while lettuce leads the list of vegetables with potential concerns.

Concern about pesticide residue was highest in 1989 at the time of the controversy over use of the growth regulator Alar on apples. Over time, confidence in the safety of produce and belief in the health-enhancing value of produce increased due to concerted educational efforts by the produce industry and health professionals. Some supermarkets advertise use of a certification system to verify that produce meets legal pesticide residue minimums or contains no residues detectable by test sensitivity. Many supermarkets also offer organic produce.

9.3.1 Organic Food

The organic market grew from $2.3 billion in 1994 and $6.7 billion in 2000 to $9.7 billion in 2002. Organic foods are in food-service operations and mainstay supermarkets as well as specialized markets. In 2003, 51 percent of U.S. women indicate they have seen the U.S. Department of Agriculture organic seal where they do most of their shopping (Burfields, 2003). The largest organic market is fresh produce (NBJ, 2001). Correspondingly, 92 percent of organic users have purchased fruits and vegetables (Hartman & Group, 2001). Price is the greatest barrier to purchasing organic produce, with 63 percent associating higher costs with organic produce compared to conventional produce (Nelson, 2002).

Consumers with children under 18 years of age are less likely to buy organic produce than those without young children. Consumers in the West are more likely to buy organic produce than other regions, and households with incomes of $75,000 or more have the highest likelihood to purchase organic produce (Produce Merchandising Staff, 2002).

The Hartman Group (Hartman & Group, 2001) found that consumers who select organic products do so for health and nutrition reasons, followed by taste, belief in food safety, and environmental concerns. Those selecting organic foods believe there is a relatively high risk associated with consuming conventionally produced food. They select organic products because they believe they are "grown without pesticides," are chemical-free, and are safer for the environment (Hartman & Group, 2001; HealthFocus, 2003; NMI, 2001; Zehnder et al., 2003). Using organic food is mentioned as a practice used to maintain health by 37 percent of consumers (FMI/Prevention, 2001). When asked what attributes of organic foods are of greatest importance, 63 percent of those who select organic foods indicate "grown without pesticides" (NMI, 2001). Environmental protection is also mentioned by over half of organic consumers. The statement "Grown without pesticides" may be the most important characteristic of organic foods and has been rated very important to more consumers than "certified organic" (HealthFocus, 2003).

9.3.2 Integrated Pest Management (IPM)

Virtually every land grant university in the United States has a research group developing environmentally responsive pest control strategies. These methods include use of good insects to attack harmful ones, use of insect-resistant varieties of plants, and production management techniques. As a last resort, if pests reach an economically significant level, pesticides may be employed. Prior to selecting a pesticide, impact on the worker, the environment, and the target pest are evaluated. When consumers hear about the IPM approach, their attitudes toward farming practices and food safety are positive (Bruhn et al., 1992). Furthermore, Govindasamy and colleagues (Govindasamy and Italia, 1997) found people indicate strong support for IPM through both a high willingness-to-purchase and willingness-to-pay a premium for IPM-grown produce. Once informed about IPM, consumers surveyed were more willing to pay a premium for and more willing to switch supermarkets to obtain IPM rather than organic produce. An economic analysis of purchase intent found those with higher income, younger individuals, those who frequently purchase organic produce, and those who live in suburban areas were more likely to purchase IPM-grown produce and to pay a premium (Govindasamy and Italia, 1997).

9.3.3 Waxes

Waxing is a latent concern for many consumers. A national survey in 1995 indicated that only 35 percent of women and 43 percent of men said they definitely would eat fresh produce knowing that it is coated with an approved food-grade wax. Consumer attitudes appear to reflect a preference for natural, unmodified produce, doubt as to the long-term safety of ingesting wax, and the perception that wax affects product taste (Packer, 1995).

9.3.4 Microbiological Hazards

Consumers volunteer that the greatest threat to food safety is microbiological (Cogent Research, 2004; Research International, 2004). Outbreaks of foodborne illness have occurred where fresh produce was identified as the source of the pathogen. Cantaloupes have been the source of *Salmonella*, frozen strawberries were contaminated with hepatitis A, and lettuce, sprouts, fresh apple juice, and fresh basil were implicated in *E. coli* O157:H7 outbreaks. Consumers responded by avoiding the implicated product. After these incidences, as many as 60 percent of consumers indicated they were more concerned about bacterial contamination of fresh produce than in the previous year (Packer, 1998). Although consumers believe produce grown in the United States is safer than imported produce, the U.S. Economic Research Service notes that outbreaks occur from domestic as well as imported produce (Zepp et al., 1998). Care must be taken in production and processing to avoid or destroy potential pathogens.

9.3.5 Perception of Produce Processed for Convenience

Convenience is highly valued among today's consumers. A review of successful new product introductions confirms that foods with increased convenience are well received (Grocery Manufacturers of America, 2004). Consumers report an increased use of pretrimmed, washed, and bagged fresh produce. Overall 32 percent indicated they purchased convenience products more frequently than five years ago. More households with children purchased convenience products (31 percent), compared to households without children (25 percent) (Packer, 2004).

9.3.6 Perception of Produce Modified by Biotechnology or Genetic Engineering

U.S. consumers remain relatively uninformed about biotechnology. Only 10 percent of U.S. consumers indicate that they have heard "a lot" about the topic and only 36 percent believe products modified by biotechnology are in the supermarket (Cogent Research, 2004). Although soybeans and corn are the primary food applications of biotechnology, among consumers who think these products are in the market 44 percent mention vegetables, 20 percent corn, 18 percent fruit and 14 percent tomatoes.

Most U.S. consumers have a positive attitude toward biotechnology, with 59 percent believing biotechnology will benefit themselves and their family within the next five years (Cogent Research, 2004). Furthermore, in a 1000-person nationwide telephone survey conducted in 2004, 66 percent of consumers indicate they would purchase produce modified by biotechnology to reduce pesticide use, and 54 percent said they would purchase products modified for better taste (Cogent Research, 2004). When told that biotechnology-modified plants could be used to produce cooking oil with less saturated fat, only 15 percent indicated that the use of biotechnology would have a negative effect on their likelihood to purchase.

Few U.S. consumers perceive modifications by biotechnology as risky, and labeling food as genetically modified is not a priority. When asked in an open-ended question about food-safety concerns, only 1 percent volunteered concerns about genetically modified food. (Cogent Research, 2004) In contrast, concerns about disease/contamination and handling/preparation were mentioned by 22 percent and 29 percent, respectively. When asked if there was information not currently on a food label they would like to see added, only 1 percent asked for information indicating if the food was genetically altered (Cogent Research, 2004). When asked to select one item from a list of potential label additions, 17 percent chose labeling if the product was genetically altered, 33 percent selected if pesticides were used in production, 8 percent if the product was imported, 16 percent that they needed no additional information, and 15 percent said they did not know (Bruskin, 2001).

Public perception of potential environmental benefits and risks influences perception of the technology. When asked if a series of potential risks are very, somewhat, or not at all important, the potential for contamination of plant species by genetic transfer was considered a very important risk by 64 percent of consumers (Pew Initiative, 2002). Other potential risks and the percentage of consumers considering the risk very important include the potential to create super weeds (57 percent), to develop pesticide resistant insects (57 percent), to reduce genetic diversity (49 percent), and the potential that modified plants could harm others (48 percent).

Many consumers value potential benefits made possible through biotechnology. When specific benefits are identified, 74 percent rated cleaning toxic pollutants as very important (Pew Initiative, 2002). Other potential benefits and the percentage of consumers considering the benefit very important include reducing soil erosion (73 percent), using less fertilizer (72 percent), developing drought-resistant plants (68 percent), developing disease resistant trees (67 percent), and using less pesticide (61 percent).

Since 1997, the percentage of the public having a positive view toward biotechnology has decreased and that expressing concern has increased (Cogent Research, 2004; Research International, 2004). This change could be related to negative media coverage and the perception that potential risks were not under control. A content analysis of articles found claims of harm were expressed in 70 percent of the articles, while discussions of benefits were covered in only 30 percent of the stories (Center for Media and Public Affairs, 2000). Discussion of harm focused on environmental or human health, while benefits were generally limited to increased production. This is not likely to be an important benefit to consumers where food is abundant. In 2001, articles on Starlink corn dominated media coverage at 73 percent, while the ability to detect biotechnology components was the focus of 11 percent of the articles and labeling of biotechnology foods was discussed in 10 percent of the articles (Center for Media and Public Affairs, 2002). Negative comments exceeded benefits by an 8 : 1 ratio. Although Starlink corn led to no known human illness, the potential for an allergic response was frequently mentioned. Furthermore, the incidence illustrated that modified plant products could unexpectedly be in a wide variety of foods.

9.4 ATTITUDES TOWARD BIOTECHNOLOGY AMONG EUROPEAN CONSUMERS

Few Europeans (11 percent) consider themselves informed about biotechnology (INRA, 2002). Questions related to facts of biology indicate basic knowledge is lacking among the general population. In 1999, only 35 percent correctly responded that the following statement is false: "Ordinary tomatoes do not contain genes while genetically modified tomatoes do." Furthermore, only 42 percent recognized that eating genetically modified fruit does not change your personal genes, with 24 percent believing human genes would be changed, and the remainder were uncertain. A comparison of responses in 1996 and 1999 indicates there has been little increase in public knowledge. In fact, fewer consumers responded correctly in 1999 to the statement that eating genetically modified fruit changes human genes compared to 1996, with 42 percent correct responses in 1999 compared to 48 percent in 1996. Only about half of European consumers were aware of many of the applications of biotechnology (INRA, 2002). Slightly over half, 56 percent, were aware that genetic modification could be used to make plants resistant to insect attack. Similarly, only about half were aware of various medical benefits. A national survey in the United Kingdom found that 22 percent said there were no benefits from genetically modified food and 31 percent indicated they did not know about benefits (ABE, 2002). Furthermore, 69 percent of UK consumers indicated they did not know enough about genetically modified foods to make a decision about their use.

The European cautious approach to genetic modification is not surprising when basic knowledge about biology is lacking, and people believe that consumption of modified food can change human genetic material. Furthermore, awareness of potential benefits is limited. Nevertheless, Europeans differentiate between applications of biotechnology and express varying degrees of support depending on potential benefit. When asked to rate if a biotechnology application is useful, risky, or should be encouraged, medical application and environmental remediation were considered most useful (ABE, 2002). Usefulness rating for modifying food to obtain high protein, longer shelf-life, or changed taste was lower than the other applications explored, with a usefulness rating at the midpoint. The support for environmental applications is consistent with a national survey in France. A majority of consumers (63 percent) indicated it was acceptable to grow genetically modified crops in France, if this allowed the development of more environmentally friendly practices (ABE, 2002).

Europeans considered all applications of genetic engineering somewhat risky (INRA, 2002). The applications considered most risky were food production and the cloning of animals. When asked to rate the moral acceptability of various biotechnology applications, making plants resistant to insects and modifying food to increase protein, shelf-life, or taste, were below the midpoint of moral acceptability. Those applications European consumers felt should be encouraged corresponded to perceptions of usefulness and moral acceptability. Medical and environmental applications received the highest ratings, with insect-resistant plants and food-related modifications receiving the lowest.

Because consumers respond differently to various applications of biotechnology, it is not the technology itself that is acceptable or unacceptable, but how the technology is used. In an analysis of European attitudes, Gaskell (2000) observes that as the perceived usefulness of applications declines, there is an increase in perceived risk and a decline in moral acceptability and support. Perception of the value or worth of potential benefits is key to acceptance of the application of biotechnology and the products produced.

9.5 CONSUMER HANDLING PRACTICES

Studies of consumer attitudes and self-reported behavior indicate most people handle food safely; however, members of every demographic group report mishandling, which can result in increased likelihood of foodborne illness (Redmond and Griffith, 2003). Generally, a larger percentage of people over 45 years compared to those younger than 45 years report following safe handling practices (Albrecht, 1995; Altekruse et al., 1996; Jay et al., 1999; Klontz et al., 1995; Li-Cohen and Bruhn, 2002; Williamson et al., 1992). Men are less likely than women to follow kitchen sanitation procedures (Albrecht, 1995; Altekruse et al., 1996; Jay et al., 1999; Li-Cohen and Bruhn, 2002). Those with at least some college education are less likely to follow safe handling guidelines compared to those with 12 years or less schooling (Altekruse et al., 1996; Li-Cohen and Bruhn, 2002).

Much of the safe food handling research is based upon self-reported behavior. Observation of behavior indicates that people overstate their compliance to safe handling guidelines. For example, Audits International (1999) found that 79 percent of consumers correctly identified instances in which hand-washing was necessary during food preparation. However, 20 percent of those consumers were observed to neglect hand-washing practices. Similarly, 97 percent of consumers believed that eating lettuce that had been moistened by raw poultry drippings was a "risky" food-handling practice, yet 98 percent of these consumers were observed cross-contaminating ready-to-eat foods with raw meat or raw egg during food preparation (Anderson et al., 2004).

9.5.1 Selecting Produce

Consumers avoid produce with cuts, bruises, or obvious blemishes. Although some blemishes are cosmetic, avoiding products in which the tissue is broken protects safety, as bacteria has been shown to spread more rapidly in fruit with cuts and bruises. Cross-contamination can occur in the grocery shopping cart. Although some consumers routinely separate meat and poultry from raw produce, focus group discussions indicate this does not occur with others (Li-Cohen and Bruhn, 2002).

9.5.2 Bringing Produce Home

The Fight BAC food safety guidelines emphasize the need to separate raw meat and poultry from foods to be eaten raw. However, many consumers do not realize the potential for juices to cross-contaminate occurring in the shopping bag. Less than 30 percent of consumers in a nationwide mailing indicated they ask for meat, poultry, and fish to be bagged separately from fresh produce (Li-Cohen and Bruhn, 2002). More than half of consumers surveyed indicated they had no special requirements for produce packaging.

9.5.3 Home Storage

Most consumers store produce in the refrigerator; however, some items are stored at room temperature (Li-Cohen and Bruhn, 2002). Room temperature storage is appropriate for optimum quality of tomatoes, bananas, and unripe climacteric fruits, but refrigeration lengthens the freshness and slows bacterial growth should a produce item contain harmful microorganisms. In a nationwide survey, 42 percent stored apples and 24 percent stored melons at room temperature (Li-Cohen and Bruhn, 2002).

Most consumers store fresh produce either in the refrigerator produce drawer or on a shelf. Meat/poultry juices can contaminate produce if these items are stored above other foods. There is a potential for 30 percent of consumers to either store produce wherever there is room or specifically place meat and poultry on a shelf above other food (Li-Cohen and Bruhn, 2002).

Contaminants can transfer from unclean surfaces to produce in the refrigerator. Frequent cleaning of the home refrigerator is not a universal practice. Although 50 percent of consumers indicate they clean their refrigerators at least once a month, the remainder cleaned two or three times a year, or less frequently (Li-Cohen and Bruhn, 2002).

9.5.4 Hand Washing

Although people should wash their hands before beginning food preparation, consumers do not always follow this practice. Most studies in the United States accepted that hand washing had taken place if people merely rinsed with water. Between 20 and 60 percent of consumers reported not washing their hands before starting meal preparation or after handling raw meat or poultry (Altekruse et al., 1996, 1999; Food Safety Inspection Staff, 2000; Yang et al., 1998). Regarding handling produce, almost half acknowledged that they do not always wash their hands before handling produce (Li-Cohen and Bruhn, 2002). When food handling was observed, only 45 percent of consumers attempted to wash their hands before food preparation. Of those who washed their hands, only 84 percent used soap (Anderson et al., 2004).

9.5.5 Kitchen Sanitation

Slightly more than half of consumers report washing the sink before handling fresh produce and about half wash the sink after handling (Li-Cohen and Bruhn, 2002). Most (69 percent) indicated using a cleanser or cleaning solution for washing, 40 percent used dishwashing liquid, 27 percent used bleach, and 19 percent used antibacterial soap, while 11 percent indicated they washed with water only.

Consumers do not always follow the Fight BAC recommendations to prevent cross-contamination. Between 20 and 70 percent of consumers report using the same utensils and/or cutting board unwashed when preparing raw produce after cutting raw meat or poultry (Altekruse et al., 1996, 1999; Bruhn and Schutz, 1999; Food Safety Inspection Staff, 2000; Jay et al., 1999; Klontz et al., 1995; Li-Cohen and Bruhn, 2002; Williamson et al., 1992; Yang et al., 1998).

9.5.6 Washing Produce

Consumers respond to survey questions that they wash produce with 81 percent saying they wash just before eating or cooking and 21 percent before placing in the refrigerator; 6 percent acknowledge that they seldom or never wash produce (Li-Cohen and Bruhn, 2002). The item washed least frequently is melon, with 36 percent indicating they never wash this fruit (Li-Cohen and Bruhn, 2002).

Many believe it is not necessary to wash melon because the rind is not consumed (Li-Cohen and Bruhn, 2002). Consumers have indicated that they did not think it was necessary to wash home-grown or organic produce (Li-Cohen and Bruhn, 2002). In actual practice, consumers may not wash produce, even when it will be eaten raw. Anderson et al. (2004) observed that vegetable washing was inadequate. When preparing salad, 6 of 99 subjects made no effort to clean the vegetables, 70 rinsed the lettuce, 93 rinsed the tomato, 47 rinsed the carrots, and 55 rinsed the cucumber. Rinsing time ranged from 1 s for tomatoes, cucumber, and carrot to up to 55 s for tomatoes, but averaged less than 12 s.

Consumers use different techniques to wash produce. Relatively efficient methods, such as peeling, rubbing with hands, scrubbing with a brush, and washing under running water were all used. Washing under running water was most common (Li-Cohen and Bruhn, 2002). As many as 20 percent, however, soak produce in a container, a method not recommended because contamination can spread to other produce items. As many as 4 percent wash produce with dish detergent. This method is not recommended because residue can remain on the product (FDA Talk Paper, 2001; Food Safety Inspection Staff, 2000).

9.5.7 Storage of Leftovers

Most consumers recognize that cut fruit should be refrigerated; however, some volunteered that they stored cut melons at room temperature. When responding to the development of a safe handling brochure, consumers advised that the words "always" be added to the advice to refrigerate leftovers immediately. The participants knew people who stored melons and other cut fruit at room temperature (Li-Cohen and Bruhn, 2002).

9.6 SUMMARY: STAYING HEALTHY, EATING HEALTHY

Research on health-related behavior indicates that people make rational decisions when they are aware and have some knowledge of health problems (McIntosh et al., 1994). Acquisition of knowledge alone does not automatically produce compliant behavior (Ajzen, 1991). It is not uncommon to make judgments based upon optimistic bias (Frewer et al., 1994). The risks of foodborne illness are considered unlikely, and therefore the practices to increase safety seem too much trouble. Convincing consumers to change unsafe handling practices may be difficult, because fruits and vegetables are not commonly associated with foodborne illness. About one-third of those responding to a mail questionnaire on safe handling indicated they were not interested in receiving information on safe handling (Li-Cohen and Bruhn, 2002). Therefore, it is important to both acknowledge that fruits and vegetables promote health and are tasty, and that they must be handled appropriately to avoid illness.

Educational materials must be presented through convenient sources. Those willing to receive information indicated they preferred brochures in supermarkets (54 percent) or information on individual produce containers (46 percent),

(Li-Cohen and Bruhn, 2002). Safe handling guidelines must be practical, easy to incorporate into daily activities, and presented simply. Consumers indicate they prefer pictures and a minimum of words and want to know the reason for any recommendation (Li-Cohen et al., 2002). Consumers suggest that safe handling be available in English and other languages. Consumers advise that media representatives should be encouraged to repeat the safe handling steps and advertise the availability of guidelines on the Web. Information should also be targeted toward children as part of school studies and with special activities. For example, safe handling should be a required component of any school gardening program.

Recommendations on safe handling of produce should remind consumers that eating fruits and vegetables is healthy, but care should be taken to clean produce and minimize any risks that may be present. Consumers believe produce is safe and are reluctant to take extraordinary steps to enhance safety further (Li-Cohen et al., 2002). Additionally, people say they prefer not to purchase special cleaning solutions, but would consider using materials already in the home, such as vinegar or a scrub brush. People prefer as many pictures as possible and most acknowledge that they will just skim text. Recommendations then should be concise, practical, and use materials readily available at home.

Box 9.1 presents guidelines refined through consumer focus groups (Li-Cohen et al., 2002), peer reviewed and available at http://anrcatalog.ucdavis.edu/InOrder/Shop/Shop.asp (type "Safe Handling" in the search box).

BOX 9.1 SAFE HANDLING OF FRUITS AND VEGETABLES

Eating a variety of fruits and vegetables is healthy, but care must be taken to be sure fruits and vegetables do not become contaminated with harmful bacteria. In the United States, one out of four people suffers from foodborne illness each year. Some of these illnesses have been traced to eating raw fruits or vegetables.

Everyone is at risk for foodborne illness, but people who are younger than 5, older than 50, diabetic, take antibiotics or antacids, and whose immunity is compromised are at higher risk.

Bacteria are everywhere. Harmful bacteria may be on fruits and vegetables, hands, and kitchen counters and sinks, even when they look, feel, or smell clean.

This publication provides guidelines for protecting you from harmful bacteria.

AT THE SUPERMARKET

- In the grocery cart, **separate** fruits and vegetables from meat, poultry, and fish to avoid cross-contamination.
- When bagging fresh fruits and vegetables to take home from the supermarket, put fresh produce and meat, poultry, and fish in **separate bags**.

HOME STORAGE

- All cut or prepared fruits and vegetables should be stored in the refrigerator along with many types of whole fruits and vegetables.
- When using the refrigerator, place produce in the produce drawer or on a refrigerator shelf.
- Store meat, poultry, and fish in the clean meat drawer or on a tray on the bottom shelf **below** other refrigerated foods. This prevents meat, poultry, or fish juices from dripping on other foods.

PREPARE THE KITCHEN

- Clean the sink with hot, soapy water or cleanser **before** and **after** washing and preparing fresh fruits and vegetables.
- Always wash cutting boards and preparation areas before and after food preparation. Wash preparation areas and utensils especially well after preparing meat, poultry, or fish and before preparing foods that will be eaten without cooking.
- If possible, use one cutting board and preparation area for fruits and vegetables and a **different** cutting board and preparation area for meat, poultry, and fish.
- Always wash knives after cutting meat, poultry, or fish with hot soapy water before cutting fresh fruits and vegetables; or, use different knives for cutting meat products and fresh produce.
- Washing with soap or detergent removes soil and food, but it removes only some bacteria. **For additional safety, always sanitize cutting boards and food preparation** areas after cutting meat, poultry, or fish, or any produce item with visible dirt or that grows on or in the ground. Sanitize by one of the following methods:
    - Pour boiling water over the clean wood or plastic cutting boards for 20 seconds.
    - Rinse clean wood plastic cutting boards with a solution of 1 teaspoon chlorine bleach in 1 quart (4 cups) of water.
    - Place plastic cutting boards in the dishwasher and run, using normal cleaning cycle.

WASH YOUR HANDS

Always wash hands with hot, soapy water for at least 20 seconds before and after handling fresh fruits and vegetables.

WASH ALL FRUITS AND VEGETABLES

- Always wash fruits and vegetables, including those that are organically grown, come from a farmer's market, or were grown in your own garden.
- Wash fruits and vegetables just before cooking or eating.
- Wash under running water.
- When possible, scrub fruits and vegetables with a clean scrub brush or with hands.
- For melons, scrub the rind with a brush under running water before cutting or peeling. This removes bacteria before it is spread by the knife when slicing. Sanitize the brush by putting it in the dishwasher, placing it in boiling water for 20 seconds, or rinsing it in a bleach solution of 1 teaspoon chlorine bleach in 1 quart (4 cups) of water.
- Dry fruits and vegetables with **disposable** paper towels.
- Do not use antibacterial soaps or dish detergents to wash fruits and vegetables because soap or detergent residues can remain on the produce. The FDA has not evaluated the safety of the residues that remain on the produce. However, the effectiveness of these washes is not currently standardized.
- Soaking fruits and vegetables in water is not recommended because of the potential for cross-contamination.
- Remove outer green leaves from items like lettuce or cauliflower **before** washing. Trim the hull or stem from items like tomatoes, strawberries, and peppers **after** washing.
- Ready-to-eat, prewashed, bagged produce can be used without further washing if it has been kept refrigerated and is used by the "use-by" date. If desired, produce can be washed again under running water.
- Precut or prewashed produce sold in open bags or containers should always be washed under running water before using.

REFRIGERATE ALL LEFTOVERS

- Peel leftover melons and store fruits in the refrigerator.
- Store all cut produce in a clean container in the refrigerator.

FOR MORE INFORMATION: USEFUL WEB SITES

- USDA/FDA Foodborne Illness Education Information center: http://www.nal.usda.gov/foodborne/index.html
- U.S. FDA/Center for Food Safety and Applied Nutrition: http://vm.cfsan.fda.gov/list.html
- Gateway to Government Food Safety Information: http://www.foodsafety.gov

REFERENCES

ABE (2002) Public attitudes to agricultural biotechnology. Available at http://www.ABEurope.info. Accessed February 8, 2006.

I. Ajzen, The theory of planned behaviour. *Organ. Behav. Hum. Decis. Proc.*, 50, 179–211 (1991).

J.A. Albrecht, Food safety knowledge and practices of consumers in the U.S.A. *J. Cons. Stud. Home Econ.*, 19, 119–134 (1995).

S.F. Altekruse, et al., Consumer knowledge of foodborne microbial hazards and food-handling practices. *J. Food Prot.*, 59, 287–294 (1996).

S.F. Altekruse, et al., A multi-state survey of consumer food handling and food-consumption practices. *Am. J. Prevent. Med.* 16, 216–221 (1999).

J.B. Anderson, et al., A camera's view of consumer food-handling behaviors. *Am. Diet. Assoc.*, 104, 186–191 (2004).

Audits International, Consumer food handling practices, 1999. Available at www.foodonline.com/storefonts/audits.html. accessed February 9, 2006.

C. Bruhn, Consumer attitude toward locally grown produce. *Calif. Agric.*, 46, 13 (1992).

C.M. Bruhn, Consumer and retailer satisfaction with the quality and size of California peaches and nectarines. *J. Food Qual.*, 18, 241–256 (1995).

C. Bruhn and H. Schutz, Consumer food safety knowledge and practices. *J. Food Safety*, 19, 73–87 (1999).

C. Bruhn, et al., Consumer response to information on integrated pest management. *J. Food Safety*, 12, 315–326 (1992).

Bruskin, National opinion polls on labeling of genetically modified foods. Center for Science and the Public Interest, 2001. Available at http://www.cspinet.org/new/poll_ge-foods.html. Accessed February 8, 2006.

T. Burfields, Customers accepting organics, 2003. Available at www.thepacker.com/icms/_dtaa2/content/2003–16620–290.asp. Accessed February 8, 2006.

Center for Media and Public Affairs, Food for Thought III. Reporting on diet nutrition and food safety news. International Food Information Council, www.IFIC.org, 2000.

Center for Media and Public Affairs, Food For Thought IV. Reporting on diet nutrition and food safety news. International Food Information Council, www.IFIC.org, 2002.

Cogent Research, U.S. Consumer attitudes toward biotechnology. International Food Information Center, www.IFIC.org, 2004.

Economic Research Service, Food Consumption. ers.usda.gov/data/food comsumption, 2005.

FDA Talk Paper, FDA advises consumers about fresh produce safety. FDA, 2001. Available at http://vm.cfsan.fda.gov/~lrd/tpproduce.html. Accessed February 8, 2006.

FMI/Prevention, "Reaching Out to the Whole Health Consumer," in *Trends in the Supermarket. Shopping for Health.* Food Marketing Inst., Washington, DC, 2001.

S.B. Foerster, et al., California dietary practices survey; focus on fruits and vegetables trends among adults, 1989–1997. California Department of Health Services Public Health Institute, Sacramento, CA, 1998, p. 41.

Food Safety Inspection Staff and U.S. Department of Agriculture, Focus groups shed light on consumers' food safety knowledge. *The Food Safety Educator*, 5, 1–8 (2000).

L. Frewer, et al., The interrelationship between perceived knowledge, control and risk associated with a range of food-related hazards targeted at the individual, other people and society. *J. Food Safety*, 14, 19–40 (1994).

R. Govindasamy, et al., Quality of agricultural produce: consumer preferences and perceptions. Rutgers University, New Jersey, Vol. 3, Nos 2–3, 1997.

G. Gaskell, Agricultural biotechnology and public attitudes in the European Union, 2000. Available at http://www.agbioforum.org/. Accessed February 8, 2006.

R. Govindasamy and J. Italia, Consumer response to integrated pest management and organic agriculture: an econometric analysis. Rutgers University, 1997. Available at http://aesop.rutgers.edu/~agecon/pub_index.htm. Accessed February 8, 2006.

Grocery Manufacturers of America, Convenience benefits continue to make lifting easier, 2004. www.gmabrands.org.

D. Hales, "What America Really Eats," in *Sacramento Bee*, Parade section of the newspaper Sacramento, California, 2004, pp. 6–7.

Hartman & Group, *Healthy Living: Organic and Natural Products Organic Lifestyle Study.* Spring. Bellevue, Washington, 2001.

HealthFocus, *HealthFocus Trends Survey.* Health Focus, Atlanta GA, 2003.

INRA, Eurobarometer 52.1 The Europeans and Biotechnology, 2002. Available at http://europa.eu.int/comm/public_opinion/archives/eb/ebs_134_en.pdf. Accessed February 8, 2006.

S.L. Jay, et al., A national Australian food safety telephone survey. *J. Food Prot.*, 62, 921–928 (1999).

A.A. Kader, "Quality and Safety Factors: Definition and Evaluation for Fresh Horticultural Crops," in A.A. Kades, Ed., *Postharvest Technology of Horticultural Crops*, 3rd edn., University of California Agricultural and Natural Resources, Oakland, CA, 2002, pp. 279–285.

K.C. Klontz, et al., Prevalence of selected food consumption and prevalence behaviors associated with increased risks of food-borne disease. *J. Food Prot.*, 58, 927–930, (1995).

A.E. Li-Cohen and C.M. Bruhn, Safety of consumer handling of fresh produce from the time of purchase to the plate: a comprehensive consumer survey. *J. Food Prot.*, 65, 1287–1296, (2002).

A.E. Li-Cohen, et al., Refining consumer safe handling educational materials through focus groups. *Dairy, Food Environ. San.*, 22, 539–551 (2002).

W. Lockeretz, Urban consumer's attitudes toward locally grown produce. *Am. J. Altern. Agric.*, 1, 83–88 (1986).

W.A. McIntosh, et al., Perceptions of risks of eating undercooked meat and willingness to change cooking practices. *Appetite*, 22, 83–96 (1994).

NBJ, Organic foods report 2001. *Nutrition Business J.*, San Diego, CA, 2001.

A. Nelson, Nonbuyers of organics prove to be a hard sell. *The Packer Online*, September 12, 2002.

A. Nelson, Nutrition news incites consumer purchase. *The Packer Online.* Available at www.thepacker.com/icms/_dtaa2/content/wrapper.asp?alink=2004–1411, 2004.

NMI, *Health and Wellness Trends Report.* Natural Marketing Inst., Harleysville, PA, 2001.

Packer, "Convenience Craze," in *The Packer Fresh Trends 1995.* Vol. 101, Vance Publishers, Lenexa, KS, 1995, pp. 56–63.

Packer, "Catching the Shopper's Eye," in *Fresh Trends.* Vol. 102, Vance Publishers, Lenexa, KS, 1996a, pp. 50–52.

Packer, "Turning Consumers Off," in *Fresh Trends.* Vol. 102, Vance Publishers, Lenexa, KS, 1996b, pp. 84–86.

Packer, "Microbes Grab the Spotlight," in *Fresh Trends.* Vol. 104, Vance Publishers, Lenexa, KS, 1998, pp. 20–26.

Packer, "Brands," in *Fresh Trends.* Vol. 106, Vance Publishers, Lenexa, KS, 2000, pp. 24–32.

Packer, "Eye Appeal Influences Shoppers," in *Fresh Trends.* Vol. 107, Vance Publishers, Lenexa, KS, 2001a, pp. 38–42.

Packer, "Fruits of Concern, Igniting Fears, Food Safety First," in *Fresh Trends.* Vol. 107, Vance Publishers, Lenexa, KS, 2001b, pp. 62–65.

Packer, "By the Numbers," in *Fresh Trends.* Vol. 110, Vance Publishers, Lenexa, KS, 2004, pp. 7–14.

Pew Initiative, Environmental Savior or Saboteur? Debating the Impacts of Genetic Engineering, 2002. Available at http://www.pewtrusts.com. Accessed February 8, 2006.

Produce Merchandising Staff, More Consumers are buying organic produce. *The Packer Online*, 2002. www.thepacker.com/icms/_data2/content/2002-162738_733.asp. Accessed February 8, 2006.

E.C. Redmond and C.J. Griffith, Consumer food handling in the home: A review of food safety studies. *J. Food Protect.*, 66, 130–161 (2003).

Research International, *Trends in the Supermarket.* Food Marketing Institute, Washington, DC, 2000, p. 90.

Research International, *Trends Consumer Attitudes and the Supermarket.* Research International, Washington, DC, 2004.

D. Williamson, et al., Correlating food safety knowledge with home food-preparation practices. *Food Technol.*, 46, 94–100 (1992).

S. Yang, et al., Multi-state surveillance for food-handling, preparation, and consumption behaviors associated with foodborne diseases: 1995 and 1996 BRFSS food-safety questions. *Morb. Mortal Wkly Rep.*, 47SS, 33–57 (1998).

G. Zehnder, et al., As assessment of consumer preferences for IPM and organically grown produce. *J. Extension*, 41, 1–5 (2003).

G. Zepp, et al., "Food Safety and Fresh Fruits and Vegetables: Is There a Difference Between Imported and Domesticated Produced Products," in *Vegetables and Specialties/VGS-274/April.* USDA Economic Research Services, Washington, DC, 1998, pp. 23–28.

CHAPTER 10

The Economics of Food Safety and Produce: The Case of Green Onions and Hepatitis A Outbreaks

LINDA CALVIN*

U.S. Department of Agriculture's Economic Research Service, Washington, DC, USA

BELEM AVENDAÑO

Department of Economics of the Universidad Autónoma de Baja California, Mexico

RITA SCHWENTESIUS

Center for Economic, Social, and Technological Research on Agroindustry and World Agriculture, Universidad Autónoma de Chapingo, Chapingo, Mexico

10.1 INTRODUCTION

In fall of 2003, large outbreaks of hepatitis A in the United States were associated with consumption of green onions from Mexico. Despite nearly a decade of industry

*The views expressed here are those of the authors and may not be attributed to the Economic Research Service or the U.S. Department of Agriculture.

Microbial Hazard Identification in Fresh Fruit and Vegetables, Edited by Jennylynd James

and U.S. Food and Drug Administration (FDA) activities targeted at reducing microbial contamination of fresh produce at the grower level, outbreaks of foodborne illness, though infrequent, still occur. The FDA acknowledges that it is not possible to guarantee food safety in terms of microbial contamination with current technology. Even growers with the best food-safety practices may still have contaminated product – all sources of risk cannot be controlled. However, there is concern that some growers are lagging behind government and consumer expectations in adopting safer practices.

The recent outbreaks of hepatitis A demonstrate one of the most important challenges in food safety. Incentives for individual growers to adopt stronger food-safety programs are increasing, but are still not adequate to entice all growers to upgrade their practices. Many Mexican growers had adopted safer practices before the outbreaks. As long as some growers do not adopt safer practices, all growers face the economic consequences of an outbreak.

Using the 2003 outbreaks of hepatitis A as an example, this chapter reviews the economics of food safety, particularly the economics of adopting better food-safety practices at the grower level. In an integrated North American industry, the same economic forces affect both U.S. and Mexican growers. The first section discusses the cost/benefit analysis growers must undertake when deciding how much to invest in food-safety practices. Peculiarities of the market for food-safety characteristics demonstrate some of the obstacles to increasing food safety to socially desirable levels. The story of the green onion case begins with a description of the U.S. and Mexican green onion industry. Earlier research provides a unique perspective on adoption of food-safety practices in the Mexican green onion industry before the food safety crisis (Avendaño, 2004). This is followed by an account of the outbreaks. The next section discusses the market impacts of the outbreaks. Interviews with a limited number of Mexican growers after the outbreaks provide the basis for an analysis of how the impact varied across farmers by the level of food-safety practices already adopted. After the chaos of the outbreaks subsided, a wide range of market participants and government agencies took actions to reduce the probability of future outbreaks by promoting adoption of better food-safety practices.

10.2 BENEFITS AND COSTS OF ADOPTING BETTER FOOD-SAFETY PRACTICES

When individual growers choose whether to adopt additional food-safety practices, they weigh their private benefits and costs. The discussion assumes food-safety practices are not mandated by law and growers' choices are not restricted. Mandatory food-safety standards will be addressed below. Typically, growers adopting a new production practice expect to either receive a higher price for a higher quality good, reduce risk, or lower their costs of production.

In the case of adopting food-safety practices, growers do not receive a higher price. As a result, some growers choose not to adopt safer practices. However, other benefits may influence growers' decisions to adopt better food-safety practices. These benefits are mostly related to risk – the reduction in probability of unpleasant events, such as a catastrophic drop in sales if contaminated produce is traced to their operations, damage to reputations, lawsuits, and so on. These benefits only accrue in the event of an outbreak. Until an outbreak occurs, growers may think that the probability of ever experiencing the benefits is very low. Afterwards, growers may revise their estimates of these benefits. A more immediate benefit of adoption of better food-safety practices is that many retailers and food-service buyers now require third-party audits of food-safety practices as a condition of purchase. Having higher food-safety standards gives growers broader market access.

Weighing against potential benefits are the costs of adopting new food-safety standards, which are immediate and, often, large. Costs may include investment in new infrastructure such as water purification plants, training for workers to improve hygiene in the fields, upgrades to recordkeeping systems, and use of third-party audits for compliance with good agricultural practices (GAPs) in the fields, and good manufacturing practices (GMPs) in packinghouses. GAPs are voluntary guidelines for minimizing the risk of microbial contamination in produce. The FDA published the guidelines for GAPs in 1998. It recommends, but does not require, GMPs for firms packing raw, intact fruit and vegetables.

Growers adopt new food-safety practices if expected benefits exceed expected costs. However, not all growers will necessarily make the same decision with respect to adopting more food-safety practices. Even among growers of the same crop, benefit–cost analyses upon which decisions are based can vary depending on characteristics of the grower and the operation.

Do the decisions individual growers make about food-safety practices ensure the level of food safety desired by consumers and society at large? Possibly not (Caswell and Mojduszka, 1996; Mitchell, 2003). Markets do not always work smoothly for all goods. Private decisions by growers may not be socially optimal because of imperfect information and negative externalities. Imperfect information, which exists when buyers and sellers cannot identify certain characteristics of a product, may reduce the incentives to adopt new food-safety practices by hindering the development of different prices for different levels of food safety. Box 10.1, "Imperfect Information and Prices for Safer Food," provides a more detailed explanation of this phenomenon.

BOX 10.1 IMPERFECT INFORMATION AND PRICES FOR SAFER FOOD

If the price of produce varied with its safety, there would be more incentive for growers to provide safer food. Other observable characteristic of produce (size, color, etc.) frequently receive price premiums. As long as buyers and sellers can clearly identify the different qualities of produce, growers will produce the variety of qualities that consumers want to buy.

In most cases, it is difficult to tell if produce is contaminated until a consumer gets sick. Assume there are two types of produce – high-safety and low-safety. When buyers and sellers cannot identify the two types of produce, economists say the market is characterized by "imperfect information." Consumers may not be willing to purchase an item if they think the probability of contamination is relatively high and they have no information on which items may be safer than others. When either the seller or buyer has the advantage of having more information than the other about the food-safety characteristics of a product, economists call this "asymmetric information." A grower could deliberately underinvest in food safety, compared with other growers, to reduce costs and then sell this produce, which is more likely to be contaminated, to unsuspecting consumers. Imperfect and asymmetric information hinder the efficient operation of markets (Akerlof, 1970).

Consumers would prefer more information before they purchase produce. Market participants try to provide more information on food-safety characteristics of their produce, but this is particularly difficult because there are often no good tests for the presence of microbial contamination. One increasingly important strategy is for growers to use third-party audits to verify their compliance with GAPs and GMPs. A successful audit verifying compliance with GAP and GMP principles does not guarantee food safety – an audit is an informed opinion on the state of farm operations at a particular point in time. It is, however, an important first step in improving food safety and signaling to buyers that certain practices are in place conforming to U.S. Food and Drug Administration (FDA) guidelines. The FDA is concerned that some outbreaks have been traced back to firms that have successfully completed third-party audits (Calvin, 2003). Firms with third-party audits can have very different commitments to food safety.

Traceback to the grower is another mechanism that provides information about the food-safety level of produce. However, it only provides information after the outbreak occurs (Golan et al., 2004a). A complete traceback to the grower of the contaminated product is not always possible. As the traceability process improves, poor production practices will become increasingly difficult to hide. If the grower of the contaminated product is identified, that firm will suffer a loss of reputation and commercial buyers may shift to other suppliers. Other growers will take note of the negative economic impact and perhaps reassess their own need for more food-safety practices.

Despite more information about the probability of safety, a market for produce of various food safety characteristics has not developed. Growers with sophisticated food-safety programs do not receive a price premium for their products. Advertising product from one producer as being safer may be a risky strategy for a retailer or food-service firm, because, in most cases, it is not actually possible to guarantee food safety. Also, advertising a particular source of product as being safer than others may provide consumers with information that undermines their confidence in the product in general, regardless of the source (Golan et al., 2004b). For example, suppose a retailer advertises a produce item from a particular source as having less probability of some type of contamination than other sources. If consumers never knew that kind of contamination was possible on produce, this information may make them less willing to buy that type of produce, regardless of safety claims.

Negative externalities also affect the incentives to adopt additional food-safety practices. Negative externalities exist when one party's production or consumption choices have a negative impact on another party's well-being without a cost to the injuring party. If the injuring party had to pay the injured, the former might change behavior to account for the additional costs. Society as a whole may demand more food safety than consumers in the grocery store or food-service outlet. Of course, in the event of a large outbreak of foodborne illness, consumers are on the frontline, facing health problems and medical bills, lost days of work, and so on. But everyone along the marketing chain associated with the contaminated produce will face potential costs. Even those not directly associated with contaminated produce may suffer. For example, if a foodborne illness is traced to a particular product, but not a particular grower, all producers of that food item may feel the full effect of decreased demand. The Centers for Disease Control and Prevention (CDC) and FDA incur substantial costs in tracing an outbreak back to the contaminated product. They conduct farm and packinghouse investigations and review inspection results. Some level of government often ends up paying for many of the medical costs incurred in dealing with an outbreak. In their private benefit–cost analyses, growers do not consider the benefits and costs that might accrue to others if food safety were improved and may therefore provide less food safety than society desires.

When there are outbreaks of foodborne illness, other groups in the produce industry, marketing chain, or government facing increased costs may try to impose new rules on growers to encourage or force them to implement food-safety measures more in line with society's total demand for food safety. For example, grower organizations may put into place voluntary or mandatory practices. This would reduce the negative impact of one producer with contaminated produce on other producers of the same product. Retailers and food-service buyers may require growers to obtain third-party audits showing compliance with GAPs and GMPs to reduce the chance that their businesses will be associated with an outbreak. Governments may also impose higher standards on producers.

10.3 U.S. AND MEXICAN GREEN ONION INDUSTRY

Mexico is the dominant force in North American green onion production. Most green onions consumed in the United States come from Mexico. In 1978 and 1979, all (or most) shipments were from the United States – shipment data do not always capture small amounts of trade (Fig. 10.1). In 1980, shipments of green onions began entering from Mexico, and this production eventually replaced winter production in Arizona and parts of California. By 1986, Mexican shipments exceeded U.S. shipments and this trend has continued ever since. Total shipments of green onions in the U.S. market (from Mexican and U.S. sources) are growing, up 48 percent from 1990 to 2003.

As a labor-intensive crop, green onions are cheaper to grow in Mexico than in the United States. Green onions, like other crops that are hand bunched, such as

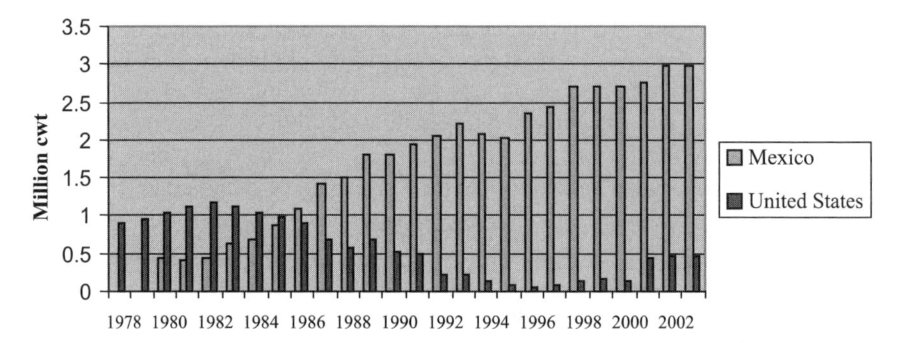

Figure 10.1 Annual shipments of green onions to the U.S. market. (*Source*: Fresh fruit and vegetable shipments, by commodities, states and months, Agricultural Marketing Service, USDA.)

radishes, involve more hand labor in the harvesting and packing process than most fruit and vegetables. Each person that handles green onions potentially increases the probability of microbial contamination. For example, in a typical operation, as many as nine different people might touch a green onion. If the harvesting and packing plant operations ever become more mechanized, costs might be reduced sufficiently that production would return to the United States.

The North American green onion industry is highly integrated. Buyers demand green onions on a year-round basis, and shippers source from both Mexico and the United States depending on season and availability. In 2003, shipment data shows that 87 percent of the supply available in the United States came from Mexico. Shipment data do not pick up production in all states, so it undercounts the importance of U.S. production. In 2003, only production from California, South Carolina, Texas, and Arizona was included. California accounted for 68 percent of U.S. shipments. Summer production in other states is not captured by these statistics. U.S. shipments are largest during the summer months and Mexican shipments peak during the winter (Fig. 10.2).

Originally, the Mexican green onion export industry produced only during the winter in Mexicali, Baja California, and San Luís Río Colorado in the adjoining state of Sonora (Fig. 10.3). The demand for year-round supplies has led some growers to produce all year. Summer production is located in the cooler western coastal range in areas such as Ojos Negros, Valle de la Trinidad, El Cóndor, and Valle de Guadalupe. The summer production areas are more isolated, with poorer infrastructure, which may be more challenging from a food-safety perspective.

At the time of the hepatitis A outbreaks in the United States in fall 2003, there were 26 green onion growers in Mexicali and San Luís Río Colorado. Eight of these growers also had summer operations in the mountains. One additional grower planted green onions only in the summer production region. Summer shipments (mid-May through early October) to the United States accounted for 27 percent of total calendar year shipments in 2002 (the last full year before the outbreak). The output of the four firms named by the FDA as being the source of the contaminated produce represented a relatively small share of the area's

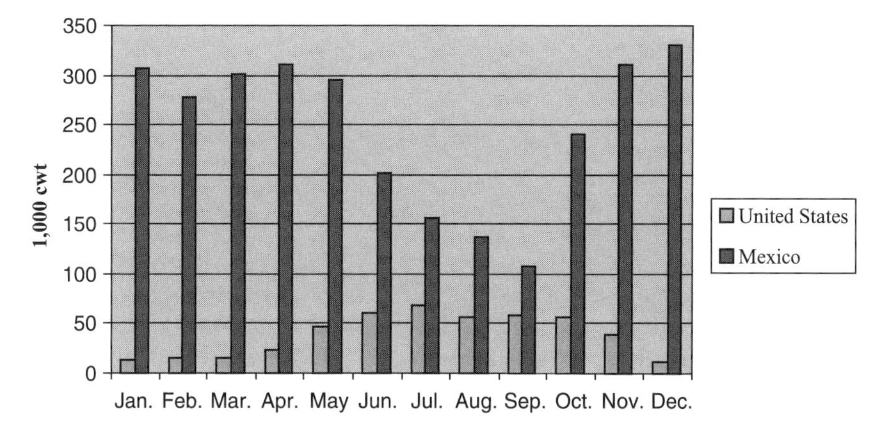

Figure 10.2 Monthly shipments of green onions, 2002. (*Source*: Fresh fruit and vegetable shipments, by commodities, states and months, Agricultural Marketing Service, USDA.)

summer production of green onions and an even smaller share of total winter and summer production. However, the problems of these firms affected the whole industry.

Shippers in the United States market Mexican green onions. Shippers and their suppliers typically develop close relationships, whether the suppliers are domestic or foreign. Often, Mexican firms grow for a U.S. firm that demands the same standards from their domestic and foreign suppliers. A shipper may require that a grower have a third-party audit verifying compliance with GAPs or GMPs. A wide range of individual food-safety programs exists on both sides of the border.

Before the 2003 outbreak of hepatitis A in the United States, many growers in Mexico already used third-party certification for complying with GAPs and GMPs. Despite survey results suggesting that most growers have an interest in food safety, a lack of concern by only a few growers can affect the entire industry. Table 10.1 shows results from a 2002 survey of GAP use by horticultural producers

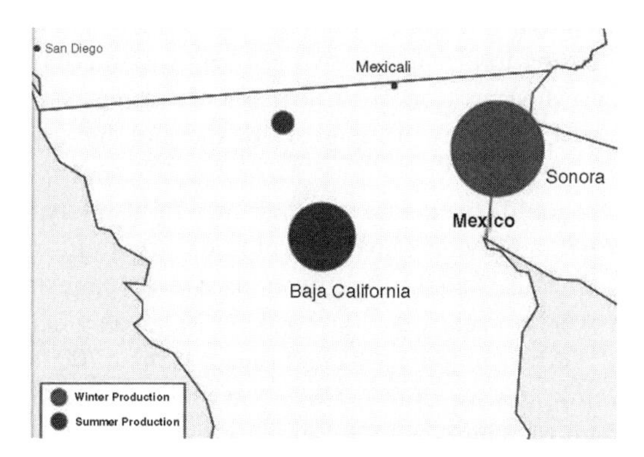

Figure 10.3 Mexican production areas for green onion exports. *Source*: Economic Research Service. U.S. Department of Agriculture.

TABLE 10.1 Grower Use of GAPs in the Mexicali Valley, Mexico, 2002

Item	Number of Growers
Level of GAP Compliance	
GAP compliant	3
In process of becoming GAP compliant	2
No GAP program	2
Source of Motivation for Adopting GAPs	
Own initiative	5
Shipper demands	1
Own initiative and shipper demands	1
Most Important Reason for Adopting GAPs	
Maintain market access	3
Produce a safer product	2
Receive higher price	1
Other	1

Source: Avendaño, 2004.

in the Mexicali Valley, Mexico (parts of Baja California and Sonora). Results for use of GMPs were similar. Although the survey's sample size is small, the results provide some insight into why growers adopt better food-safety practices. Firms in the sample grew green onions and exported to the United States. Growers were questioned about their operations in the Mexicali and San Luís Río Colorado areas. Firms were selected based on proportional random sampling with respect to farm size.

Three of the seven respondents were completely GAP compliant. Mexico is the North American market leader in green onions, with many dynamic and sophisticated producers. For some growers, adopting GAPs and GMPs was part of a natural progression to provide a differentiated product – organic certification, pesticide-residue-free certification, and GAP and GMP certification. One grower expected the next step to be applying Hazard Analysis Critical Control Point analysis to his packinghouse. According to the survey, two growers were in the process of becoming GAP compliant. Most growers plan investments in new food-safety practices over a period of several years. Two growers did not have any type of GAP certification.

Although growers could conceivably do their own food-safety testing, third-party audits of a GAP program provide a level of additional credibility. All five producers with some level of GAP use had third-party audits. Developing stronger food-safety programs is expensive. In recent years the costs for Mexican green onion growers to become GAP and GMP compliant have ranged from $700,000 to $2,500,000 (Avendaño and Schwentesius, 2005). Costs vary by size of operation, whether firms have their own packing shed, and environmental conditions. For example, growers with access only to open-ditch irrigation water may face higher costs than those with access to deep well water to achieve the same level of expected food safety.

Almost all Mexican growers (five of seven) in the survey stated that the decision to adopt better food-safety standards was, or would be, a result of their own initiative, and not due to the requirements of their shipper. All seven growers responded regardless of their compliance with GAPs. One grower reported that shipper demands were, or would be, an important factor in decision making. Another said both factors played, or would play, a role in the adoption decision.

Growers cited different reasons for adopting GAPs. Three of seven growers adopted more food-safety practices to maintain market access in an environment characterized by growing consumer concern about food safety. Certain U.S. and Canadian buyers require that products meet high standards for food safety, and growers had to adopt GAPs to compete for those sales. Many growers say that, in the past, buyers were concerned about food safety, but food safety was not necessarily their top concern when buying green onions. Since the outbreak, buyers are, for the time being at least, requiring GAPs and GMPs. About a quarter of the Mexican growers sell to the United Kingdom, a very demanding market in terms of food safety, in addition to the United States and Canada. Mexican growers selling to the U.K. market must also comply with EurepGAPs, the private European version of GAPs. The core food-safety components of GAPs and EurepGAPs are very similar, but EurepGAPs also address other issues, such as environmental quality and worker welfare. Growers typically produce to one set of standards that will meet the needs of all their buyers.

Two producers said the most important reason for adopting new practices was to ensure production of safer food. Safer practices enable firms to minimize risk in their operations. Outbreaks traced to growers lead to catastrophic loss of sales and reputation. Food safety is a particularly important issue for firms that have brandname recognition. The produce industry is not typically associated with strong brandname recognition, but fresh-cut products such as bagged salads and bagged baby carrots are an exception. The green onion industry sells fresh-cut green onions. Most of the industry's products are targeted at the food-service industry and involve washed, chopped, or diced green onions packaged in consumer-ready bags. A fresh-cut component to an industry promotes adoption of more food-safety practices.

Although one grower responding to the survey cited the prospect of a higher price as the most important reason for adopting new food safety practices, the general view in the industry is that compliance with GAPs increases marketing opportunities rather than price.

10.4 THE HEPATITIS A OUTBREAKS

On November 15, 2003, the FDA announced that hepatitis A outbreaks in September in Tennessee, North Carolina, and Georgia were associated with raw or undercooked green onions (U.S. FDA, 2003a). At that time, the FDA reported that the green onions in the Tennessee case "appeared" to be from Mexico. One person in Tennessee died (The Packer, 2003a). On November 20, 2003, the FDA announced that green onions from Mexico were implicated in the Tennessee and Georgia outbreaks (U.S. FDA, 2003b). The FDA never determined the source of the green onions associated with the outbreak in North Carolina. In late October and early November, before the FDA's first announcement regarding contaminated green onions, another very large outbreak of hepatitis A occurred in Pennsylvania among diners at one restaurant. Over 500 people contracted hepatitis A and three died (Dato et al., 2003). On November 21, the FDA announced that this outbreak was also associated with green onions from Mexico and named the four firms that grew the product associated with the outbreak (U.S. FDA, 2003c). Identification of the four firms was based on epidemiological and traceback evidence.

Hepatitis A is a liver disease caused by the hepatitis A virus. In most cases, the symptoms are mild (jaundice, fatigue, abdominal pain, loss of appetite, nausea, diarrhea, and fever). Most people recover fully and some never even know they have the virus. The disease occasionally can be severe, particularly for people with liver disease. The virus is transmitted by the fecal–oral route. Produce can become contaminated when a person who has hepatitis A, or whose hands are contaminated with the virus, comes into contact with the produce, or when the produce is exposed to water contaminated with the hepatitis A virus.

Contaminated green onions have been implicated in previous foodborne illness outbreaks in the United States. In 1999, the FDA began testing a sample of domestic and imported produce for three microbial pathogens (but not hepatitis A). Contamination was found on both domestic and imported green onions (U.S. FDA, 2001a, b). Green onions may also be particularly susceptible to contamination because "plant surfaces are particularly complex or adherent to viral or fecal particles" (Dato et al., 2003).

The CDC and FDA investigated the cause of the 2003 hepatitis A outbreaks. First, the CDC looked for the product that was most likely to be the source of the contamination. In each case, the contaminated product was consumed in a restaurant (U.S. FDA, 2003a). Local health officials, who must first determine whether the contamination occurred at the point of service in their jurisdiction, decided that the original contamination occurred at some point before the green onions arrived at the restaurants. In the Pennsylvania case, restaurant workers also became ill at the same time as the patrons, implying they were not the original source of contamination. However, officials noted that the storage practices used at the restaurant in Pennsylvania could have contributed to intermingling of uncontaminated and contaminated green onions, which may explain the extent of the outbreak (Dato et al., 2003).

Once local officials determined the green onions were not contaminated in the restaurants, the FDA became involved in the traceback to determine the origin of the contaminated product. The FDA believed the green onions originated in Mexico and were contaminated there too. It is difficult to pin down exactly where the produce became contaminated – at the farm, packing shed, or in the distribution chain as the produce made its way into the U.S. food distribution system. However, the hepatitis A virus sequences from the outbreaks traced to Mexico were identical or very similar to sequences of sick people living along the U.S.–Mexican border or returning from visits to Mexico (U.S. FDA, 2003d).

The FDA named four growers in Mexico as being associated with the outbreaks and issued an import alert, ordering border inspectors to reject all shipments of green onions from these firms. The four firms named by the FDA as being associated with the outbreak did not have third-party audits of GAPs for their summer operations, although one did for its winter operation. As often happens, many of those most hurt by the news of contaminated product were not necessarily those who may actually have been responsible for the problem. The fields associated with the contaminated green onions only produced in the summer season (mid-May through early October) and were not in production when the adverse publicity broke in November. With an incubation period of up to 50 days, green onions associated with illness in September and early November would have been harvested from July through early September (assuming seven weeks for incubation and one week for packing and shipping). In November, winter production was in full swing in Mexicali, Baja California, and San Luís Río Colorado in the adjoining state of Sonora. Unfortunately, consumers did not distinguish between the two groups of producers, much to the dismay of winter producers without any connection to the implicated summer production.

Mexican officials inspected the firms named by the FDA. Because the FDA could not provide any corroborating physical evidence, Mexico did not accept its claims that the green onions were contaminated on the four farms. The Mexican Secretary of Health, however, temporarily suspended operations at one of the four farms for unsanitary conditions.

During the first week of December 2003, CDC and FDA officials joined with Mexican officials in an investigation of green onion farms in Baja California and Sonora, Mexico. Following an outbreak of foodborne illness, the FDA and CDC will often go to a foreign country, if invited, to participate in an investigation to try to determine the causes of the contamination and measures to prevent a recurrence. Of the four firms that were implicated in the traceback, none was growing in the summer production area, three were producing green onions in the winter production area, and one was growing other produce items but not green onions. As a result, U.S. officials did not necessarily expect to find the point of contamination. The mission was more broad-based – to identify conditions that could promote contamination.

On December 9, the FDA issued a press release outlining some of its preliminary findings from the trip to Mexico. It did not find evidence of hepatitis A on the four farms. It did, however, identify issues of concern that could have played a role in the spread of the disease, including poor sanitation and inadequate hand-washing facilities. The FDA also raised questions about worker health and hygiene, and the quality of water used in the fields, packinghouses, and the making of ice (U.S. FDA, 2003d). It also noted that many firms were in the process of, or had just completed, improvements to their water systems and other facilities.

As is common with produce, there is no reliable test for the presence of hepatitis A on green onions. Microbial contamination is usually low-level and sporadic, making it hard to detect. The FDA does routine testing for pesticide residues, but these are much easier to detect; if one piece of produce shows evidence of pesticides, all produce from the same field is likely to have the same problem. Because of the difficulty of testing for microbial contamination, the FDA cannot depend on tests conducted at the border and instead relies on promoting GAPs and GMPs to reduce the incidence of microbial contamination. Also there is rarely any physical evidence in a foodborne illness outbreak. By the time people begin to show symptoms of illness, the contaminated food has usually been consumed or discarded. In the case of hepatitis A, which has a long incubation period, there is even less likelihood of existence of produce that can be tested for the presence of contamination.

The United States and Mexico disagreed in this case about the level of evidence needed to implicate the four Mexican firms. The FDA had epidemiological evidence and traceback records associating the outbreaks with green onions shipped from the four farms in Mexico and determined that conditions there were consistent with potential hepatitis A contamination at the farm level. The FDA considered this information to be adequate for making decisions. Mexico argued that the FDA did not have physical evidence linking the contamination to the farms and that the green onions could have been contaminated at some point in the U.S.-based marketing chain. This analysis examines the impact of the FDA's announcement that the contaminated green onions originated in Mexico. It does not depend on where the green onions were actually contaminated, which is beyond the scope of economics.

Because of the difficulty of finding physical evidence of contaminated produce, the FDA sometimes bases decisions on epidemiological and traceback evidence alone. For example, in 1996, the FDA used these types of evidence to associate a large foodborne illness outbreak – 1465 people – in the United States and Canada with Guatemalan raspberries. After two consecutive years of outbreaks, the FDA denied all imports of Guatemalan raspberries. Issuing an import alert without physical evidence was very rare in 1997. The FDA based its decision on epidemiological and traceback evidence related to past outbreaks and its observations regarding current production practices (Calvin et al., 2002). In 1999, a handful of Guatemalan growers were allowed to begin exporting raspberries to the U.S. market. The growers had to adhere to the Model Plan of Excellence, which was a mandatory food-safety program developed by the Guatemalan government and the Guatemalan Berry Commission, in consultation with the FDA. Growers also had to pass Guatemalan government and FDA inspections before exporting the raspberries. Not until 2000, after several years of U.S. outbreaks, did the FDA actually observe the parasite *Cyclospora* on a Guatemalan raspberry (Ho et al., 2002). Waiting until 2000 before trying to resolve the contamination problem would have put more U.S. consumers at risk. Since 1997, the FDA has become less reluctant to deny imports based on epidemiological and traceback evidence alone.

10.5 ECONOMIC IMPACT OF THE OUTBREAKS

A foodborne illness outbreak can have a severe impact on shipments and prices of the food item associated with the outbreak. Free-on-board (FOB) prices peaked on Friday November 14, 2003, at $18.30 per box of medium green onions (48 bunches to a box), the day before the FDA's first announcement implicating green onions in the outbreaks (Fig. 10.4). Free-on-board price is the average, unweighted unit price received by the shipper or grower-shipper. It excludes freight and insurance costs.

The first FDA announcement stated that the green onions in one outbreak "appeared" to come from Mexico. On Thursday November 20, the FDA announced that the green onions came from Mexico. Growers think this announcement had the most significant effect on general market demand. By Friday November 21, when the FDA made its third announcement naming four Mexican growers associated with the outbreak, prices of green onions had declined to $12.43 per box. One week later, the day after Thanksgiving, prices had declined to $7.23 per box.

Amid falling demand and confusion triggered by the FDA announcements, the industry tried to right itself. Buyers called their shippers to find out what kinds of food-safety programs they and their growers had in place or to seek concrete evidence of earlier verbal assurances regarding food-safety practices. Buyers also sent out inspection teams to examine growing and packing operations. Shippers also dispatched audit teams to their growers. Shippers and growers faxed copies of their third-party audits for compliance with GAPs and GMPs, if they had audits. For some growers and buyers, the level of concern regarding food safety clearly fluctuates with events.

On November 24, 2003, the National Restaurant Association advised its members to stop using Mexican green onions or to cook them thoroughly. Thoroughly cooking green onions can reduce or eliminate the hepatitis A virus.

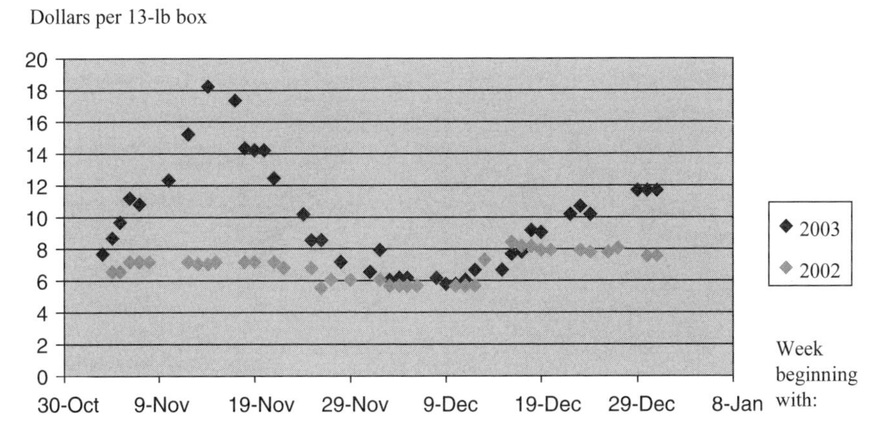

Figure 10.4 Daily green onion free-on-board prices in the United States, 2002 and 2003. (*Source*: Western melon and vegetable report, Agricultural Marketing Service, USDA.)

Several major U.S. restaurant chains announced that they were removing green onions from their menus until further notice (The Packer, 2003b). Some retailers also removed green onions from their stores. Green onions are not a menu staple in the United States and in many food preparations regular onions can be substituted for green onions. Shippers tend to disagree about the relative sensitivity of the retail and food-service markets to bad publicity.

In November and early December of 2002, the price of green onions was nearly steady during the entire period compared to more dramatic swings during the same period in 2003. The industry considered the 2003 price the day before the first FDA announcement to be extraordinarily high. Between November 14 and December 10, prices fell 72 percent as demand for green onions dropped because of food-safety concerns. Prices then rose steadily from $5.73 on December 10 to $11.73 on December 31, 2003. Consumer demand during the November and December holidays is typically high, and alternative supplies were very limited in the United States. After the first of the year, however, prices declined again and did not increase above the previous year's corresponding prices until April 2004.

Shipments of green onions from Mexico actually increased in the week following the first FDA announcement on November 15 (Fig. 10.5). Growers already had green onions packed and waiting in their coolers, and there was strong demand in the run-up to Thanksgiving. The second and third FDA announcements on November 20–21 had a more serious impact on shipments. During the next two weeks (November 23 to December 6, 2003), Mexican shipments of green onions to the United States declined 42 percent, compared with a 13 percent decline during the same time in 2002. Beginning the week of December 7, shipments began to rebound, and during the week beginning December 21, shipments were 97 percent of the volume of the same week in 2002. In the first three months of 2004, after the high holiday demand, shipments generally lagged behind the levels of the previous year.

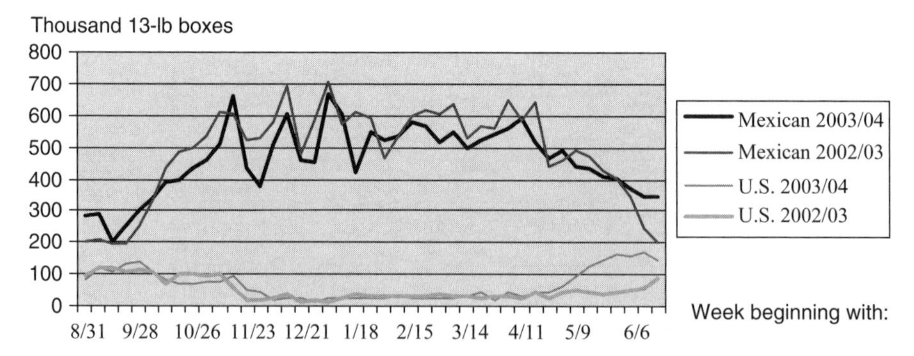

Figure 10.5 Weekly U.S. and Mexican green onion shipments, 2002/2003 and 2003/2004. (*Source*: Fresh fruits, vegetables and ornamental crops: weekly summary shipments, Agricultural Marketing Service, USDA.)

For the two-week period of November 16–29, estimated losses for the Mexican growers, considering lost sales and decreased prices on actual sales, totaled $10.5 million. Growers incurred additional losses when fields went unharvested due to low demand. In the last week of November, Mexican growers left 48 hectares of green onions unharvested. In December an additional 317 hectares were left unharvested. Green onion fields are planted every few weeks to provide a continuous supply for harvest. With the decline in demand, growers probably cancelled some planned plantings. The decline in harvest resulted in a decline in demand for labor, which had a serious impact on the local economy.

In other foodborne illness cases, markets for the foods implicated in the illness rebounded when outbreaks ceased and the problem appeared to have been resolved. But in cases with repeated outbreaks, industries associated with the contaminated product can face serious long-term impacts. Mexican cantaloupes were associated with back-to-back U.S. outbreaks of Salmonella in 2000, 2001, and 2002 (Calvin, 2003). In late 2002, the FDA put all Mexican cantaloupes under import alert and, by May 2005, only six growers could export to the United States. In 1997, the FDA put all Guatemalan raspberries under an import alert after consecutive outbreaks associated with that product. With the economic losses associated with the outbreaks and the cost of adopting more food-safety practices, the Guatemalan industry declined to a handful of producers. The few remaining growers did not export to the United States at all in the spring of 2004. The Mexican raspberry industry benefited from the problems in Guatemala and is now a major supplier of raspberries to the United States. If one production area has food-safety problems, alternative sources may become more attractive.

10.6 IMPACT OF OUTBREAKS ACROSS GROWERS

Interviews with a limited number of Mexican growers in June 2004 indicated that the impact of the hepatitis A outbreaks varied across growers of different types. Here, we discuss the impact only on Mexican growers. There were so few U.S. growers selling green onions at the time of the FDA announcements about the outbreaks that there was virtually no negative impact on the U.S. industry. These post-outbreak interviews with eight growers were independent of the 2002 pre-outbreak survey.

First, although all growers were affected by the general loss of consumer confidence in green onions and lower prices, growers with third-party audits of compliance with GAPs and GMPs had higher volumes of sales than other growers (Table 10.2). If retail and food-service buyers needed green onions, they sought growers with the best food-safety programs first, although they did not pay more for the green onions. For these growers, green onion shipments did not decrease markedly, nor were their other crops affected. Growers who were in the process of becoming GAP compliant and had audits to demonstrate their progress to date in improving food safety also fared reasonably well. Their shipments of green onions usually fell a bit and demand for some of their other crops dropped

TABLE 10.2 Impact of Food Safety Outbreaks on Mexican Growers, by GAP Status

	Impact	
GAP Status	Volume of Green Onion Sales	Demand for Other Products
GAPs	Fairly constant	None
Partial GAPs	Down a bit	Some impact
No GAPs	Down by 50 percent	Down by about 30 percent
No GAPs and named by FDA	No sales and most fields plowed under	Shippers stopped selling all or almost all products from these growers

Source: Avendaño and Calvin, 2004.

slightly. For producers who were not GAP compliant, their green onion sales declined to about half the normal volume and demand for other products sold by these firms declined by about 30 percent of their usual levels. For those growers who were not compliant with GAPs and were named by the FDA as being associated with the contaminated green onions, the impact was catastrophic. U.S. shippers did not want green onions or any other products. These growers plowed up most of their green onions and in some cases sold small amounts to the domestic Mexican market.

The market impact also depended on a grower's particular buyers. Growers could face a dramatic loss in sales if they had several big retail or food-service buyers who decided to take green onions off their shelves or menus until consumer confidence was reestablished. In interviews, growers were divided about which market, retail, or food-service, was most sensitive to the food-safety problem and their views probably depended on their particular experience with buyers. Growers not named by the FDA said the negative market impact lasted from one to four months.

Those growers with buyers from the United Kingdom were fortunate, because sales to that country appear to have been unaffected. There were no cases of hepatitis A related to green onions outside the United States, and commercial buyers in the United Kingdom who already demanded strong food-safety practices from their suppliers, appeared to be confident of the product. Although there were no cases of hepatitis A in Canada, demand in that market did fall. Officials from the Canadian Food Inspection Agency also visited Mexico in February 2004 to look at production and packing facilities, but did not issue import alerts. In practice, because all Mexican green onions going to Canada are sold by shippers located in the United States, the U.S. import alert also affected Canada with or without an official import alert. It is not clear what accounted for the difference in market response. The outbreak was probably not as well publicized in the United Kingdom as it was in the United States and Canada. Researchers have estimated the impact of positive and negative publicity in a foodborne illness outbreak associated with strawberries (Richards and Patterson, 1999). Prices responded more strongly to bad news than good news.

The impact also varied by type of packing process. Traditionally, green onions have been packed in a carton and then covered with ice to prevent them from drying out during shipment. A newer technique involves iceless packing. The green onions are packed in polyethylene bags in a cold room held at 3.3–4.5°C. This pack does not require ice during transportation, although cooling is required. The food-service industry and the U.K. market favor the iceless pack. Some buyers feel iceless green onions are less risky than green onions packed with ice, water being a potential source of contamination. Those firms with iceless production appeared to have fared a bit better in terms of sales than firms not employing this technique. Some firms expect the iceless pack trend to continue growing because of the perception by some buyers that it reduces the risk of contamination.

10.7 RESPONSE TO THE OUTBREAKS

In May 2005, three of the four Mexican growers that the FDA put on import alert were still restricted from exporting to the United States. They continue trying to resolve their food-safety issues and remove their names from the alert. Growers on import alert are in the unenviable position of having to continue summer production just to clear their reputation with the FDA and buyers. Some growers might have preferred to abandon summer production. The firm that was closed down by Mexican officials for unsanitary operations has reopened and it can ship to the Mexican market, though not to the U.S. market.

One of the four firms named by the FDA as being associated with the contaminated green onions was removed from the import alert in late October 2004. To have the alert lifted, the grower had to provide the FDA with documentation showing that he was complying with GAPs and GMPs. Then the FDA, with Mexican federal and state officials, inspected the firm to determine if the import alert could be lifted.

All green onion growers who did not already have GAPs and GMPs, not just the four growers named by the FDA, had to undertake new investments because of new buyer demands for food-safety assurances. In addition, Mexican green onion growers were unanimous in their desire for a mandatory food-safety program to ensure that a few noncompliant growers could not hurt their businesses and reputations again in the future. This was particularly important given the large investments required to improve food-safety practices. Growers requested that the Mexican government ensure mandatory compliance with GAPs and GMPs. The growers worked with the Mexican government agency charged with food safety, Servicio Nacional de Sanidad, Inocuidad y Calidad Agroalimentaria (SENASICA), and the state government of Baja California to develop a mandatory export protocol for green onions that was approved in July 2004. SENASICA is part of the Mexican equivalent of the U.S. Department of Agriculture, Secretaría de Agricultura, Ganadería, Desarollo Rural, Pesca y Alimentación (SAGARPA).

The Mexican program has its own GAPs and GMPs, based on the FDA guidelines for minimizing microbial contamination, but with more specific requirements for green onions instead of general guidelines for all produce growers. In order to export, growers are required to be certified by SENASICA as meeting the new food-safety standards. This program is completely independent of any FDA interests and is a Mexican government response to Mexican growers. The three growers who remain on import alert cannot export to the U.S. market until the FDA approves their operations, regardless of any SENASICA certification.

Growers needed to evaluate the benefits and costs of the new mandatory program to decide whether to continue exporting. Some growers face more costly investments than others because of their environmental conditions, such as dependence on open-ditch irrigation canal water. As the cost of doing business increases, not everyone will necessarily be able to participate.

Growers might still want to consider adopting additional food-safety practices above the minimum required and would have to consider the expected benefits and costs of doing so. Growers surveyed in 2002, before the outbreak, about their reasons for adopting GAPs and source of motivation for adopting GAPs, were resurveyed in 2005. All growers indicated that the most important reason for adopting GAPs in 2005 was to maintain sales to the United States, a major shift from the pre-outbreak survey results when only three producers listed that as their most important reason. Also, the demands of shippers were more important in the decision-making process than they were before the outbreak, when growers' own initiative was the most important motivator.

Growers had to apply for SENASICA certification by October 31, 2004. Anticipating difficulties for some growers with the short deadline, SENASICA required a series of audits during the summer to help growers identify and resolve weakness. By late October, 19 percent of the growers had met the food-safety requirements and received SENASICA certification (Roche, 2004). Another 38 percent had practices in place, had submitted their paperwork, and were awaiting certification. These two groups of growers, 57 percent of the firms (but a larger percent of total production), could export as far as Mexico was concerned. For many growers, however, developing adequate food-safety practices over such a short time period was not possible. By the end of October 2004, a group of small producers, 43 percent of firms, were not yet ready to apply for certification. In May 2005, 32 percent of the firms were still uncertified by SENASICA.

During the 2004/2005 season, a few firms who did not have SENASICA certification but who did have third-party audits from reputable companies, exported to the United States. This group did not include any growers on the import alert. U.S. buyers accepted third-party audits as an adequate precaution from these Mexican growers. When the green onion export program began, SENASICA did not have legal instruments to enforce a mandatory program. Although the government seeks changes in laws to give SENASICA this power, new laws are not yet available. It is not yet clear how SENASICA intends to deal with the growers exporting without their certification. The small growers have become disenchanted with the mandatory program as the difficulty and expense of receiving SENASICA certification has become more obvious. Growers who have SENASICA certification are not satisfied with the current situation.

Some of the small growers may struggle to be competitive under a mandatory program. If there are fixed costs, smaller growers will be spreading the investment over a smaller volume of output and raising their per unit costs more than larger growers. Becoming certified is generally more expensive for packinghouses than for field operations and, in the 2004/2005 season, several firms had to arrange for other firms to pack their produce when they were unable to achieve certification for their packinghouses. The packinghouse owners required field certification from these growers before packing their product.

A factor favoring a mandatory program is that Mexico is a major supplier of produce to the U.S. market, and outbreaks associated with green onions may affect consumer confidence in other Mexican produce items. SENASICA may have considered this negative externality on other producers when it agreed to design a mandatory food-safety program for green onion exporters.

There have also been activities on the U.S. side of the border in response to the outbreak. Every outbreak of foodborne illness provides the FDA and market participants with an opportunity to gain information and, if necessary, reassess their strategies to promote food safety. In January 2004, the FDA and CDC met with produce industry leaders to discuss commodity-specific GAPs that would provide guidelines tailored to individual commodities (The Packer, 2004a). The produce industry convened a meeting on June 9–10, 2004, to consider additional commodity-specific guidelines for several products, including green onions. Much of the industry debate focused on whether additional commodity-specific guidelines or just universal compliance with current GAP guidelines was needed. Industry leaders urged producers and supply chain participants to be proactive in developing and implementing appropriate, voluntary food-safety systems rather than be regulated by the government. In mid-2005, the U.S. green onion industry had nearly completed its own green onion food-safety guidelines. When they are released, there will be three major competing guidelines that apply to green onions – the FDA's general GAPs, SENASICA's specific green onion guidelines, and the U.S. green onion industry's specific guidelines. So far, the U.S. market appears to be favoring FDA's GAPs over SENASICA's guidelines. It remains to be seen how buyers will respond to the new U.S. green onion industry guidelines.

In the late spring and early summer of 2004, U.S. shipments of green onions were up substantially from the year before (Fig. 10.5). U.S. shippers might have been hedging their bets by having a bit more domestic production available. However, production costs are higher in the United States than in Mexico and many U.S. growers faced fairly low prices. Industry experts expected U.S. shipments would return to more normal levels in 2005. The impact of the outbreak on U.S. food-safety practices is uncertain. Mexican green onion producers, who face a mandatory program to export while their U.S. competitors still face only voluntary guidelines, are particularly interested in this issue. However, almost all retailers and food-service buyers now demand GAPs and GMPs for green onions, even if they did not before the outbreaks of hepatitis A in 2003. Many U.S. growers will be forced to adopt new food-safety practices because of buyer demands. These food-safety practices will be a cost of doing business, not a personal choice to distinguish the firm. These additional costs may put some U.S. producers at an economic disadvantage because they are only summer producers, growing for a short season compared with Mexican growers producing over an extended season or even on a year-round basis.

On the legal front, in late 2004, more than 300 people filed claims against the restaurant in Pennsylvania where the contaminated green onions were served and the restaurant had settled 134 claims. In turn, the restaurant sued its green onion suppliers – the U.S. firms that sold the Mexican green onions (The Packer, 2004b).

10.8 CONCLUSIONS

The example of the recent outbreaks of hepatitis A associated with green onions demonstrates that although incentives for individual growers to voluntarily adopt stronger food-safety programs have been increasing, they are not always adequate to achieve universal adoption. And one grower without an adequate food-safety program can cause financial problems for the entire industry. This analysis shows that with better traceback and third-party audits, the negative impact can vary by type of producer by identifying those with the most responsibility for an outbreak or those whose products pose more risk in the aftermath of an outbreak.

The response of Mexico and the United States to the outbreak differed. Mexico adopted a mandatory food-safety program for exporters, although in practice it has not yet been possible to enforce the program because of a lack of an appropriate legal mechanism. A mandatory program may reduce the probability of additional outbreaks, which could have more serious long-run impacts on the reputation of Mexican green onions and perhaps Mexican produce more generally. However, a mandatory program may force some producers out of the industry. The United States did not impose any mandatory changes on its domestic green onion industry. However, because buyers are reported to be demanding GAPs since the outbreak, a voluntary program may be effectively mandatory in practice. The outbreak associated with green onions demonstrates the important role of buyers in strengthening food-safety practices.

Both Mexico and the United States, however, have followed the same trend in developing crop-specific guidelines for green onions since the outbreak. In Mexico it is a government guideline, and in the United States it is an industry-developed guideline. It remains to be seen how retail and food-service buyers will respond to the proliferation of food-safety standards for green onions.

REFERENCES

G. Akerlof, The market for "lemons": quality, uncertainty, and the market mechanisms. *Quarterly J. Econ.*, 84(3), 488–500 (1970).

B. Avendaño, El impacto de la iniciativa de inocuidad alimentaria de EE.UU. en las exportaciones de hortalizas frescas de México. Unpublished Ph.D., Center for Economic, Social, and Technological Research on Agroindustry and World Agriculture, Universidad Autónoma de Chapingo, Chapingo, Mexico, 2004.

B. Avendaño and R. Schwentesius, Factores de competitividad en la producción y exportación de hortalizas: el Valle de Mexicali, Baja California, México. *Problemas del Desarrollo*, 36 (140) (2005). Instituto de Investigaciones Económicas.

L. Calvin, "Produce, Food Safety, and International Trade: Response to U.S. Foodborne Illness Outbreaks Associated with Imported Produce," in J. Buzby, Ed., *International Trade and Food Safety*. U.S. Department of Agriculture, Economic Research Service, Agricultural Economic Report Number 828, November 2003.

L. Calvin, W. Foster, L. Solorzano, J.D. Mooney, L. Flores, and V. Barrios, "Response to a Food Safety Problem in Produce: A Case Study of a Cyclosporiasis Outbreak," in B. Krissoff, M. Bohman, and J. Caswell, Eds., *Global Food Trade and Consumer Demand for Quality*. Kluwer Academic/Plenum Publishers, New York, 2002.

L. Calvin, B. Avendaño, and R. Schwentesius, *The Economics of Food Safety: Green Onions and Hepatitis A Outbreaks*. U.S. Department of Agriculture, Economic Research Service, E-Outlook VGS-305-01, December 2004.

J. Caswell and E. Mojduszka, Using informational labeling to influence the market for quality in food products. *Am. J. Agri. Econ.*, 78(5), 1248–1253 (1996).

V. Dato, A. Weltman, K. Waller, M. Ruta, A. Highbaugh-Battle, C. Hembree, S. Evenson, C. Wheeler, and T. Vogt, Hepatitis A outbreak associated with green onions at a restaurant – Monaca, Pennsylvania, 2003. *MMWR* 52(47), 1155–1157 (2003).

E. Golan, B. Krissoff, F. Kuchler, L. Calvin, K. Nelson, and G. Price, Traceability in the U.S. food supply: economic theory and industry studies. U.S. Department of Agriculture, Economic Research Service, Agricultural Economic Report Number 830, 2004a.

E. Golan, T. Roberts, E. Salay, J. Caswell, M. Ollinger, and D. Moore, Food safety innovation in the United States: evidence from the meat industry. U.S. Department of Agriculture, Economic Research Service, Agricultural Economic Report Number 831, 2004b.

A. Ho, A. Lopez, M. Eberhart, R. Levenson, B. Finkel, A. da Silva, J. Roberts, P. Orlandi, C. Johnson, and B. Herwaldt, Outbreak of cyclosporiasis associated with imported raspberries, Philadelphia, Pennsylvania, 2000. *Emerg. Infect. Dise.*, 8, 783–788 (2002).

L. Mitchell, "Economic Theory and Conceptual Relationships Between Food Safety and International Trade," in J. Buzby, Ed., *International Trade and Food Safety*. U.S. Department of Agriculture, Economic Research Service, Agricultural Economic Report Number 828, 2003.

T. Richards and P. Patterson, The economic value of public relations expenditures: food safety and the strawberry case. *J. Agric. Res. Econ.*, 24, 440–462 (1999).

R. Roche, Personal Communication, 2004. Chief of the Food Safety Program, Baja California State Committee on Plant Health.

The Packer, Onions tentatively blamed in death. Shawnee Mission, KS, *The Packer*, November 3, 2003a, p. A5.

The Packer, Green onions yanked from menus; demand falls. Shawnee Mission, KS, *The Packer*, November 24, 2003b, p. A1.

The Packer, Buyers, not growers, hold key to food safety. Shawnee Mission, KS, *The Packer*, June 21, 2004a, p. A6.

The Packer, Chi-Chi's sues 3 green onion suppliers, sells some assets to Outback chain. Shawnee Mission, KS, *The Packer*. August 9, 2004b, p. A3.

U.S. Department of Agriculture, Agricultural Marketing Service. Fresh fruit and vegetable shipments; by commodities, states, and months. Various issues.

U.S. Department of Agriculture, Agricultural Marketing Service. Fresh fruits, vegetables and ornamental crops: weekly summary shipments. Various issues.

U.S. Department of Agriculture, Agricultural Marketing Service. Western melon and vegetable report. Various issues.

U.S. Food and Drug Administration, Center for Food Safety and Applied Nutrition. FDA survey of imported fresh produce, January 18, 2001a. Available at www.cfsan.fda.gov/~dms/prodsur6.html. Accessed October 29, 2004.

U.S. Food and Drug Administration, Center for Food Safety and Applied Nutrition. Survey of domestic fresh produce: interim results, July 31, 2001b. Available at www.cfsan.fda.gov/~dms/prodsur9.html. Accessed October 29, 2004.

U.S. Food and Drug Administration, Center for Food Safety and Applied Nutrition. FDA talk paper: consumers advised that recent hepatitis A outbreaks have been associated with green onions, November 15, 2003a. Available at www.cfsan.fda.gov/~lrd/tphep-a.html. Accessed October 29, 2004.

U.S. Food and Drug Administration, Center for Food Safety and Applied Nutrition. FDA statement: the FDA today issued the following statement to clarify its actions on imported Mexican green onions (scallions), November 20, 2003b. Available at www.cfsan.fda.gov/~lrd/fphep-a.html. Accessed October 29, 2004.

U.S. Food and Drug Administration, Center for Food Safety and Applied Nutrition. FDA statement: Statement on hepatitis A, November 21, 2003c. Available at www.cfsan.fda.gov/~lrd/fphep-a2.html. Accessed October 29, 2004.

U.S. Food and Drug Administration, Center for Food Safety and Applied Nutrition. FDA statement: FDA update on recent hepatitis A outbreaks associated with green onions from Mexico, December 9, 2003d. Available at www.fda.gov/bbs/topics/NEWS/2003/NEW00993.html. Accessed October 29, 2004.

INDEX

Microbial Hazard Identification in Fresh Fruit and Vegetables, Edited by Jennylynd James